COMPUTER INFORMATION SYSTEMS DEVELOPMENT: Design and Implementation

Copyright © 1985
by South-Western Publishing Co.
Cincinnati, Ohio

ISBN: 0-538-10860-6

Library of Congress Catalog Card Number: 83-51621

1 2 3 4 5 6 7 8 K 7 6 5 4

Printed in the United States of America

COMPUTER INFORMATION SYSTEMS DEVELOPMENT:
Design and Implementation

David R. Adams
Associate Professor, Computer Information Systems
Northern Kentucky University
Highland Heights, Kentucky

Michael J. Powers
The Country Companies
Bloomington, Illinois

V. Arthur Owles
Industry Consultant
Wang Laboratories, Inc.
Lowell, Massachusetts

Published by

J86 **SOUTH-WESTERN PUBLISHING CO.**

CINCINNATI   WEST CHICAGO, ILL.   DALLAS   PELHAM MANOR, N.Y.   PALO ALTO, CALIF.

CONTENTS

PREFACE

PERSPECTIVE

This book, in part, represents an implementation of the *Model Curriculum for Undergraduate Computer Information Systems Education* of the Data Processing Management Association-Education Foundation (DPMA-EF). Specifically, the information presented in this book meets or exceeds the content called for in the suggested outline for course *CIS-5—Structured Systems Analysis and Design*. Correspondence between this book and the course specifications is assured by the fact that the text was developed under the oversight of the DPMA-EF, with content appropriateness and technical accuracy validated through independent review.

The DPMA-EF curriculum specifies a structured approach to systems development through use of structured methods within an established life cycle. The curriculum is aimed at graduating students qualified as entry-level programmer/analysts in business-oriented computer facilities.

CONTENT LEVEL

This text is designed to support an undergraduate course. It is assumed that students using this text will have completed an introductory course in computer information systems (CIS). Students should also have completed two semesters of work in structured programming designed to impart skills in the development of COBOL programs that solve business problems. Students should be familiar with the terms and techniques of program development to gain full value from a course based on this book.

In addition, this text is designed for use in a second course in systems analysis and design. Prior to use of this text, students should have completed a review of structured systems analysis work. A companion text, *Computer Information Systems Development: Analysis and Design,* by Powers, Adams, and Mills, is designed to provide a background leading up to the content of this text. However, the first two chapters of this book review the highlights of the companion text. Thus, students with sufficient background and qualifications could use this text without previous work in the field. Conversely, students who have just completed a course based on the companion work may be able to move quickly through the first two chapters to concentrate on the new material.

CONTENT HIGHLIGHTS

The book uses a basic, easily taught systems development life cycle as a framework. This life cycle divides a typical systems development project into five phases and 15 activities. This book concentrates upon the third phase of the model life cycle structure—Detailed Design and Implementation. As indicated, the first two phases are reviewed in the opening chapters. This review serves to establish a context for the main instructional content that follows. Finally, since emphasis is on design, the Installation and Review phases of the project structure are overviewed in the final chapters.

(The final phases of the project structure are encompassed in a companion course, CIS-7, entitled *Applied Software Development Project.* As the focal point of the CIS-7 course, students actually develop computerized information systems. A companion text, *Computer Information Systems Development: Management Principles and Case Study* is designed for use with this course.)

This book makes heavy use of case methods for illustration and instruction. Case citations and examples are provided at appropriate points throughout the text. In addition, an Appendix provides an opportunity for in-depth experience and systems development practice.

The Appendix reviews the principles and practice of systems development projects. Then, three separate case studies are provided as a basis for class or supplemental assignments.

The chapters of this book are divided into two categories:

- Phases and activities of the systems development life cycle.
- Skills applied in systems analysis and design.

'Activities' Chapters

These chapters deal with the individual activities of the Detailed Design and Implementation Phase of the project structure. For the other phases, activities are reviewed in single chapters. These activity chapters use standard subject headings and follow a common presentation pattern. Within each of these chapters, there are standard sections on:

- Activity Description
- Objectives
- Scope
- End Products
- The Process
- Personnel Involved
- Cumulative Project File.

For each activity, two of these areas are treated as keys to a student's understanding of the analysis process. They are the objectives of each activity and its end products. The other areas tend to be be natural consequences of these two.

'Skills' Chapters

The second series of chapters deals with the individual skills applied in analysis and design of computer information systems. These chapters cover:

- The Roots of Systems Design
- The Process of Systems Design
- The Technical Environment of Systems Design

- Application Design Strategies
- File and Database Design
- Foundations of Software Design
- Evaluating Software Design
- Software Design Strategies
- Test Specifications and Planning
- Software Testing Strategies

ACKNOWLEDGMENTS

To assure accuracy and appropriateness for the content of this text, a highly experienced, objective group of persons was asked to review the manuscript during development. The careful readings and thoughtful comments of this group represented, cumulatively, an important contribution to the soundness of this text. Their contributions are acknowledged with sincere thanks.

Donald R. Deutsch, General Electric Information Systems Co., Nashville, TN

Dr. W. Terry Hardgrave, George Mason University, Fairfax, VA

Dr. William Hetzel, Hetzel and Associates, Jacksonville, FL

ESTABLISHING DESIGN SPECIFICATIONS I

Part I establishes a framework for the main topic of this text—systems development. It is assumed that, before enrolling in the course for which this text is used, you have had coursework in the prerequisite subjects of systems analysis and design. Because students may have used varying texts in prerequisite coursework, and also because there may have been a time interval between the study of analysis and design and the present course, this book begins with a review of the principles of analysis and design.

As the title for this part of the book implies, the end result of analysis and design phases of a systems project is a set of design specifications. These design specifications, in turn, become the basis for systems development and implementation.

This part has five chapters. The first chapter reviews the phases of the systems development life cycle leading up to the beginning of systems development. The remaining four chapters recap the specific skills applied and results derived from systems analysis and design.

If you have just completed your study of systems analysis and design—and if your instructor feels you are ready to move immediately into developmental topics—you may be instructed to begin your work in Part II of this book. If so, the content of Part I will still be valuable—possibly necessary—as reference material.

1

1 THE ROOTS OF SYSTEMS DESIGN

LEARNING OBJECTIVES

On completing reading and other learning assignments for this chapter, you should be able to:

☐ Explain the main differences between systems analysis and systems design.

☐ Give the rationale for adopting a structured methodology for development of computer information systems.

☐ Describe and list typical phases of the systems development life cycle (SDLC).

☐ Describe the transition from systems analysis phases of the SDLC to design phases.

☐ Explain how the documentation from systems analysis becomes input to systems design.

☐ Explain how software design differs from systems design.

BACKGROUND

This text is designed to support continuing study in computer information systems development. It is assumed that the student using this text has done prior work in the areas of identifying system needs, defining those needs, and developing user-oriented systems designs. The background and skills appropriate to this introductory-level work are covered in a companion text, *Computer Information Systems Development: Analysis and Design*, by Powers, Adams, and Mills.

The major portions of this text deal with the design and development that take place after nontechnical, user-oriented specifications have been developed. The specifications for what a system is to do are, in general, the products of *systems analysis. Systems design*, then, encompasses the formulation of the technical plans and methods to implement the specified system.

Establishing a Working Structure

The development of a computer information system is so large an undertaking that it is virtually impossible to understand all aspects of a problem at the outset. Some sort of structured approach is necessary. A proven, practical methodology is needed. A standard method for dealing with complexities or uncertainties lies in breaking jobs down into a series of individually definable activities. Although the overall system to be built may be complex and filled with uncertainties, the subparts are small enough to be grasped and managed separately. The structure that results establishes a series of orderly steps that organize and monitor the total effort—a *project*.

If an organization develops systems on a continuing basis, it is desirable to have standard project structures. Thus, systems development projects can be compared with one another. Further, people can be assigned to projects with the assurance that they will understand—and be able to complete—their assignments. For this reason, most computer information systems organizations establish a standard project structure, typically known as a *systems development life cycle (SDLC)*.

THE SYSTEMS DEVELOPMENT LIFE CYCLE

A systems development life cycle, basically, is a formalized description of a project structure for the development of computer information systems. Any large organization that is involved continuously in the development of computer information systems will have established some kind of life cycle. The specific steps within any given life cycle will vary among organizations. However, the underlying principle remains the same: It is necessary both to establish a fundamental structure for the management of projects and also to apply structures that are comparable among projects. This comparability, in turn, makes it possible to train people for effective work on multiple projects and also to establish budgets and controls because the project structure is known and uniform.

Phases

A typical systems development life cycle consists of standard *phases* involving measured amounts of activity. A flowchart for a typical life cycle composed of five stages with interim management decision points is shown in Figure 1-1. The phases within this particular life cycle structure are:

- Investigation
- Analysis and general design
- Detailed design and implementation
- Installation
- Review.

This breakdown centers upon the structure of phases that divide the work incrementally and establish interim checkpoints for management review and commitment. A different look at the same project structure, focusing upon the flow of work within the project and the results delivered, is shown in the data flow diagram of Figure 1-2.

The phase structures illustrated in Figures 1-1 and 1-2 represent the major breakdowns of the overall project. Phases, in effect, are the portions of a project that are carried out between the milestones represented by management decisions for the allocation of funds and the setting of priorities among projects.

Activities

Phases, in turn, are broken down into smaller units of work, known as *activities*. An activity, basically, is a subset of a phase. In the life cycle used to illustrate this book and its companion work, a sequence of 15 activities is used, as illustrated in the Gantt chart shown in Figure 1-3. Note that the identified activities are structured within phases. In other words, activities break down phases into still more manageable units of work.

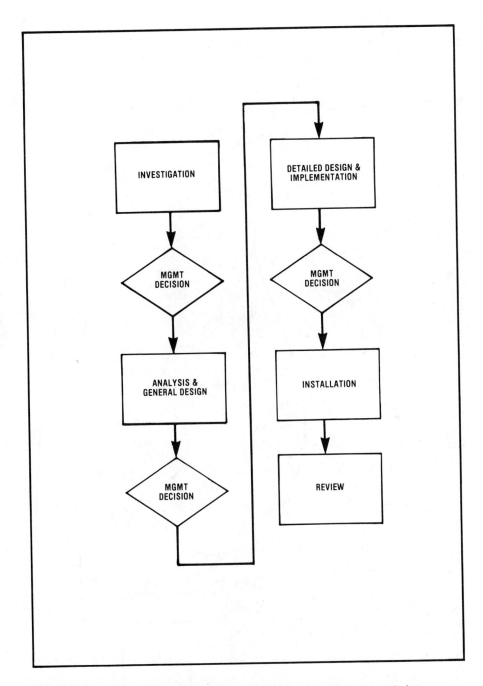

Figure 1-1. The systems development life cycle—a control-oriented view.

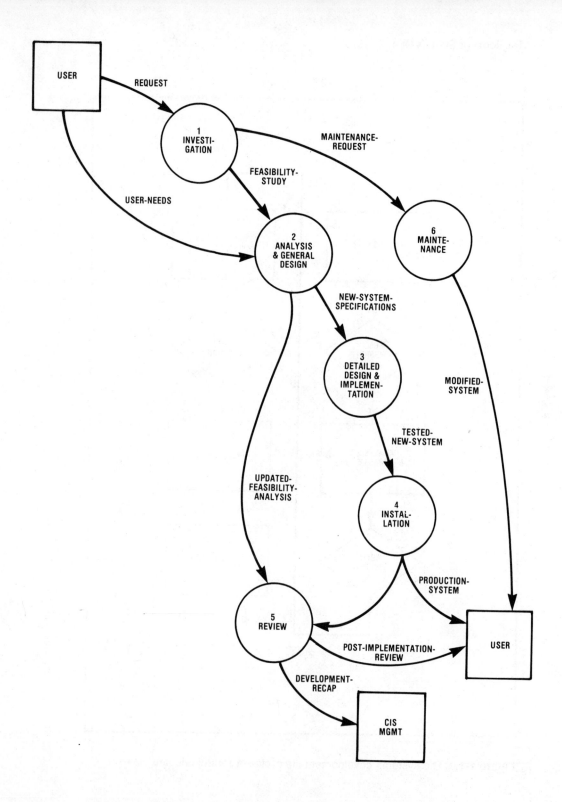

Figure 1-2. The systems development life cycle—a process view.

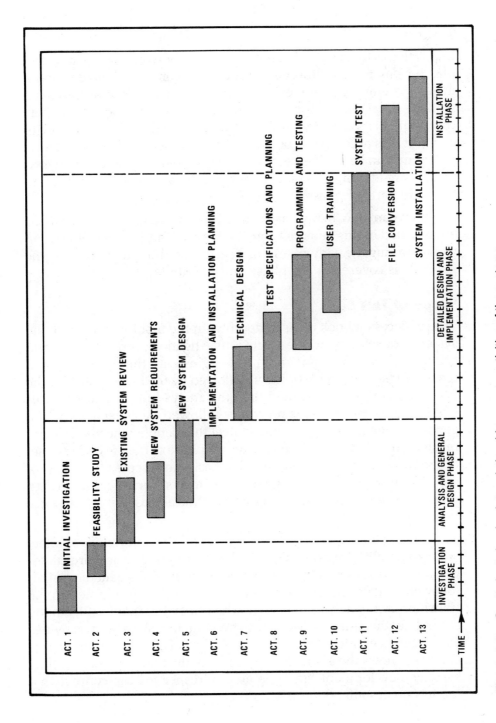

Figure 1-3. Gantt chart showing interrelationships among activities of the systems development life cycle.

Tasks

In practice, activities still can represent units of work that are too complex for day-to-day supervision and management. Consider that a major systems development project can involve several thousand working days. During the course of the project, scores of people may contribute to its completion. To assure control throughout the life of a project, it is necessary to break down efforts so that the work of each individual can be checked every few days. To do this, individual *tasks* usually are identified. Tasks, in effect, are parts of activities that, in turn, are parts of phases.

There are too many tasks in a typical project life cycle to be enumerated in this book. Rather, in this book and in the companion work covering the early phases of a systems development project, the life cycle is covered at the phase and activity levels.

Scope of This Book

The predecessor book deals in depth with the first two phases of the systems development life cycle—investigation and analysis and general design. The companion text begins with the processing of a request for systems development, followed by the investigation of the feasibility of that request, and then the analysis of present practices and identified needs. This process is followed by development of a general design for a new system to meet the identified needs. The end result of the analysis and general design phase of the life cycle is a set of specifications for a new system that has been accepted by both prospective users and CIS designers. The analysis and general design phase also determines economic and technical feasibility of the proposed system.

Thus, at the conclusion of the second phase, analysis is complete and a general technical design has been proposed. The third phase begins with a reexamination and modification of this general design, followed by detailed technical design. This book concentrates on the transition from analysis to design—the transition, in effect, from the second to the third phase—and then on the skills and activities associated with detailed technical design. The remainder of this chapter reviews the efforts and practices of the first two phases of the systems development life cycle as a refresher for the content that follows on systems design.

NATURE AND PURPOSE OF THE SDLC

The methodology of the systems development life cycle is a basic management tool for organizing the work of large projects and the efforts of large CIS departments. However, the SDLC should not be considered a "straightjacket." In its basic structure, it is independent of the development techniques employed.

Systems development life cycles vary by organization and depend, in part, on management style. The phases and activities of a life cycle should be driven by clearly stated objectives, not by lengthy lists of tasks. The particular task lists that are derived from those objectives will vary greatly depending on the nature of the project and on the development tools and methodologies employed. For example, two useful techniques—*prototyping* and *version installation*—are covered in the predecessor text. Use of such techniques significantly modifies the tasks performed in the various life cycle phases and activities.

The phases of the life cycle presented in this and the predecessor text are outlined briefly below.

THE INVESTIGATION PHASE

The investigation phase of the life cycle establishes a framework for receiving, reacting to, and evaluating requests for the development of computer information systems. The two activities within this phase include, first, an initial investigation to determine the size and scope of the request. Then, the second activity, a feasibility study, involves a combined evaluation by the requester and experienced systems analysts. This evaluation results in a report indicating whether the systems request is feasible technically and economically, and a recommendation on whether the system should be scheduled for development.

A data flow diagram showing the processing, flow of information, and transformations of information associated with the investigation phase of the systems development life cycle is shown in Figure 1-4. Process 1 corresponds with Activity 1: Initial Investigation. The remaining processes indicate the major steps that occur during Activity 2: Feasibility Study.

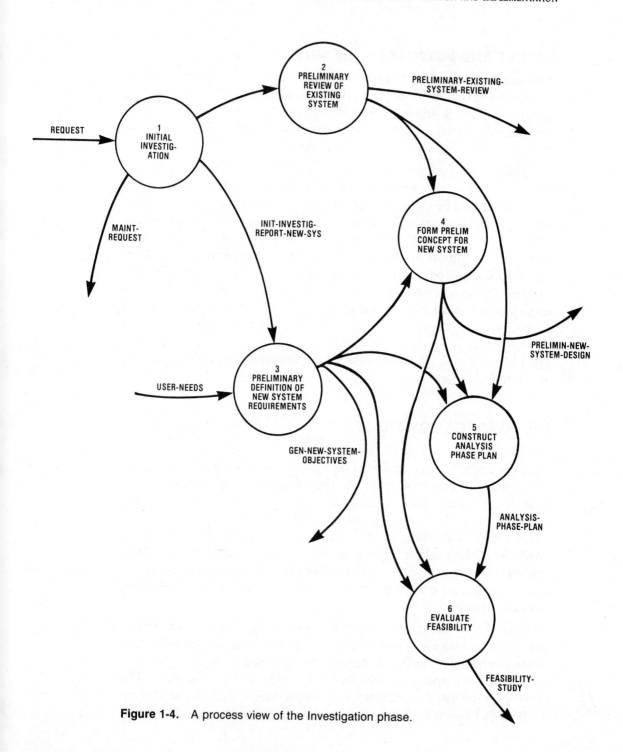

Figure 1-4.   A process view of the Investigation phase.

The feasibility study sets the stage for evaluating progress and performance later in the project. That is, the recommendations and forecasts contained in the feasibility study are refined during succeeding phases involving analysis, design, and implementation. Ultimately, a review at the conclusion of the project evaluates the recommendation of the feasibility report. That is, the feasibility report becomes a standard: The designed, implemented, and installed system should do at least as well as the initial feasibility study indicated it would.

Further, as work progresses on a system, the feasibility study provides a framework, or a stabilizing force for the project. As users and CIS personnel learn more about a system under development, there is a temptation to expand its scope, to add refinements to the procedures, or to produce a result that is more "elegant" than the initial interpretations indicated. Such tendencies must be evaluated and moderated as a project develops. Without some standard such as the feasibility report to provide a framework, it would become extremely difficult to keep a project on track and avoid overruns in both time and money.

ANALYSIS AND GENERAL DESIGN PHASE

This phase represents a major segment of a systems development project, perhaps 40 or 50 percent. Activities include in-depth studies of existing systems, specification of requirements for the new system from the users' point of view, and design of the new system at a general level. The design documentation becomes a basis for the transition from analysis to design within the overall project makeup.

At the conclusion of the analysis and general design phase, user management has seen and approved a model of the processing sequences of the new system, the outputs to be produced, and the inputs anticipated. Thus, the user has previewed and "bought into" the makeup of the new system and has approved both the expenses for development and the procedures to be followed when the new system is implemented. This general design is the basis for updating the feasibility analysis done during the first phase. In addition, enough design must be done during this phase so that technical members of the project team can be sure that they can deliver the specified results.

Figure 1-5 is a data flow diagram of the analysis and general design phase. Processes 1 and 2 correspond with Activity 3: Existing System Review. Processes 3 through 5 correspond with Activity 4: New System Requirements. Processes 6 to 8 cover Activity 5: New System Design.

Following approval of the user specification, the project goes through a transition from emphasis upon systems analysis to a concentration on systems design. This transition accepts and applies user specification documentation as the major input for design efforts.

DETAILED DESIGN AND IMPLEMENTATION PHASE

Most of the computer-oriented work of systems development takes place during the *detailed design and implementation phase.* Hardware and software specifications, possibly begun in the previous phase, are refined. Programming plans are established. Programs actually are written and tested. A core group of users is trained. Then, with user participation, the system goes through testing that is extensive enough to result either in acceptance or in specifications for further modification. When users are ready to accept the system on the basis of this testing, the steering committee is asked for approval to proceed with the installation that will begin the useful life of the new system. The approval, at this point, is based chiefly on user confirmation of acceptability of the design and operating features of the new system.

INSTALLATION PHASE

The main achievement of the *installation phase* is conversion from the existing procedures to the new ones. The remaining users are trained. The old system is phased out. At the conclusion of this phase, the new system has been implemented and is in ongoing use.

This is the point in the life cycle when the impact of change is felt fully by the organization and its people. Thus, the work of this phase can involve considerable human discomfort and a myriad of organizational problems. These problems can be both extensive and serious. However, advance planning and sensitivity in execution of plans can avoid or minimize problems.

Depending on the nature of the system, extensive user training may be required during this phase. It may be necessary or beneficial

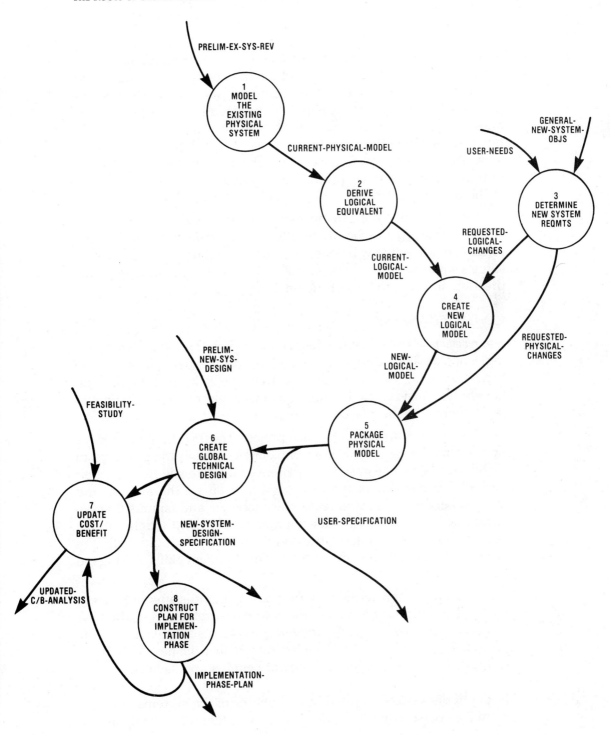

Figure 1-5. Process view of the Analysis and General Design phase of the systems development life cycle.

to conduct special demonstrations, briefings, and continued consultations that help users to understand the full potential of their system. This is especially true if a system includes MIS or DSS features. The prospective management potential of these tools will have been identified during systems analysis. Once these tools are actually available, however, extended potential values may be uncovered—values not anticipated during development. In such situations, an investment of time in user training may produce high yields by enhancing the value of the new system.

Note that no specific management review is called for at the end of this phase. At this point, results speak for themselves. The implemented system now belongs to its users, and user acceptance constitutes some measure of success. This success will be evaluated more precisely in the activities of the ensuing phase.

REVIEW PHASE

The *review phase* of a systems development project is devoted to learning. Considerable effort and money have been expended to develop a new system. Considerable experience has been gained, and, hopefully, considerable learning has taken place. Now it is time to review what has been accomplished.

Two reviews are useful for each project. The first should take place shortly after the system has been implemented, while the project team is still together and can share experiences that are still fresh in their minds. The purpose of this review is to evaluate the process and methods employed and to recap the successes and failures that occurred during the systems development project. Although nobody enjoys discussing failures, this type of review should help the organization to improve the systems development skills it brings to future projects.

A second review takes place perhaps six months after implementation. The purpose of this review is to measure results of the new system to see how they compare with projections at the outset of the project. The emphasis is on determining whether the new system has fulfilled, in actual use, its promises of benefits and savings.

In-depth discussions of the first two phases of the systems development life cycle occupy the majority of the content for the companion, predecessor text. Most of the remainder of this book, then, is devoted to the general design tasks that begin late in the second phase and the more detailed technical design activities associated with the third phase of the life cycle—detailed design and implementation. In addition, this book also covers, at an overview level, the final two phases of the life cycle—installation and review.

In other words, this is, in large part, a book about the transition from user specifications developed during systems analysis into the design of a new system that implements those specifications.

DESIGN INPUTS

Two major outputs of the analysis and general design phase become inputs to detailed systems design:

- User specification for the new system
- New system design specification.

User Specification

The *user specification* describes and documents all of the logical processing functions for the new system. Included are one or more physical models—data flow diagrams that represent the system to be developed from the user's perspective. These models encompass the human-machine boundaries at which people interact with computerized portions of the system. These boundaries denote both the batch and on-line portions of the system as well as the business cycles for batch processing. Also included in the user design specification are rough layout forms for the inputs and end products that users anticipate.

New System Design Specification

The *new system design specification* is, in effect, the cumulative documentation delivered as an end result of the work in the analysis and general design phase. This product is simply an extension of the user specification. Sufficient general design work is included to update and improve the accuracy of the initial feasibility evaluation of the system. At this point, for example, additional data on facilities, equipment, and computer processing capabilities are needed so that technical requirements and developmental costs can be estimated more closely than was possible previously.

The material added to the user specification to arrive at the new system design specification represents the beginning of systems design efforts. To illustrate the relationship between these two documents, the portions of the descriptions below in *italics* indicate content revisions or substantial additions. The documentation of the new system design specification includes:

- Overview narrative. This document is a project rationale and impact statement that addresses three areas. The first covers the goals and objectives of the organization and provides a basis against which system requests are evaluated. The second describes the system's purpose, goals, and objectives, as well as the basic, logical functions that the system must perform. The third is an overview statement of changes to be made between the existing system and the new one.

- System function. A system function, written primarily for the user's benefit, is a concise—but processing-free—description of what the system will accomplish. It is, in effect, a black-box description of the computer portion of the system.

- Processing. Processing descriptions include a context diagram and a hierarchical set of data flow diagrams. Diagram 0 should identify the major subsystems. Lower-level diagrams should show physical packaging from the user's perspective. Differentiations should exist among manual and computer processing, batch and on-line processing, timing cycles, and performance requirements. *Computer processing should be defined to the job-stream level and should be documented using systems flowcharts.* (A *job stream* is a sequence of programs, or *job steps,* comprising a single processing job.) *The distribution versus centralization of computer processing must be documented. In the case of a distributed network, data communication design must be included.*

- Data dictionary. This document supports the data flow diagrams.

- Outputs to the user. This section consists of an index sheet listing all outputs. The index sheet is followed by an output documentation sheet and a rough format for each output. *Both user-oriented outputs and outputs to support security functions are included.*

- Inputs to the system. This section consists of an index sheet listing all inputs. The index sheet is followed by an input documentation sheet and a rough format for each input. *As with outputs, security- and control-related inputs also are included.*

- User interface with the system. Routine aspects of how user personnel work within the system and how user personnel interface with the computerized portions of the system are contained in the definition of manual processes. This section may include rough outlines of human-machine interactive conversations. Also included are explanations of the impact upon job descriptions and the number of positions in the user area. *User responsibilities as they relate to security and control processing are added.*

- *Data files. This document describes requirements in terms of files, rather than data stores, highlighting the transition from analysis to general design. File access methods and storage media are specified, along with approximate quantity of stored data and anticipated growth.*

- *Performance criteria. These descriptions of expectations from the new system are critical for both computer and manual processing. Included are required response times, anticipated volumes of transactions, and other performance data.*

- *Security and control. Measures for security and control are discussed as they apply to hardware, computer processing, and manual processing.*

- Policy considerations. This section lists any relevant policy decisions that have not yet been made.

- *Computer operations interfaces with the new system. These specifications are still at a general level and are not detailed operating instructions. However, descriptions should include hardware, data communication requirements, timing, projected volumes, impact on existing operations, backup and record retention requirements, and recovery procedures.*

In addition, there are two possible, or optional, items of documentation that may be appended to the new system design specification:

- The project team may recommend use of packaged application software. If so, the new system design specification will include a description of the software package and recommendations for its adaptation and use.

- If the proposed system involves significant hardware and/or systems software changes, technical specifications will be prepared and incorporated in the documentation delivered at the end of the analysis and general design phase.

Examples of Design Inputs

To visualize new system design specification documents in their role as working tools for systems design, consider the example of the water billing system that is the focal point of discussion in the companion text. The situation, based on an actual systems development project, involves a request handled by the computer information systems department at Central City, a community with a population of approximately 75,000.

Central City operates its own water utility that serves approximately 20,000 homes and businesses. Billing is semi-monthly, with some 25 percent of customers receiving bills on a cycled basis every two weeks. Thus, each cycle involves the processing of approximately 5,000 bills.

The request for systems development services, in this instance, is initiated by the Midstate Sanitary District, which provides sewage services for Central City and some of its surrounding communities. In the past, the sanitary district has had no billing problem. The district had the power simply to tax its customers on the basis of property values. However, a new ruling by a federal agency has disallowed this practice. The sanitary district, along with other sanitary districts across the country, has been told that charges to customers must be based upon the amount of service used. A ruling has come down that water usage can be equated to sanitation services, since the two services are, of necessity, proportionate. Thus, the sanitary district has approached the city with a request that its new billing system be added to the existing water billing system.

A subsequent feasibility study established that it was not possible simply to add sanitary district billing to the existing water billing system. The water billing system had been computerized for some time. However, there was not enough capacity in the existing system, in terms of both file size and processing time, to handle the expanded work load that would result. Thus, it was determined that, if the two billing jobs were to be combined, a new system would have to be developed.

A context diagram for this system is presented in Figure 1-6. This context diagram establishes the scope of the system and provides a basis for additional data flow analysis and for the presentation of processing specifications that come out of the analysis and general design phase.

For example, Figure 1-7 is a Diagram 0 for the water billing system. In a single document, this diagram presents an overview of the entire system. All processing requirements are summarized through the identification of 10 processes in this data flow diagram and the associated data flow arrows, data stores, and external entities that interact with the processing functions.

A Diagram 0 like the one in Figure 1-7 serves as the basis for additional analysis and detailing of the data flows, processing, files, and human-machine interactions required to implement the new system. That is, each of the bubbles in a Diagram 0 can be broken down and represented in additional levels of data flow diagrams. To illustrate, consider just two of the bubbles shown in Figure 1-7. Process 3, APPLY NEW READINGS, accepts reading data, updates the customer master file, and triggers special bills. In any processing function with this level of complexity, you would know, immediately, that additional levels of detailing would be required. The same is true for the processing represented by bubble 5, PREPARE BILL.

To illustrate how high-level data flow diagrams are *partitioned* or *decomposed* as part of the modeling process, consider Diagram 5 in the water billing system, illustrated in Figure 1-8. This data flow diagram represents a further detailing of bubble 5 in the Diagram 0. Figure 1-8 breaks down the PREPARE BILL processing function to show a partitioned version of what happens when bills are processed. Note that separate procedures are identified for producing normal, incycle bills and for special bills that fall out of the regular cycles for specific customers.

In this example, Diagram 5 is still at too high a level to serve as a basis for the technical design of the new system. Therefore, the process is partitioned even further, as shown in Diagram 5.2 in Figure 1-9. This diagram includes separate bubbles for all of the processing

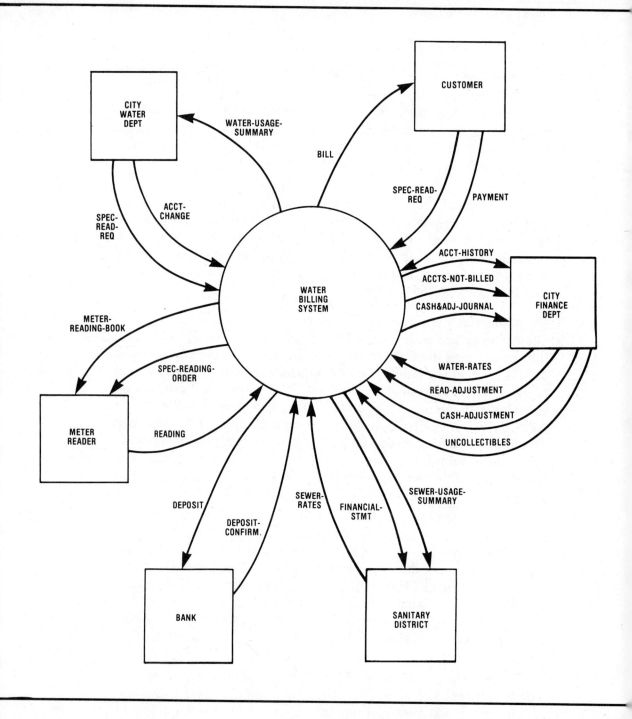

Figure 1-6. Context diagram for a new Central City water billing system.

DIAGRAM 0

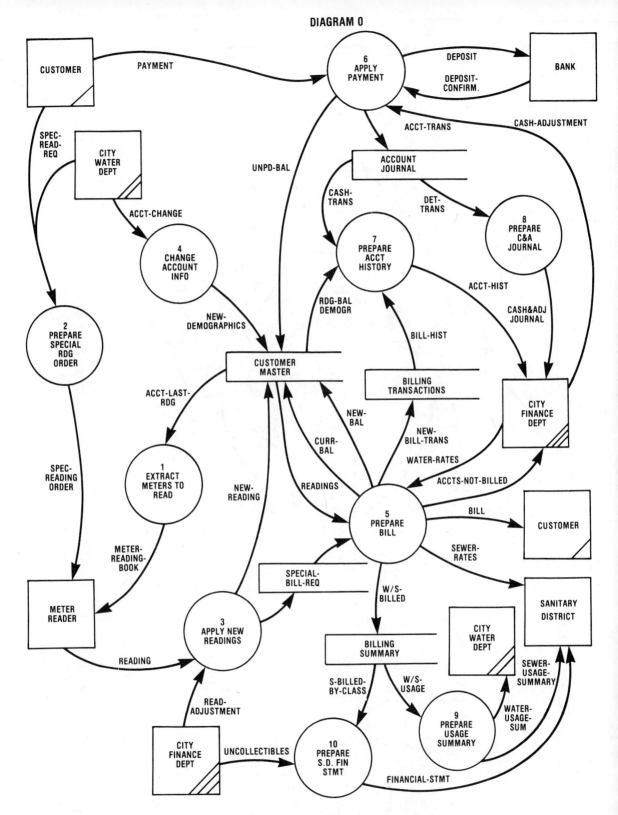

Figure 1-7. Diagram 0 for a new Central City water billing system.

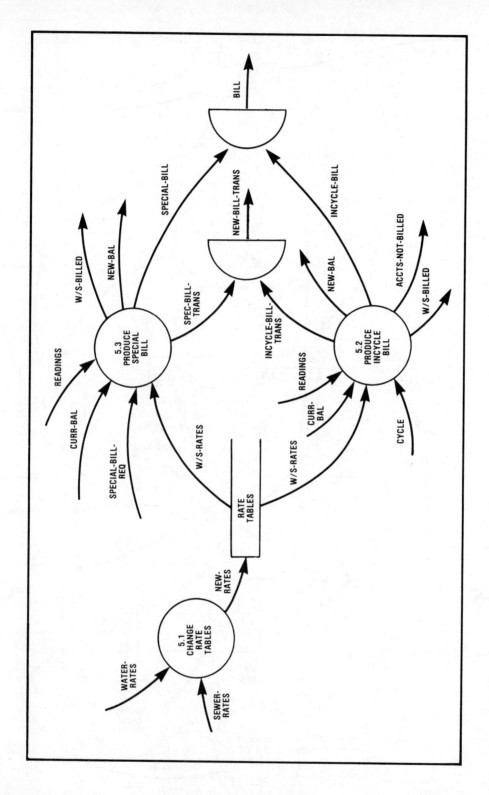

Figure 1-8. Diagram 5—PREPARE BILL—corresponding with the Diagram 0 in Figure 1-7.

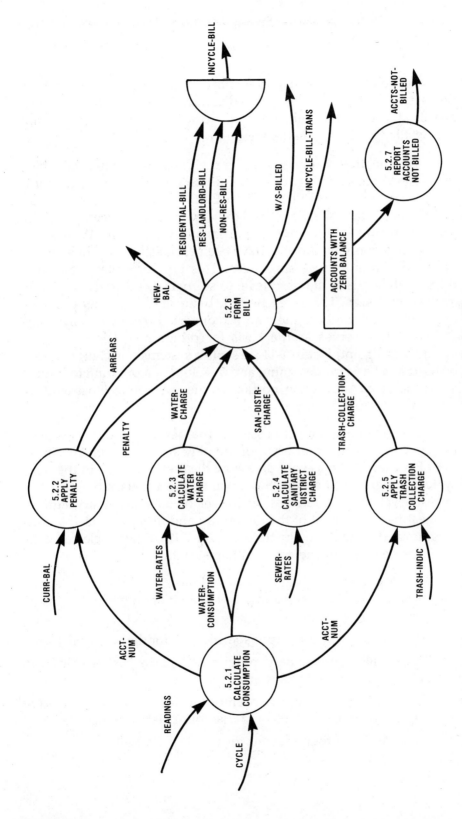

Figure 1-9. Diagram 5.2—PRODUCE INCYCLE BILL—corresponding with the Diagram 5 in Figure 1-8.

steps needed to calculate components of a finished water-sanitation bill. At this level, the data flow diagram and supporting process specifications are detailed sufficiently to serve as a suitable input for the detailed design of the new system.

As still another example, consider the Diagram 3 for the water billing system, shown in Figure 1-10. This data flow diagram partitions the process for the APPLY NEW READINGS function shown in Diagram 0 (Figure 1-7). Note that boundary lines have been drawn to separate manual and computer portions of the system. Within the computer portion, boundaries have been established to delineate the batch and on-line processing areas. This type of systems analysis documentation is at the right level to support direct transition from analysis to design. That is, each of the delineated computer processing areas, batch and on-line, can be converted readily to a systems flowchart that serves as the basis for program development. To illustrate this point, Figure 1-11 presents a systems flowchart that can be used as a basis for designing and developing a program to implement the batch processing portion of the subsystem illustrated in Diagram 3 (Figure 1-10).

Supplementary input to assist in the detailed design of a new system comes from other areas of the new system design specification. For example, for every data flow diagram included in the new system design specification, there will be process narratives for the lowest level process bubbles and a set of data dictionary entries delineating the content of the data flows and files to be used by the diagrammed system. To illustrate, Figure 1-12 contains selected data dictionary entries for the water billing system.

Also important as inputs to detailed design are the layouts and specifications for the outputs and inputs that are devised as part of the user specification. To illustrate:

- Figure 1-13 shows an output specification for the water bill form.
- Figure 1-14 is a preliminary, rough sketch of the postcard bill form to be produced by the system.
- Figure 1-15 is an output specification for the terminal screen display to be made available for account status references.
- Figure 1-16 is a rough layout for the display itself.

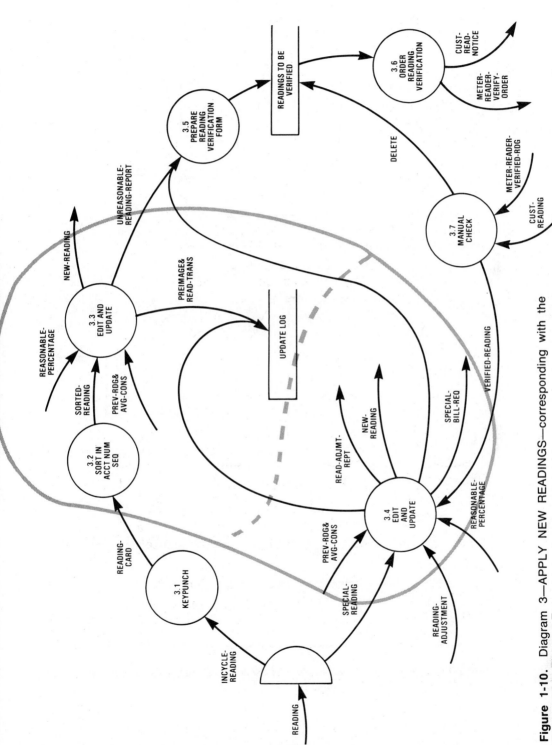

Figure 1-10. Diagram 3—APPLY NEW READINGS—corresponding with the Diagram 0 in Figure 1-7.

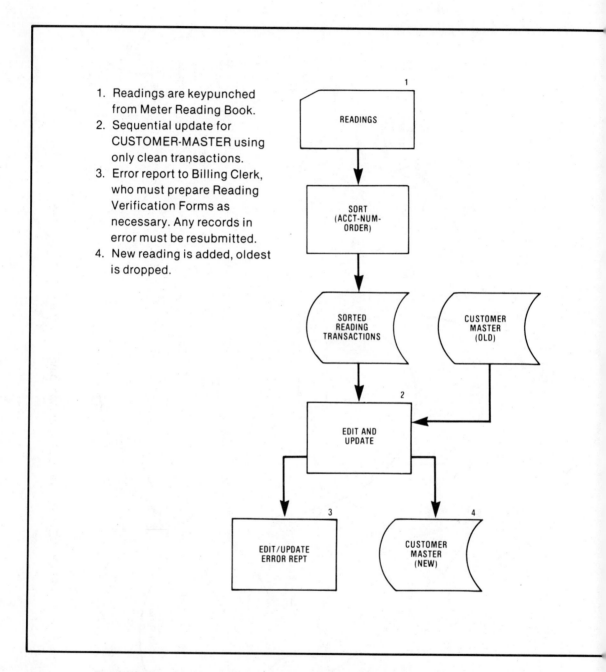

1. Readings are keypunched from Meter Reading Book.
2. Sequential update for CUSTOMER-MASTER using only clean transactions.
3. Error report to Billing Clerk, who must prepare Reading Verification Forms as necessary. Any records in error must be resubmitted.
4. New reading is added, oldest is dropped.

Figure 1-11.   Systems flowchart for APPLY INCYCLE READINGS job stream.

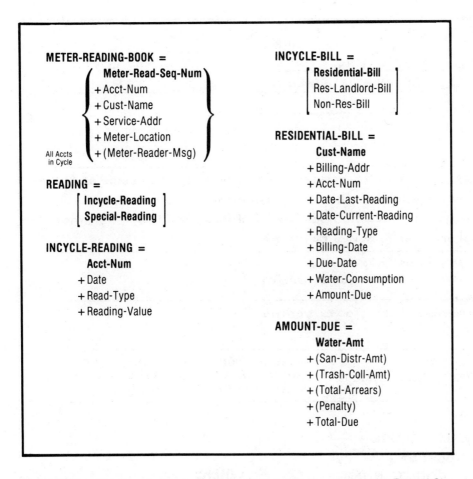

Figure 1-12. Notations for the content of selected data flows of the new Central City water billing system.

OUTPUT

System: **WATER BILLING**   Date prepared: **NOV 15, 1983**

Output name: **RESIDENTIAL-BILLING**   Prepared by: **JRP**

Output number/id: _____

Design documentation attached: **X** Rough format   ____ Detailed layout

Output media: **X** Printed Report   ____ Screen Display

PURPOSE/USE: Sent to residential customers who reside at the service address. (Bills for residential service sent to landlords -- see RESIDENTIAL-LANDLORD-BILL). Customer returns stub with payment.

ORGANIZATION (order, level of detail, totals):
Mail route order

DISTRIBUTION/ACCESS: To customer

SIZE/VOLUME:
5000-7000/cycle

FREQUENCY: Four cycles - billing run every two weeks. (Individual customer receives bill bimonthly)

DATA CONTENT:

CUST-NAME
+ ADDRESS
+ ACCOUNT-NUM
+ DATE-LAST-READING
+ DATE-CURR-READING
+ READING-TYPE
+ BILLING-DATE
+ DUE-DATE
+ WATER-CONSUMPTION
+ AMOUNT-DUE

WHERE

AMOUNT DUE =

    WATER-AMT
+ (SAN-DISTR-AMT)
+ (TRASH-COLL-AMT)
+ (ARREARS)
+ (PENALTY)
+ TOTAL-DUE

Figure 1-13.   Form specifying residential bill output.

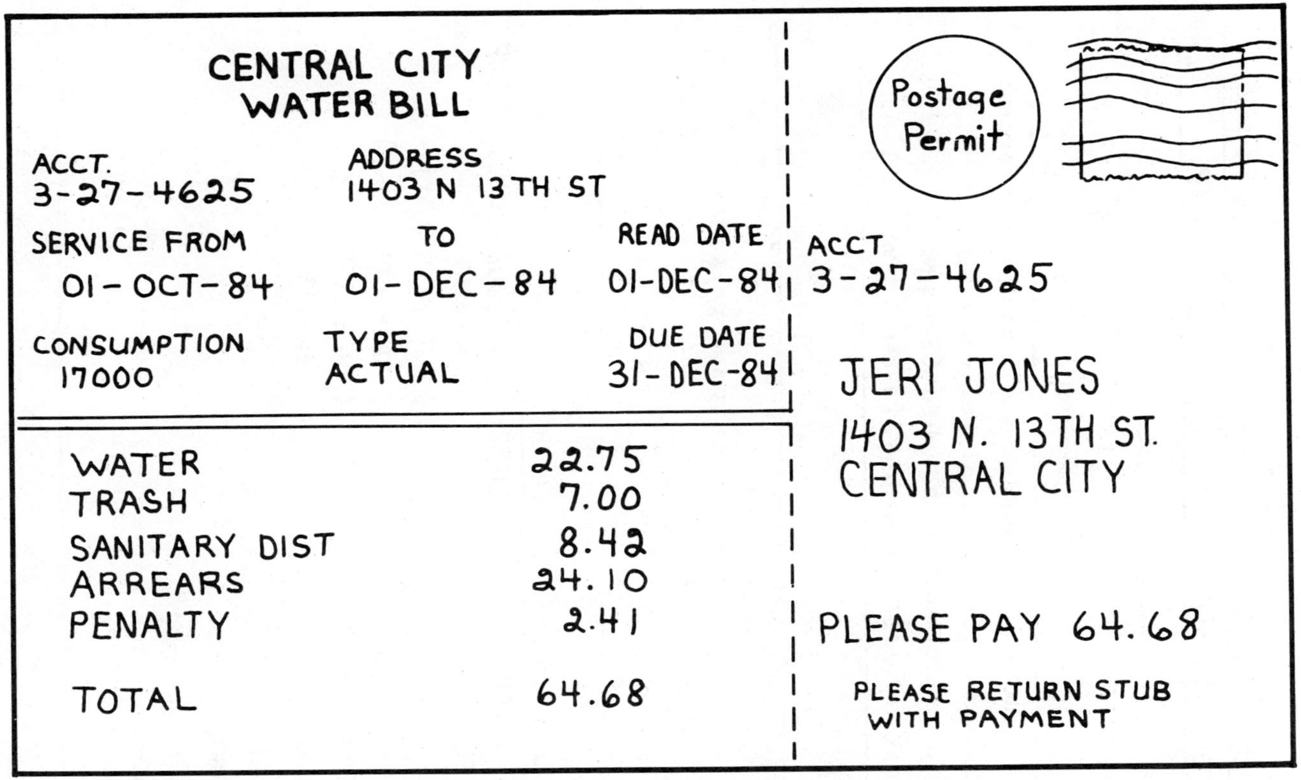

Figure 1-14. Rough format for residential bill.

OUTPUT

System: __Water Billing__   Date prepared: __Nov 15, 1983__

Output name: __ACCT-HISTORY__   Prepared by: __JRP__

Output number/id: _____

Design documentation attached: __X__ Rough format   _____ Detailed layout

Output media: _____ Printed Report   __X__ Screen Display

PURPOSE/USE: This is an on-line query capability—to access the transactions on an individual account. ACCOUNT—NUM must be supplied. (Reading/Cash adjustments are included as Reading/Payment transactions.

ORGANIZATION (order, level of detail, totals):
   N/A

DISTRIBUTION/ACCESS:
   Restricted to city finance dept. personnel

SIZE/VOLUME: N/A   FREQUENCY: N/A

DATA CONTENT:

```
   ACCOUNT-NUM              WHERE
+  CUST-NAME                CURR-AMOUNT-DUE = AMOUNT-DUE
+  ADDRESS                     = WATER-AMT
+  CURR-AMOUNT-DUE              + (SAN-DISTR-AMT)
     ⎰ READING-DATE  ⎱         + (TRASH-COLL-AMT)
+    ⎨ + READING-TYPE ⎬        + (ARREARS)
   6 ⎱ + READING-VALUE⎰        + (PENALTY)
     ⎰ BILLING-DATE  ⎱         + TOTAL-DUE
+    ⎨              ⎬
   6 ⎱ + AMOUNT-DUE ⎰
     ⎰ PAYMENT-DATE  ⎱
+    ⎨ + PAYMENT-AMT ⎬
   6 ⎱ + PAYMENT-TYPE⎰
```

Figure 1-15. Output form for account history screen.

INDIVIDUAL ACCOUNT HISTORY

ACCT: 3-27-4625   ADDR: 1403 N 13TH ST

NAME: JERI JONES   CENTRAL CITY

CURRENT DUE: $64.68

READINGS

DATE	TYPE	VALUE	DATE	TYPE	VALUE
01-APR-84	ACT	8044573	01-JUN-84	EST	8044597
01-AUG-84	ACT	8044650	15-SEP-84	ADJ	-33
01-OCT-84	EST	8044712	01-DEC-84	ACT	8044729

BILLINGS

DATE	WATER	SANITARY	TRASH	ARREARS	PEN.	TOTAL
10-FEB-84	10.26	2.44	7.00	0	0	19.70
11-APR-84	14.50	3.20	12.00	0	2.50	32.20
10-JUN-84	35.42	12.85	7.00	0	0	55.27
9-AUG-84	42.25	15.60	7.00	0	0	64.85
10-OCT-84	14.05	3.05	7.00	0	0	24.10
10-DEC-84	22.75	8.42	7.00	24.10	2.41	64.68

PAYMENTS

DATE	TYPE	AMT	DATE	TYPE	AMT
31-DEC-83	P	24.83	1-MAR-84	P	19.70
27-APR-84	P	29.70	17-MAY-84	ADJ	+ 2.50
25-JUN-84	P	55.27	30-AUG-84	P	64.85

Figure 1-16. Rough format for account history output screen.

As a final example of analysis documentation that serves as design input, consider Figure 1-17. This contains descriptions of all of the records to be used in the files that will support the water billing system. This type of notation represents an ideal level of documentation for input to detailed design.

Logical data analysis inputs.   The data flow documentation described above is effective in expressing systems analyses and new system designs in terms that are comprehensible to both users and systems analysts. Basically, the tools are graphic and provide a model representation of both existing and proposed new systems. However, data flow diagrams tend to be insufficiently specific in representation of data files or data-based elements that will be required to support the new system. Rather, data flow diagrams concentrate upon data transformations. Part of the design principle of data flow diagram preparation lies in "starving" a process by providing only the data needed to support each individual function. At some point before the detailed design of a system begins, therefore, it is necessary to provide a coherent, complete picture of what the files or data-based elements will look like. In other words, a *logical data structure* must be developed to model, or represent, the files to be used by the new system. Two criteria are applied in the development of logical data structures:

- *Simplicity* is an important aspect of logical data structures. Simplicity, in this regard, implies that all components of data stores be presented as fixed-length records and that these records be accessed only by primary keys.

- The components of data stores that support systems should be *nonredundant*. That is, to the extent possible, a given data element or attribute should appear only once within a file structure. This requirement applies to the total collection of files that model the data structure of an organization. Implied further is the principle that any value or attribute that can be derived from existing data elements is not retained in the file but is derived as needed.

These criteria are met through a process known as *normalization* of data stores. Normalization, in turn, is carried out in three steps, or stages, as illustrated through an example provided in Figure 1-18. For the purposes of illustration, this figure demonstrates normalization of a single

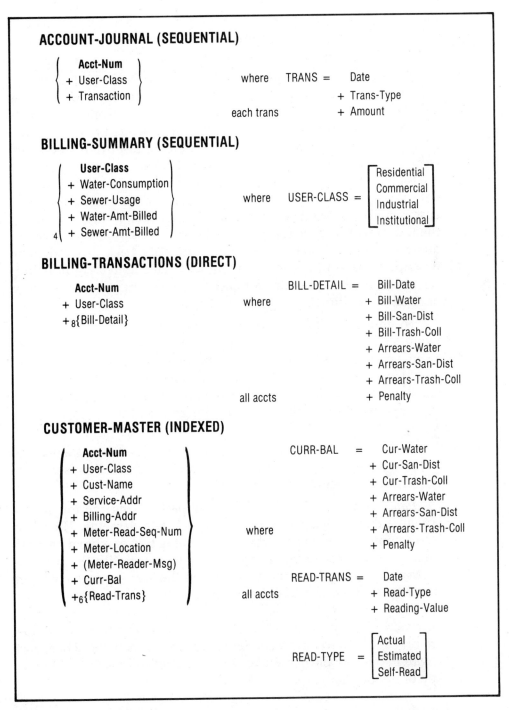

Figure 1-17. Definitions for the master files and access methods for the new Central City water billing system.

file. In practice, all of the files that comprise the data resources of an organization would be scrutinized and processed in the manner shown in this simplified example.

The listing of data elements in the topmost group of Figure 1-18 represents input. This list of all attributes within a record serves as a basis for an ORDER-FILE. The order record for any given business would list all of the fields, or data elements, needed to fill an order from a customer, withdraw required merchandise from stock, and issue an invoice to the customer. The fields shown within braces in the topmost group of items are iterated. That is, this entire group of data fields, or elements, is repeated for each item ordered by a customer. The ordered item is known as a *line-item*. For each line-item, then, data required include the listed items, including item number, quantity ordered, quantity shipped, quantity back ordered, item description, unit price, and item amount. This item amount field is calculated by the computer for each line-item entry processed. It is derived by multiplying quantity ordered by unit price.

To implement the structure shown at the top of Figure 1-18, it would be necessary to create a file with variable-length records—an arrangement that would be hard to manage. Variable-length records would be needed to accommodate repetition of *segments* of multiple line-items for individual orders. Remember, to achieve simplicity, file structure should be built from fixed-length records.

The three-step process for normalization involves deriving a *first normal form* from the original record format. This derivation is done in the leftmost column in Figure 1-18.

First normal form is developed by partitioning the initial data structure that contains repeating groups of data elements. The object is to accomplish the same purpose by forming two or more data structures that do not have repeating groups. To accomplish this, the first normal form listing in Figure 1-18 removes the iterating segments from the order record and establishes them as a separate file. In doing this, the key of the original file is *concatenated,* or added to, the key for the newly created file. In the example, order number and item number are concatenated to form a combined key for the new file. Thus, each record in the new file has a unique key that is tied back to the primary key of the original file.

ORDER-FILE = {Order-Record}
Order-Record = Order-No. +
Order-Date +
Account-No. +
{ Item-No. +
Qty-Ordered +
Qty-Shipped +
Qty-Backordered +
Item-Description +
Unit-Price +
Item-Amount }

ORDER-FILE = {Order-Record}
Order-Record = Order-No. +
Order-Date +
Account-Number

ORDER-ITEM-FILE = {Order-Item-Record}
Order-Item-Record = Order-No. + Item-No. +
Qty-Ordered +
Qty-Shipped +
Qty-Backordered +
Item-Description +
Unit-Price +
Item-Amount

A. FIRST NORMAL FORM

ORDER-FILE = {Order-Record}
Order-Record = Order-No. +
Order-Date +
Account-No.

ORDER-ITEM-FILE = {Order-Item-Record}
Order-Item-Record = Order-No. + Item-No. +
Qty-Ordered +
Qty-Shipped +
Qty-Backordered +
Item-Amount

ITEM-FILE = {Item-Record}
Item-Record = Item-No. +
Item-Description +
Unit-Price

B. SECOND NORMAL FORM

ORDER-FILE = {Order-Record}
Order-Record = Order-No. +
Order-Date +
Account-No.

ORDER-ITEM-FILE = {Order-Item-Record}
Order-Item-Record = Order-No. + Item-No. +
Qty-Ordered +
Qty-Shipped

ITEM-FILE = {Item-Record}
Item-Record = Item-No. +
Item-Description +
Unit-Price

C. THIRD NORMAL FORM

Figure 1-18. Normalization of data stores.

The second step in normalization is to focus upon and analyze the newly created files that have concatenated keys. The purpose of deriving the *second normal form* is to make sure that each non-key data element within a structure is functionally dependent upon the fully concatenated key. In this example, a separate file is constructed to encompass a series of records based upon item number. The reason is that these data elements are independent of the specific order being processed. These fields, or attributes, can be identified through reference to only a part of the concatenated key. For instance, the order number is not needed to access item number and description data, nor is order number needed to determine the price for a specific item. Rather, the item description and unit price fields are defined by the item number alone. Therefore, these fields are broken out into a separate file.

As a final step in normalization, a search is made for redundancies among identified fields. These redundancies then can be eliminated as the step toward meeting one of the criteria for establishing efficient file structures. This elimination of redundancies puts the file into *third normal form*.

In Figure 1-18, note that the fields for QTY-BACKORDERED and ITEM-AMOUNT have been eliminated. Data in these fields can be derived from other fields that already exist within each line-item record. For example, quantity back ordered can be derived by subtracting the quantity shipped from the quantity ordered. Similarly, the item amount field can be derived by multiplying the quantity ordered field by the unit price.

In third normal form, then, the content of the original file has been both simplified and rendered nonredundant. This process serves to establish clear-cut access patterns to any of the data within files about inventory items or orders, as shown in Figure 1-19. This file shows a clearly defined access path among three simple files. These files make it possible to access data elements according to their content and function, without having to search through large, complex records containing iterative record segments and redundant fields.

Once a normalized file structure has been developed, some fine tuning may be needed to adjust to the practicalities of the processing for a

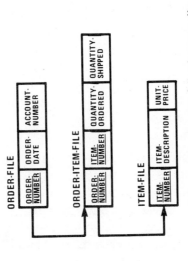

ORDER-FILE

| ORDER-NUMBER | ORDER-DATE | ACCOUNT-NUMBER | ITEM-NUMBER | QUANTITY-ORDERED | QUANTITY-SHIPPED | QUANTITY-BACKORDERED | ITEM-DESCRIPTION | UNIT-PRICE | ITEM-AMOUNT | ITEM-NUMBER | . . . | ITEM-NUMBER | . . . |

A. Logical file structure prior to normalization.

ORDER-FILE

| ORDER-NUMBER | ORDER-DATE | ACCOUNT-NUMBER |

ORDER-ITEM-FILE

| ORDER-NUMBER | ITEM-NUMBER | QUANTITY-ORDERED | QUANTITY-SHIPPED |

ITEM-FILE

| ITEM-NUMBER | ITEM-DESCRIPTION | UNIT-PRICE |

B. Logical file structure following normalization.

Figure 1-19. Data access diagrams describing file structures before and after normalization.

given application. For example, rather than computing back-ordered amounts each time a reference is made, a company experiencing a high level of such inquiries might prefer to carry this item in the file, even though it is redundant.

Once a normalized, optimized file structure has been established, the data flow diagram for the new logical system that has been approved for development would be mapped against this data structure. The documentation that results then would become the major input to the detailed technical design for the new system.

TRANSITION INTO DESIGN

Nominally, the second phase of the systems development life cycle is dedicated to systems analysis activities in which users and analysts interact closely. According to formal definitions, the activities of the analysis and general design phase are nontechnical. Nominally also, activities become highly technical during the detailed design and implementation phase of the life cycle—the point at which technical designers and programmers begin to put together the computer portions of the new system.

In practice, the boundary is not all that firm. Unavoidably, involvement in detailed design must begin in the second phase of the life cycle. Also unavoidably, any design effort that takes place during analysis and general design should be reviewed and, quite probably, revised at the beginning of the third phase.

As explained above, some design effort is necessary in the analysis activities as a basis for reappraisal of the feasibility estimates for the project. Thus, although formal descriptions indicate that design takes place in the third phase, the truth is that a number of transitional efforts cross the boundaries between the analysis and detailed design phases of a project. The types of design efforts begun in systems analysis can include:

- Controls are added to the system models. In this sense, controls encompass all measures or processing functions established specifically to assure accuracy, completeness, reliability, and quality of results from a system. A systems analyst is, under this approach, responsible for assuring that all needed controls will be built into the system, particularly at interface points between human and computer operations.

- The *data base* for the new system, including definition of the fields and records to support the application, is documented during the final stages of systems analysis, as illustrated in Figure 1-17. This documentation, in turn, serves as the basis for building master files and defining transaction files to be incorporated in the new system.

- The model for the new system is evaluated to assure that the design can be implemented on the current or proposed hardware configuration. The transition, in this area, forms a bridge between the physical model of the new system with the technical specifications for the system and the programs that will implement it. In part, these activities involve designing hardware configurations, data communication networks, and computer jobs and job streams, and deriving systems flowcharts using the data flow diagrams produced during systems analysis.

- If application software packages are to be used, the general descriptions prepared during systems analysis are enhanced with additional technical details. This is necessary to determine the level of modification and/or additional documentation needed to implement the application package in the specific installation in which it will run.

Note that the reference to the data base above has two separate words. This term is used throughout this book to refer to all of the data and information resources of an organization, collectively. On the other hand, the term *database* is used to refer to a specific, structured data organization scheme that serves as an underpinning for an extensive series of applications. The organization of and access to records within a database is controlled through specialized software known as a *database management system (DBMS)*. If the computer installation on which the new system will be implemented uses database management software, the transition between analysis and design will include activities aimed at verifying that the data structures established for the new system are compatible with database requirements.

The extent of transition activities from analysis into design varies widely with individual projects. The point is that, for systems and software design to take place, certain types of documentation are required. To the extent that this documentation is prepared as a by-product of analysis, the amount of work needed to initiate technical and detailed design is diminished. Conversely, the lower the level of design content in analysis activities, the greater the effort will be at the startup of detailed design. At best, this transitional stage is a variable that has to be evaluated in each individual case, with specific activities and levels of effort adjusted accordingly.

Illustration of Transition into Design

To show the kinds of thinking and results that are produced during the transitional stages from analysis to design, consider the series of three illustrations beginning with Figure 1-20. Together, these illustrations form an example of a data flow diagram for a hypothetical new system. For the purposes of illustration, the processes and other elements of this data flow diagram have been left unlabeled.

Design efforts applied to a diagram like the one in Figure 1-20 are directed toward identifying the portions of the system that will be computerized. Then, for the computerized portions of the system, timing considerations are established. That is, the procedures to be performed daily, weekly, and monthly are delineated. After that, boundaries are established to determine the on-line and batch areas of the computer processing system. All of these breakdowns are illustrated in the annotations that have been added to the basic data flow diagram in Figure 1-21.

Continuing the transition from analysis into design, discrete portions of the subdivided data flow diagram then are used as a basis for designing individual job streams and documenting job streams using systems flowcharts. This documentation, then, will serve as the basis for program design and development. Figure 1-22, depicting the weekly processing cycle in Figure 1-21, illustrates this transition.

Note that the weekly batch processing procedures in the data flow diagram have been divided into three processing areas. In one instance, two processing bubbles have been included within a single functional area of the data flow diagram. These areas of the data flow diagram, in turn, have been translated directly into job steps within the systems flowchart. To restate the definition, a job step is one of two or more programs and data to be executed in sequence. In an actual situation, this identification of individual job steps is a critical design activity. Guidelines for packaging job steps are covered in Chapters 2 and 4.

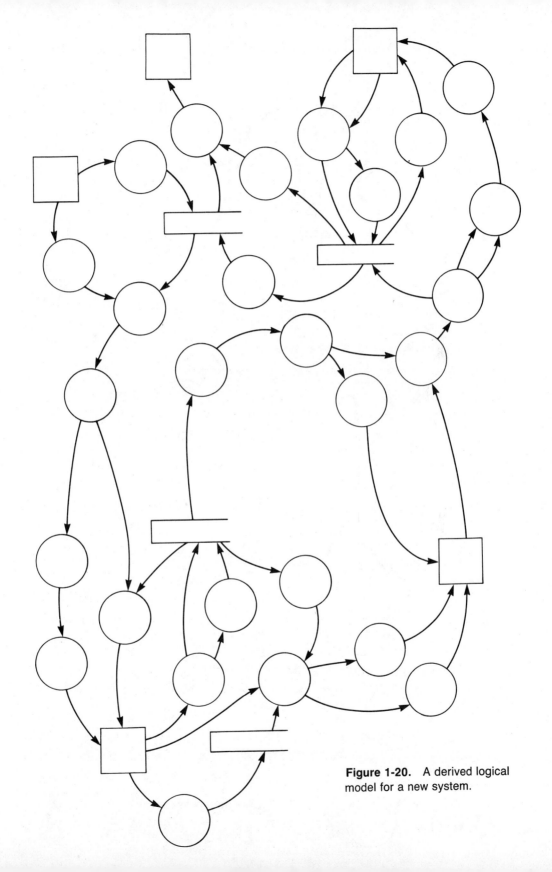

Figure 1-20. A derived logical model for a new system.

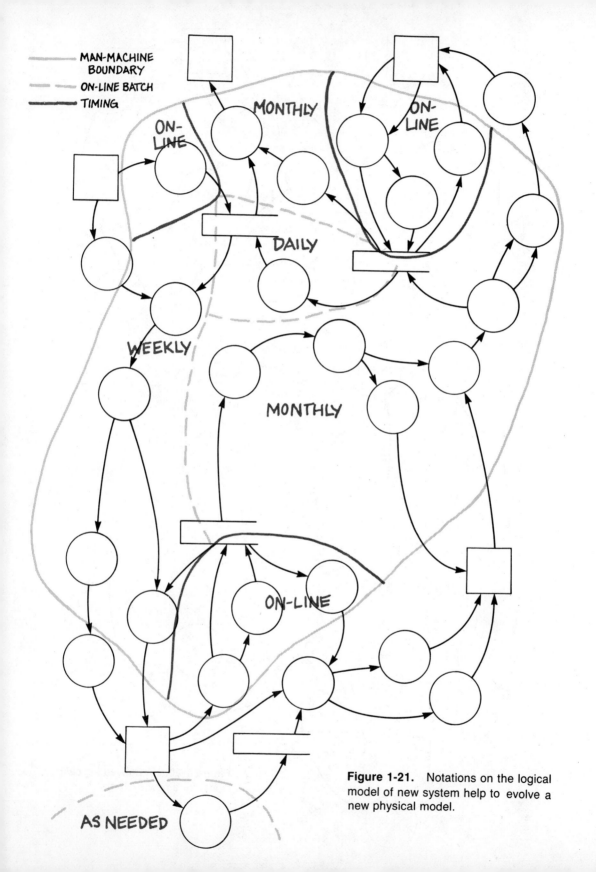

MAN-MACHINE BOUNDARY
ON-LINE BATCH
TIMING

MONTHLY

ON-LINE

ON-LINE

DAILY

WEEKLY

MONTHLY

ON-LINE

AS NEEDED

Figure 1-21. Notations on the logical model of new system help to evolve a new physical model.

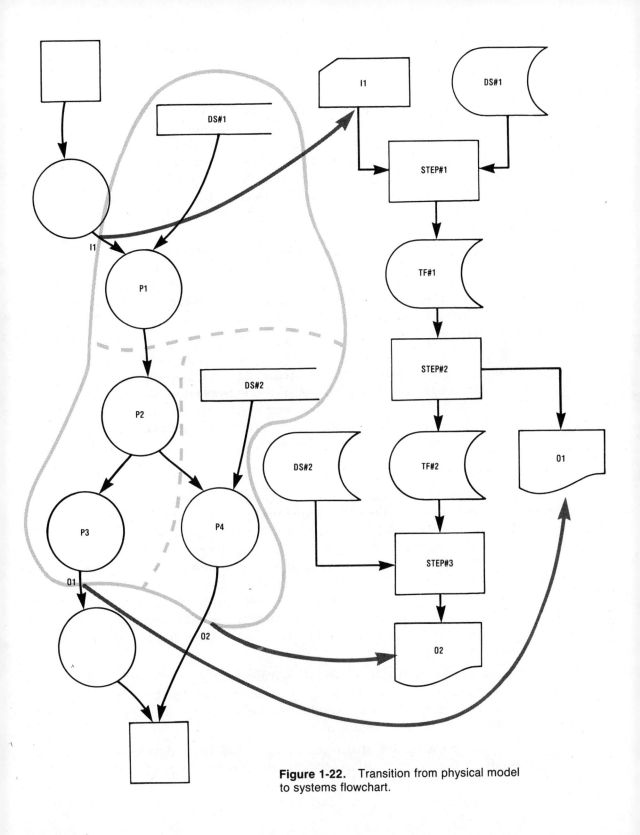

Figure 1-22. Transition from physical model to systems flowchart.

Another feature of the transition can be seen in considering what happens at junctures between human and machine processing. At each point of connection between procedures handled by people and those processed on computers, either an input or an output is generated. Thus, the segment of the data flow diagram in Figure 1-22 shows three points at which data flows cross the human-machine boundary. These are represented, in turn, by one input and two outputs.

Finally, note that, each time a data flow crosses an internal boundary within the computerized portion of the system, temporary files are created to communicate data from one job step to another. The example in Figure 1-22 has two data flows that cross processing boundaries. These data flows are represented in the systems flowchart by two temporary files.

Figure 1-22 is, of course, a simplified example developed specifically to illustrate what happens. Actual design procedures will, of necessity, be more complex. However, this example does illustrate the point that detailed design of a computer system begins with a transition from the products of analysis into documents that form the foundation for the design of the new system.

LEVELS OF DESIGN

The design of computer information systems, in itself, is a process with enough complexities to require that a perspective be established. The remainder of this book deals primarily with events and techniques for the development of designs for the computer processing portion of information systems. The general concept followed throughout a systems development life cycle is that large, complex jobs are subdivided into manageable segments and that a project is completed by following a series of steps moving to successive levels of detail. These principles hold true in the design area just as they do for systems analysis.

General design and detailed design also represent separate areas of discussion within this book. Systems design and data base design are covered in the remaining chapters of the first part of the book. Part II deals with detailed technical design.

This presentation structure, in part, has been developed to establish the needed differentiation between the areas of systems design and software design. In general, students and practitioners alike have little difficulty in understanding the differences in levels between the design of overall systems and designs for inputs, outputs, or data communications networks. These latter areas of design are clearly more detailed and technical. However, there seems to be a general temptation to think of software design as being akin to systems design. This is not the case. Software design is, in common with the other areas of detailed design, an outgrowth of systems design. Software design exists at the same level within a systems development effort as the other detailed design functions. The work performed and the results delivered differ for systems design and software design.

Systems Design

By nature, systems design retains a flavor of user orientation. That is, there is the feeling, in designing the overall system for computer processing, that emphasis is upon the solving of a user-stated problem. In practice, there will be great variations among separate systems development projects in the extent or methodologies of systems design. There is a tendency to grope or probe for the best way of proceeding, a tendency resulting mainly from a current lack of established standards or methodologies. Standard approaches are difficult to develop because systems design is tied so closely to the application itself, and applications vary widely. However, general techniques for systems design are evolving, and standard approaches are likely to gain wide acceptance in the future.

Concerns associated with systems design include:

- Human-machine interfaces must be identified. A determination must be made about how much of a given system is to be computerized and how many functions are to be handled on a manual basis.

- Boundaries must be established for batch and on-line procedures.

- Understandings must be established and rules set down about the processing cycles and schedules for identified portions of the system under development.

- Basic decisions must be made about what portions of a given system ought to be centralized or distributed, both for data resources and for processing functions.

- Basic computer processing, at the level of job steps, must be designed.

Systems design results.   The results produced by systems design are at a relatively high level. These results include:

- Overall computer hardware configuration and network requirements.
- Job streams encompassing major programs and data files are identified.
- If appropriate, application software packages are selected and the needed interfaces between these packages and the other elements of the system are devised.
- Any additional hardware or systems software requirements needed to support the application under development are identified.

Timing of systems design.   On most projects, systems design is started during the analysis and general design phase. Therefore, systems design overlaps with systems analysis. This overlap is necessary because of the requirement to update the feasibility evaluation of a project before proceeding with the extensive (and expensive) activities of the detailed design and implementation phase.

Systems design also represents a starting point for the activities in the detailed design and implementation phase. Specifically, Activity 7: Technical Design, carried out at the beginning of the phase, starts with the review, updating, and completion of systems design specifications.

Software Design

As compared with systems design, software design is far more detailed. Software design is technically oriented rather than user-oriented. Where systems design remains, at this writing, relatively unstructured, software design tends to be consistent and predictable. Regardless of the application being developed, uniform methods and techniques can be applied in identifying and designing program modules. Thus, software or program designers can follow a predictable pattern while systems designers are, in comparison, groping through individual situations to develop a solution for each. This difference is highlighted by the fact that fairly reliable sets of standards exist not only for the design of programs but also for evaluating the quality of program design and for testing program reliability.

Results of software design.   Software design results in highly detailed designs ready for use in the coding of programs that will implement the new system. Standard tools, such as structure charts and pseudocoding, are readily usable in guiding the transition from systems flowcharts to detailed designs useful as a basis for program development.

Timing of software design.   Software design is held until the third phase of the systems development life cycle, detailed design and implementation. The work, done as part of Activity 7: Technical Design, follows the completion of system design specifications. It is necessary to have these *global*, or system-wide, design specifications in place before beginning software design. In other words, no design steps can be taken until systems specifications are made final.

Summary

The specifications for what a system is to do are the products of systems analysis. Systems design encompasses the formulation of the technical plans and methods to implement the specified system.

A standard method for dealing with complexities or uncertainties lies in breaking jobs down to a series of individually definable activities. Within a computer information system organization, it is desirable to have standard project structures so that systems development projects can be compared with one another and so that job assignments have similarities from one job to another. Most computer information systems organizations establish a standard project structure known as a systems development life cycle.

A systems development life cycle is a formalized description of a project structure for the development of computer information systems. Though the underlying principle and structure of such life cycles is basic, the specific tasks involved will vary greatly among organizations and among specific projects.

Though not universal, a typical systems development life cycle consists of standard phases that include: investigation, analysis and general design, detailed design and implementation, installation, and review.

Systems documentation techniques that are necessary in building understanding of a given system include data flow diagrams and systems flowcharts. A data flow diagram concentrates upon the flow of information through a system and the transformations that take place as the data and information are processed. A flowchart presents an overall picture of the application of controls to a sequence of events that runs from a known starting point to a projected conclusion.

Phases of the SDLC are broken down into smaller units of work known as activities. To break down activities even further, tasks are established that define assignments in increments of a few working days each.

The result of the investigation phase is a report indicating whether the systems request is feasible technically and economically and providing a recommendation on whether the system should be scheduled for development.

The end result of the analysis and general design phase of the life cycle is a set of specifications for a new system that has been accepted as meeting the needs of prospective users for that system. Analysis is completed at the conclusion of the second phase, as documentation delineates the performance of the new system in user terms. Technical, or detailed, design of the new system is set to begin.

Specifications and design documentation developed during the analysis and general design phase become the input to detailed design. At the conclusion of the analysis and general design phase, user management has seen and approved a model of the processing sequences of the new system, the outputs to be produced, and the inputs anticipated.

Included in the documentation that forms the input to technical design is logical data analysis. Files or data-based elements are modeled as a logical data structure. The criteria applied in the development of logical data structures include simplicity and nonredundancy. These criteria, in turn, are met through the process of normalization. Normalization results in data structures that make it possible to access data elements according to their content and function, without having to search through large, complex records containing iterative record segments and redundant fields.

The extent of transition activities and related documentation required from analysis into design varies widely with individual projects.

At a first level within the area of computer processing design, two broad functions are systems design, which is general, and detailed technical design, which has several subparts.

Key Terms

1. systems analysis
2. systems design
3. project
4. systems development life cycle (SDLC)
5. phase
6. activity
7. task
8. prototyping
9. version installation
10. detailed design and implementation phase
11. installation phase
12. review phase
13. user specification
14. new system design specification
15. job stream
16. job step
17. partition
18. decompose
19. logical data structure
20. simplicity
21. nonredundant
22. normalization
23. line-item
24. segment
25. first normal form
26. concatenate
27. second normal form
28. third normal form
29. data base
30. database
31. database management system (DBMS)
32. global

Review/Discussion Questions

1. What are the main differences between systems analysis and systems design?

2. What are two reasons for establishing a structured methodology for CIS development?

3. What is the systems development life cycle and what five phases may it include?

4. What is the goal of the investigation phase of the SDLC?

5. What activities make up the analysis and general design phase of the SDLC?

6. What two major outputs of the analysis and general design phase become inputs to systems design?

7. What constitutes the user specification for a new or revised system?

8. Of what elements is a new system design specification composed?

9. What documentation may be prepared for input to systems design?

10. What design efforts begun in systems analysis are continued in the transition from systems analysis to systems design?

11. How do systems design and detailed technical design phases differ in levels of design?

12. What are some overall characteristics of systems design?

13. What are some characteristics of software design and how does it differ from systems design?

14. What steps are taken to prepare a Diagram 0 from a context diagram for a proposed system?

Project Assignments

The Appendix presents three case studies that can be used in support of project assignments. All three scenarios provide functional system specifications that would result from systems analysis. Included are data flow diagrams, data dictionaries, and process narratives that would result from systems analysis. Assignments that relate to the content of this chapter will be found in *PART 1: Analysis and General Design Review.*

2 THE PROCESS OF SYSTEMS DESIGN

LEARNING OBJECTIVES

On completing reading and other learning assignments for this chapter, you should be able to:

☐ Describe the process and end products of systems design.

☐ Explain the role of systems flowcharts as systems design tools.

☐ Tell what special considerations are involved in the design of on-line systems.

☐ Tell what alternatives must be identified and reviewed during the systems design process.

☐ Describe the role of the systems designer in reviewing and challenging system specifications.

FOCUS OF SYSTEMS DESIGN

Systems design may be thought of as a sequence of stepping-stones. The foundation for systems design is built upon the outputs of systems analysis. Systems analysis produces a model for a new system expressed in terms of data flows, transformations, and stores. As part of the same set of documentation, systems analysis also produces a statement of the goals and objectives of the organization for which the system is being developed and, specifically, for the individual system itself.

As described in the previous chapter, physical characteristics for the new system—at least to a level of user interaction—are added to the specification late in the systems analysis process. All aspects of the system that are designated to be computerized are addressed during systems design and may be considered technically as a *black box*. This computerized system is defined primarily by the outputs produced and the inputs from which those outputs will be derived. The processing has been specified by the portion of the leveled set of data flow diagrams that have been packaged for automation. Supporting documentation includes:

- The data dictionary, which defines all data elements and structures contained in data flows or data stores.

- The process specifications, which provide the policies and rules governing the data transformation.

The new system specification tells the designer *what* the system should do. However, at the time systems design begins, little attention has been paid to *how* these specifications will be achieved. During analysis, the primary objectives were to strip away the physical characteristics of how the system's capabilities were delivered and to develop an understanding of the logical or essential aspects of the system. Systems design takes this logical statement of the new system and adds those physical features that make the system work.

More specifically, the purpose of systems design is to derive the overall computer systems architecture that defines both the technical and software environments. Decisions on computer hardware configurations and communication methods are made. Custom development of software and the availability of commercial packages are evaluated with respect to software needs. Processing is assigned to specific computer jobs, sequences of job steps, and programs. Data bases and files are defined based upon the data stores in the system model and upon job processing boundaries.

Systems design, then, produces a set of specifications identifying the technical environment in which computer programs will interact with input, storage, and output files, and in which sequence and at what times to perform system processing.

To accomplish these results, the systems designer follows a set of orderly procedures and applies a logical thought process to the transformation from general specifications to specific procedures. This chapter describes the process of systems design at an overview level.

Two succeeding chapters, then, deal with two segments of systems design—the technical environment and application design. The discussion is within a context, remember, that assigns to systems design the function of defining the context for detailed design. Detailed design is carried out in a series of separate efforts covered in later chapters of this book.

Systems design is an interim procedure because the full detailing of the design does not take place at this point. Rather, a single black box known as a system is broken down into a set of interrelated black boxes, or subsystems. One of the things accomplished in this breakdown of black boxes is that the overall system becomes more understandable. Further, each subsystem is easier to implement with detailed designs than a larger one; the larger system would require complex operations and programs that are difficult to manage. Thus, systems design produces the overall structure for how the system will work, in preparation for detailed design. The process subdivides large systems into a series of smaller units that can be dealt with at lower, more technical levels. A primary tool for documenting the results of systems design is the *systems flowchart*.

THE SYSTEMS FLOWCHART

A systems flowchart has a limited, defined role in the design process. Specifically, a systems flowchart is a way to represent a design of the computerized part of the system in terms of sequences of black-box processing steps and controls among those black boxes. A systems flowchart is an overview document. The detailing of what goes on inside the black boxes is done with the aid of software design tools that are identified and delineated further in several later chapters.

In addition to showing processing sequences, systems flowcharting techniques also are valuable as a means of representing the configurations that result from the design of the hardware and communication components of the system.

The Language of Flowcharts

Flowcharting serves as a vehicle of communication that can be used by anyone interested in a system. Flowcharting can serve as a universal language, largely because of its simplicity. The entire language of flowcharting can be expressed in a few basic symbols, as illustrated in Figure 2-1. As shown in this illustration, the flowcharting language

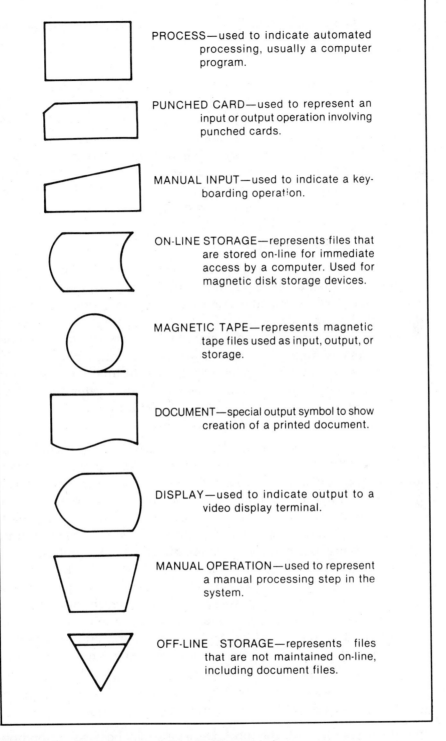

PROCESS—used to indicate automated processing, usually a computer program.

PUNCHED CARD—used to represent an input or output operation involving punched cards.

MANUAL INPUT—used to indicate a keyboarding operation.

ON-LINE STORAGE—represents files that are stored on-line for immediate access by a computer. Used for magnetic disk storage devices.

MAGNETIC TAPE—represents magnetic tape files used as input, output, or storage.

DOCUMENT—special output symbol to show creation of a printed document.

DISPLAY—used to indicate output to a video display terminal.

MANUAL OPERATION—used to represent a manual processing step in the system.

OFF-LINE STORAGE—represents files that are not maintained on-line, including document files.

Figure 2-1. Common systems flowcharting symbols.

is designed to communicate only down to the black-box level in describing systems. The rectangle representing a computer job or processing unit, in effect, becomes a black box. The remaining symbols describe peripherals of the system, or system inputs and outputs. It is the black box that, at later stages, will be subdivided using software design techniques into a hierarchy of less comprehensive black boxes. This subdivision, or partitioning, may be represented in a structure chart. The resulting black boxes, or modules, and their supporting documentation are used in writing actual programs.

Delineation of Black Boxes

One of the important skills of systems design, then, lies in analyzing data flow documentation, grouping, or packaging, processes into processing units, or *jobs,* and allocating files. These design decisions are documented using linear systems flowcharts that are suitable for a succeeding transition into detailed design. Systems design starts a model of the processing that is expressed using data flow diagrams. Systems designers create systems flowcharts that, later, technical designers use to create detailed designs of programs, hardware, systems software, and other requirements. Thus, systems flowcharts are techniques for documenting the decisions of systems designers.

Documentation of systems design comes with the preparation of systems flowcharts identifying the programs, data files, as well as input and output features of computer jobs. In this sense, a job is a single processing function that is controlled by a single computer program interacting with one or more input, storage, and/or output files. A computer job may involve on-line, interactive processing or off-line, batch processing. Batch jobs usually are divided into separate *job steps.* A *job stream,* then, is a sequence of job steps. Determining the processing scope of computer jobs and job steps is the crux of systems design. Yet, there are no hard-and-fast rules for deciding which computer programs should be defined to interact with input and output files. Rather, the designer is guided by general criteria and judgmental factors that often are influenced by intuition, experience, education, and creative insights.

The discussions below identify some of the main criteria, especially for batch jobs, that should guide design efforts.

Run times required for the job.   Decisions may be based upon how large or extensive an individual processing function may be, as determined by its projected *run time.* For example, suppose a designer, looking at the specification for a process, determines that it will take a full four hours to complete a given processing run on the available computer. Such determinations are based upon interpretation of the design specifications for the new system. The designer determines the size of the file to be processed, the amount of processing to be applied to each record, the output functions, and the corresponding time requirements.

Based on these factors, the designer estimates run time. An interruption in processing such a single-batch job could mean that all of the time and effort applied up to the interruption is wasted. When processing is resumed, the task might have to be restarted from the very beginning.

One of the objectives of systems design is to keep run times for tasks within manageable boundaries. Also, excessive time should not need to be spent on reruns of jobs that are inadvertently interrupted by system failures. For example, the designer might deem it wiser to establish, say, four one-hour job tasks, creating files of intermediate results, than one task requiring four hours to complete. Under those circumstances, systems failures should cause reruns of no more than one hour. This rerun time is additional to the time needed to complete running the job. A single task that failed near the end would require almost four hours to rerun.

Volume requirements for the job.   In some batch cases, it may not be practical, or preferable, to design computer jobs for processing complete file content on any one run. On the one hand, volume demands may contribute directly to excessive run times, as described above. In addition, processing large volumes of data during single runs may degrade processing control capabilities. For example, it is common practice to establish processing checkpoints through the use of batch control totals. Record counts, hash totals, and other batch totals are developed for each batch of transaction records. During processing runs, these totals are compared with check totals developed by the program performing the processing. Discrepancies require resolution of problems, then reruns of affected input batches. If many batches of transactions are submitted for processing within a single job run, correction and resubmission of erroneous batches may become time-consuming and tie up substantial amounts of computer time.

Definitions of job steps, then, may have to be coordinated with other system tasks to maintain manageable processing volumes. Batch control requirements, for instance, may provide criteria for identifying tasks. Other volume-related criteria may include data entry procedures, billing cycle requirements, seasonal transaction volume expectations, and device capacity limitations.

Volume requirements for on-line processing usually are stated in terms of some number of transactions within a given time period. In such cases, the processing time per transaction and equipment limitations are the major factors.

Equipment limitations. Constraints can be placed upon systems design by the capabilities or limitations of units of equipment. For example, if a large report is to be printed, a designer may want to set up a separate program to handle this task rather than perform printing as part of a file update run. Separate runs would make it possible to use special programs that operate the printer as *background* work while other, more time-critical, processing continues.

In other cases, the limited capacity of storage devices may constrain the amount of processing that can be accomplished on any computer run. If, for example, a large file requires sorting as part of a file update run, sufficient disk work space must be available to carry out the sort. If available disk space is not adequate to handle such large sorting requirements, it may be necessary to consider a slower tape sort or partition the file into two or more logically related subfiles and sort the subfiles separately into intermediate files processed in separate job steps.

For on-line processing, constraints usually result from data communication speeds, CPU memory size, and data access times for storage devices. All these factors affect the response time to a user's processing request. Thus, equipment limitations have an effect on both the equipment configuration design and the software design.

Many such device-related constraints can develop. General design specifications rarely can be elaborated routinely into ideal technical designs. Hardware barriers often require ''shoehorning'' an application into a given physical environment.

Availability of utility software. If special systems software or utility programs are available, the flowcharts may be structured to take advantage of their capabilities. For example, in designing for the classic job of creating a transaction file to be used in updating a master file, the sorting of records is a requirement for sequential batch updating. In most computer installations, this sort would be handled through utility software provided by the computer manufacturer. Thus, the job stream creates a transaction file as the transactions are entered and recorded. This serial file of transactions would be input to the utility and sorted to create the sequential transaction file as output. Then, this sorted file would be processed against the master file.

This approach would create a job stream with three separate job steps—a data entry and serial file creation job step, a sort process that builds the transaction file, and an update job step to apply the transactions against the master file. In practice, all three processes could be combined into a single job. However, the design and processing complications of combining the processes, along with the availability of a sort utility, suggest a three-step approach as most appropriate.

The availability of database, data management, and report-writing software on large numbers of modern computer systems has made it possible to develop many traditional data processing applications without ever writing original programs. If specialized processing applications are not needed to meet user requirements, job steps and/or processing tasks can be defined by available processing utilities.

Availability of library routines. Over time, a CIS installation will have developed a library of processing routines that can be drawn upon in developing new systems. Special-purpose statistical routines, mathematical functions, and other application-dependent processing modules may have been cataloged within source statement, subprogram, and core-image libraries. If these routines were developed originally with an eye toward general usefulness and adaptability to future system requirements, the routines may be usable as building blocks within other systems. An awareness of existing routines is useful for discovering processing requirements compatible with existing software; such routines can become templates, or design shortcuts, for new applications.

Experience with existing systems. Among the valuable discoveries made by systems developers as they gain experience in designing and implementing systems is the awareness of the great similarities among apparently different systems. User processing requirements may differ according to application. However, at a logical, essential level, these differences often are only variations on the same processing theme. For example, virtually all data processing systems are composed of the same basic set of generic program types. These basic types often include such program functions as data entry, file creation, file editing, file updating, file maintenance, report writing, and inquiry processing. These generic program types are fundamental building blocks for many systems. Variations in the systems commonly relate only to the particular attributes of the data to be processed rather than to the types of processing activities required.

This commonality is not particularly surprising, since most information systems share limited objectives: to maintain and report on the status of organizational entities. In performing these common functions, systems place information in files, keep the information up-to-date by applying ongoing transactions against the files, and report on the content of those information files.

Therefore, the designer applies these insights in searching the data flow diagrams for processing requirements that can be packaged into standard software components.

Existence of system boundaries.   There may be natural boundaries within systems and subsystems that define jobs and job steps. Recall that processing cycle requirements often make it necessary to establish temporary, transitional files that are used to pass information from one batch processing function to another.

For example, consider the case in which a serial transaction file is created in an on-line mode from several remote sites but will be collected and used to update a master file in batch mode. Transactions are logged as they occur. After this file is built, it is applied against the master file on a batch basis. Thus, a processing boundary exists between the data collection function and the file updating function.

Most of the important boundaries within a system are documented during the analysis and general design phase. Within the scope of automation, data flow diagrams show these boundaries as interfaces between batch and on-line processing and between batch timing requirements or cycles. Also, different output and reporting requirements that service users within the organization often will establish processing boundaries. For any of these reasons, it may be necessary to package the system according to service demands rather than for technical efficiency.

Flowchart Development Examples

To illustrate how systems flowcharts document the design of a batch processing job and are derived from data flow documentation, consider Figure 2-2. The user's timing requirements have been added to Figure 1-8, PREPARE BILL, of the Central City water billing system. Initially, the special billing process will be considered as a semi-monthly process. Later, the effect of alternative timings on the design will be discussed. Additionally, Figure 2-3 reflects the job step packaging of Diagram 5.2 for this batch job. A similar packaging would exist for Diagram 5.3.

The resulting systems flowchart appears in Figure 2-4. Several considerations have gone into this design. Although the calculation processes for incycle and special bills have many similarities, the processes have been packaged into different job steps because different data drive each. An incycle bill is generated for each account whose billing cycle matches the input cycle. Calculations are based upon the current reading stored, if the reading date is on or after the input billing date. The readings for these accounts are updated by a separate file maintenance job stream that is run as meter readings are received, batched, and input. This processing requires sequential processing of the CUSTOMER-MASTER file (or a subset of it if an alternate index access path by cycle is available and feasible).

The creation of special bills, by contrast, depends upon the account numbers in SPECIAL-BILL-REQ. This file is processed sequentially, and the matching CUSTOMER-MASTER is accessed directly. To retrieve the date of bill preparation within a single job step would add unneeded complexity. The concern over the apparent duplication of program code that ultimately will be written can be addressed through the structured design of the software in each job step. The commonly needed functions, such as CALCULATE WATER CHARGE, can be provided by a single, well-defined routine that is invoked in each job step. The strategies and design evaluations that govern this approach are discussed further in later chapters.

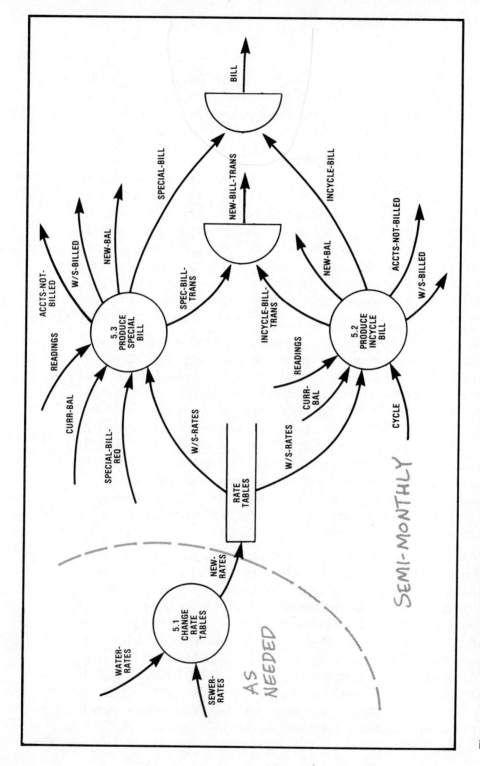

Figure 2-2. Diagram 5—PREPARE BILL—with timing considerations added.

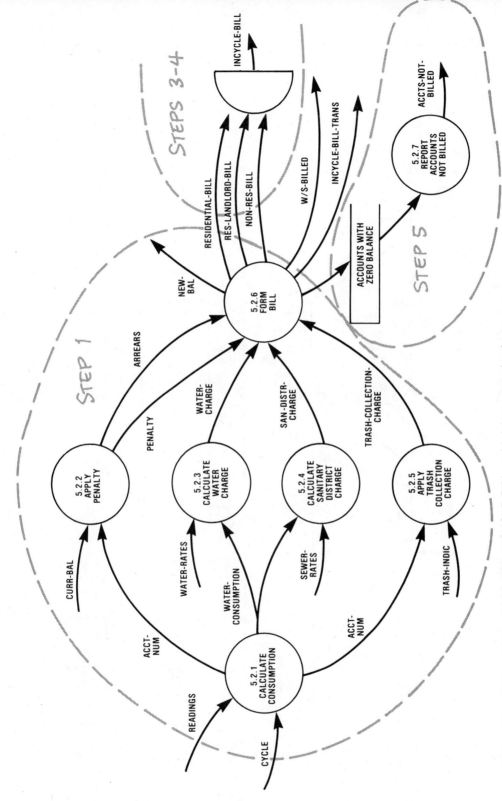

Figure 2-3. Diagram 5.2—PRODUCE INCYCLE BILL—with job step packaging.

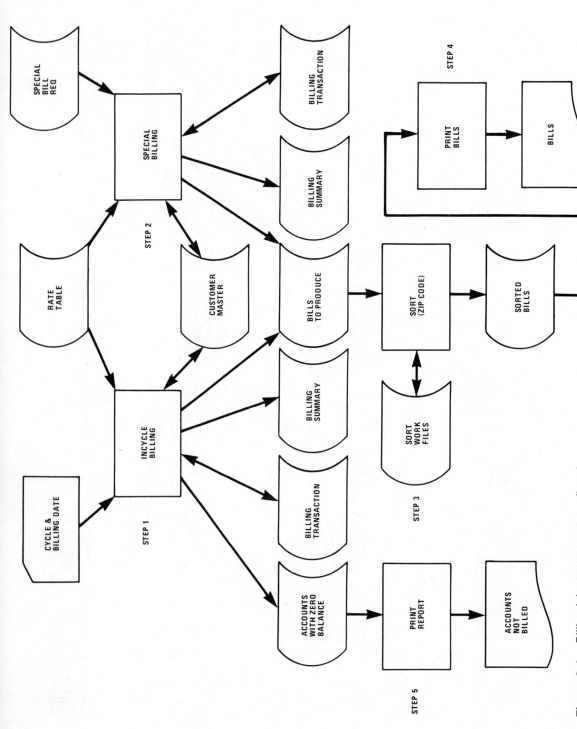

Figure 2-4. Billing job stream systems flowchart.

The preparation of the actual bills has been packaged into a separate job step because both Step 1, INCYCLE BILLING, and Step 2, SPECIAL BILLING, will generate bills. An additional basis for this design decision is that a financial savings on postage can be realized if the bills are presorted by ZIP code before mailing. An intermediate sequential file, BILLS-TO-PRODUCE, receives data from each job step. The sort utility uses this composite file of all data for bills as input and sorts it by ZIP code. This step is followed by a report writer job step that actually prints the bills for mailing.

In producing the ACCOUNTS NOT BILLED report, an intermediate file with accounts having zero balances is created. The report then can be produced through a relatively straightforward report-writer step.

Of course, different timing considerations for producing SPECIAL-BILLs would change the design. The primary effect would be to create two systems flowcharts and job streams that differ only in the first step, one with INCYCLE BILLING and the other with SPECIAL BILLING. Such an approach may be more likely because the special requests may need servicing on a more frequent basis.

An additional consideration in the design of this job is the backup, or copying, of the data currently on the computer files before the update is run. All of the files have been shown as on-line disk files. One means for backup would be to copy all such files to another portion of reserved on-line storage or to off-line media such as tape. This design decision usually is based on equipment limitations and off-site security needs. Backup procedures often occur either before updating begins or at its completion. Additionally, input transaction data must be retained. Alternatively, a log of pre- and post-update copies of a record may be created to provide file backup. Various strategies are covered later. At this point, though, it is sufficient to recognize that backup capabilities must be added to the systems flowchart in Figure 2-4.

Batch systems usually are initiated by a user in the form of a job request to a systems scheduler or administrator who manages the running of the particular job. A computer operator then initiates the job by either entering a card deck containing procedure or job control language statements into the card reader or executing a procedure through an operator's console or terminal. For a specific request, the operator also supplies the needed parameter values where required. Because operator intervention and multiple peripherals may be required, the systems flowchart is useful for operations documentation. Its overview perspective of the job provides valuable information.

SPECIAL ON-LINE CONSIDERATIONS

The value of systems flowcharts is less dramatic and apparent in documenting the design of on-line processing. In batch systems, separate programs exist for all of the separate job steps. Usually, interfaces between these programs are defined by intermediate files. As described above, natural processing boundaries exist between processing functions within a batch system. Within on-line systems, on the other hand, several processes are encompassed within a single processing procedure, or program. Processing boundaries do not show up as transitional files or differences in timing requirements. Typically, somewhat independent, smaller-scoped groups of processes are brought together into an on-line job. Thus, the single processing box shown in a systems flowchart for an on-line system may imply several processing steps—from data entry and editing through updating, retrieval, and display.

For example, consider Process 3.4 in Figure 1-10, EDIT AND UP-DATE for SPECIAL-READINGS or READING-ADJUSTMENT, which is repeated here as Figure 2-5. As indicated, this process is designated to be on-line. The systems flowchart for this job is shown in Figure 2-6. Unlike the batch processing of the incycle readings, the on-line job handles only a single reading and does so at the time the terminal operator inputs the reading. Immediate editing of input data occurs; detection of errors is displayed for operator action; and records are updated with accepted data. The results of a single such transaction then are available to the next person accessing the updated record. In this example, the job appears to contain only a single processing transaction. A more detailed review of Process 3.4 may find that the processing of SPECIAL-READING, READING-ADJUSTMENT, and VERIFIED-READING may differ from one another. The user's decision to specify these processes as on-line takes into consideration that each involves special handling by people and has a requirement of timeliness for its execution.

Another alternative to the design for preparing bills in Figure 2-4 may be to package the production of special bills, Process 3.3, with

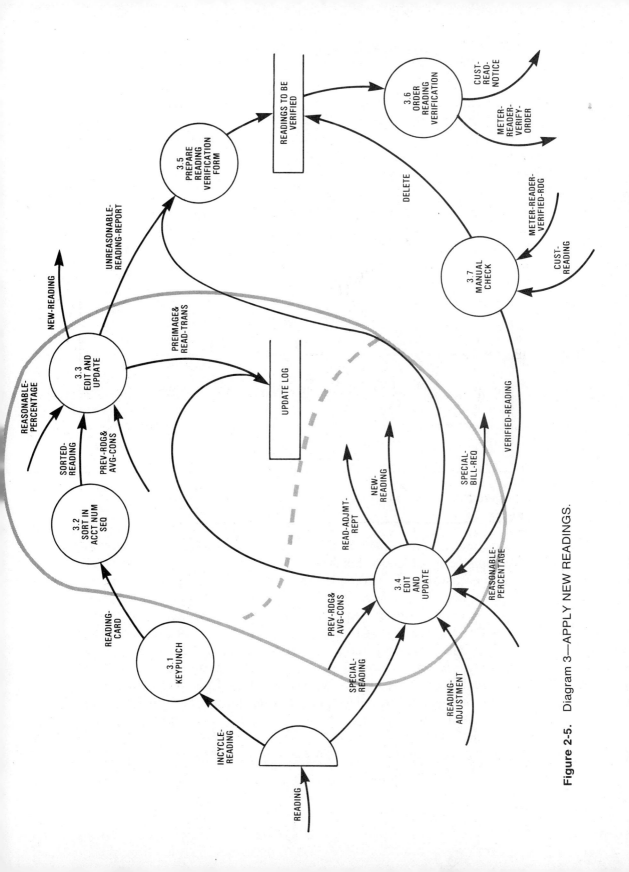

Figure 2-5. Diagram 3—APPLY NEW READINGS.

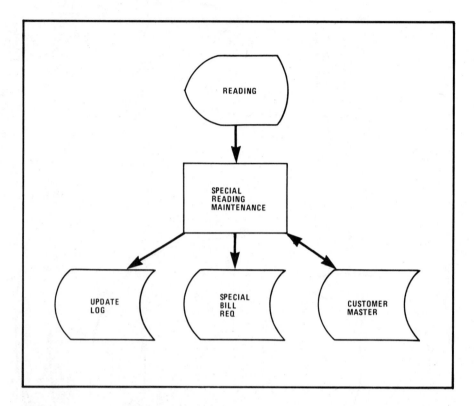

Figure 2-6. Systems flowchart for EDIT AND UPDATE, Process 3.4.

the above on-line capabilities. As a result, a special transaction can be completed in which the bill is produced and can be mailed or handed to the customer for immediate payment. This approach requires special design considerations for both software and hardware. Earlier, well-structured design required that commonly needed functions be used by the batch jobs that prepare incycle and special bills. With special bills being prepared as part of an on-line job, the need for these routines remains. Additional design consideration must be given to the usability of the routines in both on-line and batch environments. To accomplish this, these routines must perform single functions. The routines should not depend upon the physical environment in which they are used but only upon the data they receive. These necessary characteristics of the software are discussed more completely in later chapters of this text.

To produce these special bills, a printer that would be available to print, as opposed to display, output data would be dedicated to the on-line system. This printer would be considered to be another device in the network. For flowcharting purposes, both processes would be represented by a single box. A printed output, BILL, would be added to Figure 2-6 to document this design.

The typical approach in designing the on-line component of a CIS system is to group several transactions or business tasks together under an "umbrella." Each transaction is based upon some event that takes place in the business: A customer closes an account; an adjustment is made to an incorrect reading; the rate per unit for water for commercial users is changed. The transactions are disjointed from one another except for a possible tie to a common data base of information. The umbrella often is implemented through a menu program. That is, the on-line system displays a list of processing options from which the terminal operator can choose. Figure 2-7 is an example of a menu for the water billing system. Each option reflects a transaction or business function. Through the keyboard, the operator selects the processing function that is needed. After this entry, control is passed to a program that manages or controls the processing associated with the selection: requesting additional input data, editing data, acting upon data, manipulating retrieved data, or displaying a response.

Several menus may appear in an on-line system, with each corresponding to a particular grouping of functions. These groups usually reflect security considerations as to who is allowed access to what functions. For example, a water billing service clerk would have access to a menu such as that in Figure 2-7 but not to one that contained the function from Process 5.1, CHANGE RATE TABLES. The menu program, whose display is shown in Figure 2-7, and all of the programs that implement the options presented require careful design. The objective is to make implementation easier and to ensure maintainability and flexibility. A software design strategy, called transaction analysis, that facilitates achieving these qualities is discussed later.

Backup and recovery of data in an on-line system requires more sophisticated approaches than in batch processing. Because the system is needed on a continuing basis for sending and receiving timely information, special concern must be given to such problems

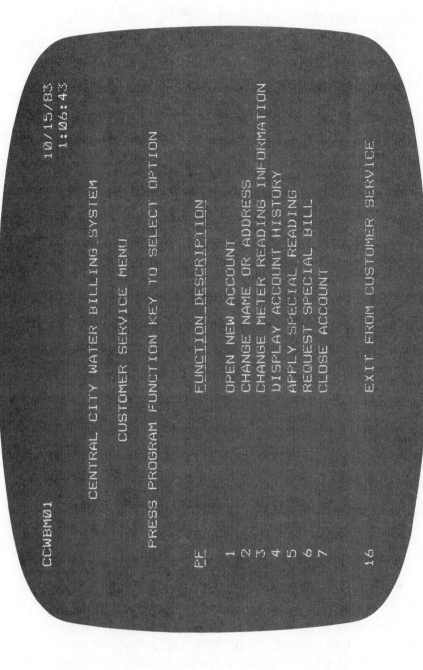

Figure 2-7. Water billing system service menu.

as addressing program failure, interruption of power, and loss of communications. Design approaches that deal with such contingencies are discussed later.

When documenting on-line systems, all of the above considerations are condensed into a single box on a systems flowchart. Additional documentation of these design decisions is necessary both to support the feasibility determined in Activity 5 and to serve as a basis for detailed design in the next phase. To document the user's interaction with the system, rough screen formats such as Figure 2-7 can be used to describe a ''conversation'' with the on-line system.

Additionally, attention must be given to the internal structure of the on-line job. The scope and organization of processing is documented with a structure chart. For on-line systems, delineation of functions often occurs with the preparation of structure charts, rather than systems flowcharts. If a flowchart shows the entire system as a single processing symbol, a structure chart details the separate procedures. A structure chart, or hierarchy chart, is used to illustrate the processing functions within a system and to describe the relationships among those functions. The structure chart presents a top-down view of the organization of software components of the system, in succeeding levels of detail. The top-most level of the structure chart, corresponding with the systems flowchart process symbol, indicates the overall function of the system to be developed. Additional levels present the different processing functions that must be performed. Structure charts can be developed to extreme levels of detail, with each processing function implemented as a single, low-level computer instruction. However, such detailing usually is avoided at the design stage. Rather, only the major processing functions and the data interfaces between those functions are indicated. Later, during program design, these structure charts are expanded to lower levels to include the computer processing operations that will implement the functions. Structure charts are used throughout software design but are especially useful at this stage with on-line systems.

SHOWING HARDWARE CONFIGURATIONS

The same, basic flowchart symbols also can be used for representations of computer equipment configurations. For this purpose, the rectangle, or box, is used to show a computer processor. All of the other symbols show peripheral equipment units. When the symbols are joined by straight lines, a direct connection is indicated between the units. When a jagged line is used, a data communication link is indicated that may be identified further according to class of service (dial-up, leased line, or special data service).

To illustrate, Figure 2-8 shows a portion of the hardware configuration that could be used to support a student records system. Processor rectangles in this diagram represent three levels of computers—the mainframe at the central campus, a large minicomputer at a branch campus, and a microcomputer used personally by a faculty member. Other symbols show terminals, card readers, printers, disk devices, and tape drives. The configuration diagram shows that the branch campus would maintain local files on disk devices. Student records all would be centralized on the main campus, with both disk and tape storage media available. A leased data communication line would connect the processor on the central campus directly to the processor on the branch campus. The data communication link between the faculty member's microcomputer and the processor on the branch campus would be through dial-up telephone service.

Thus, use of flowchart symbols provides an effective way to present equipment configurations, an important requirement in completing both systems and detailed designs.

THE PROCESS OF SYSTEMS DESIGN

Because systems design is complex and involves many unknowns, it is best to approach the job of devising a systems design through an orderly, structured process. A diagram covering such a process is shown in Figure 2-9. This process is presented as one method that organizes the job of systems design. This approach establishes a series of orderly steps, providing a basis for a predictable, controllable process. However, there is no claim that this is *the* way or the *only* method. Rather, the diagram in Figure 2-9 represents an attempt to replicate the logical processes followed by experienced systems designers whose work has been observed closely.

Understand the Needs

The first process step in systems design is to understand the needs to be met. The input to this process, as identified in Figure 2-9, is the

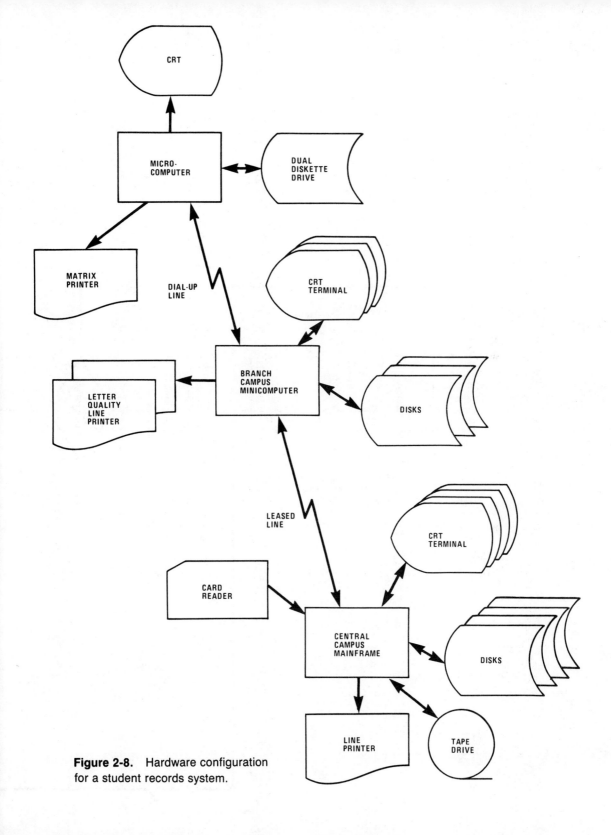

Figure 2-8. Hardware configuration for a student records system.

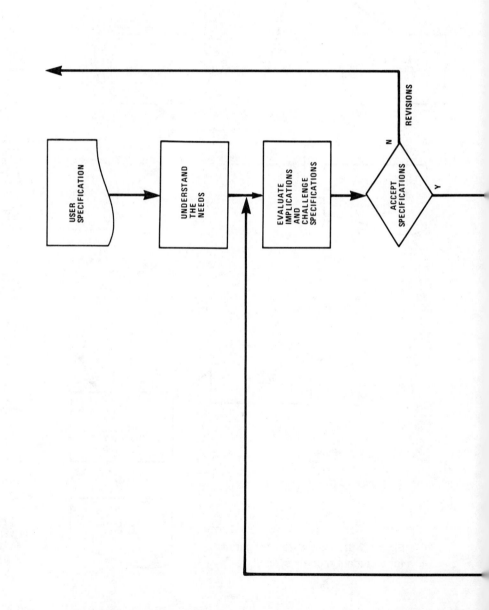

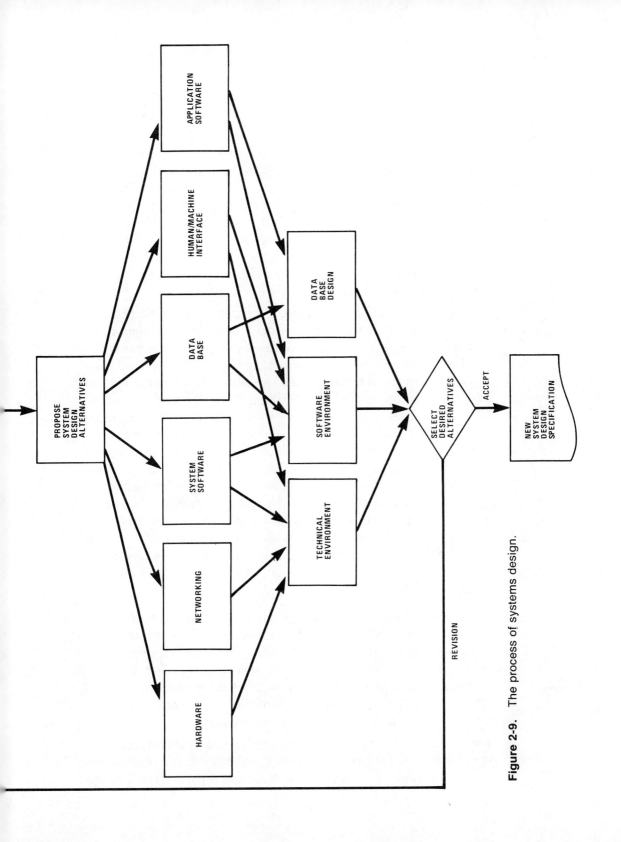

Figure 2-9. The process of systems design.

user specification—the final product of the systems analysis activities within a given project.

The transition from analysis to systems design presents an opportunity to review the system's specifications and underlying motivation, the "business case" for development. Refocusing on the total picture of the system is valuable for both team members who have recently been working on the analysis details and for those who are just joining the team as design efforts begin. Such a review reinforces a common understanding of the system needs and objectives. Additionally, it reflects the cumulative knowledge and understanding that has been developed by all those involved.

Consider the situation: Members of the project team, particularly the project team leader and key user management personnel, have worked closely together for some months, perhaps as long as six to 12 months. In the process, their knowledge has expanded and they have gained maturity in terms of their conceptual grasp of the system they are working on. It should be stressed that this understanding applies on both sides. First, computer professionals have become more sophisticated about what business their organization is in and about the specific characteristics that set their organization apart from others in the same field. Second, users also are more knowledgeable about computer capabilities, because discussions of computerized solutions and feasibility have occurred during the analysis activities. The users also will find that they even have a clearer picture of their business. These insights result from the analysis efforts to get to the logical model of the business. In developing this model, the physical aspects of how things currently are being done have been stripped away to derive an understanding of the essence of the business.

Differentiation. The uniqueness of any given organization can be analyzed and described in terms of a process known as *differentiation*. For any given organization or portion of an organization, differentiation begins by identifying the market served or specific function performed. All companies in the same line of business have commonalities, or share certain characteristics. Each organization, though, attempts to set itself apart from the competition.

For example, an organization's advertising clearly reflects an effort to appear capable of offering a unique type of product or method of service. Knowledge of the forces that mold the organization's image is important to the designer: Is the differentiation based upon a projected image of secure, solid, conservative stability? Or, is it meteoric, innovative, leading-edge, pioneering? Is it people-oriented or technology-driven? These factors are combined with economic or business forces. Growth in the scope or nature of the business or changes in corporate style may be initiated to increase revenues and market share. Other factors to be understood are avoidance of costs to improve the financial picture and improvement of service to keep customers satisfied.

The means by which an organization differentiates itself from others in the same business, the style or image that it builds, and the underlying business factors must be considered by the designer. These factors will have significant impact on the goals for a particular information system under development. The information data base, processing capabilities, and method of implementation all will be affected.

For example, in the Central City water billing system, reducing expenses is a major factor. Better access to information to provide improved customer service is a goal that requires consolidated information with flexible and easier access. In a student records system, if the administration seeks to expand the use of automation across several campuses, the designer is faced with certain challenges. Distributed resources will increase, including word processing and electronic mail. By allowing students access to course planning and evaluation capabilities, the need for security controls increases. Manual procedures become automated and require flexibility.

Differentiation also can occur for internal functions within organizations. Accounting departments, for example, choose from a series of options or alternatives in setting up the systems and procedures they use. No two accounting departments, even in companies in the same business, have procedures exactly alike. The same is true of CIS organizations.

Thus, even though a project team may have looked at other systems in the same industry or area during early phases of a project, specific differences and unique characteristics of the system under development should be recognized and accommodated. It is necessary to differentiate fine points during systems analysis. Then, in designing a specific system for a specific organization, the match will be closer to the users' expectations of the system.

Data base design.   Another facet of understanding the organization is its data base, which may be conceptualized as a corporate data model. Data base design is one of the major, critical areas of detailed design for new information systems. The data base, or the collection of files that will support a system, should constitute, in effect, a model of the organization that it will serve. Therefore, it becomes important, at this stage of a project, to enhance and affirm the characteristics, the goals, the objectives, and the working needs of the organization that will use the system. These factors, lumped together, sometimes are referred to as a *corporate culture.* A data base that models an organization should reflect its corporate culture.

Under this outlook, a data base is viewed as representing those entities, or objects, that are important to the organization and for which processing procedures are established. Customers, products, employees, raw materials, classes, orders, invoices, and so on are the business elements that define the organization. In the course of doing business, relationships are established among these entities. These relationships, in turn, are maintained and documented within computer information systems. The focal point of an information system is the collection of data representing, or modeling, the objects of importance to the organization.

During systems analysis, current processing procedures are documented and new or replacement procedures are devised. One of the primary tools used in performing this analysis is the *data flow diagram (DFD).* Data flow diagrams document the flow of data into and out of *data stores,* where data are retained, and indicate the transformations applied to the data as they move through the system. DFDs describe particular procedures for collecting and preparing data for inclusion within files and for extracting information from files for reporting. In addition, data flow diagramming helps in identifying the elements of data that must be maintained about organizational entities. Thus, in a real sense, processing procedures, as studied through DFDs, exist to service the file maintenance and reporting functions of the information system. These processing procedures support the organizational data base and the requirements for reporting on its content.

Processing procedures are transient. For example, reports change to meet needs of managers, operating personnel, and government regulations. Processing changes also occur to support changing methods of capturing and storing business information. System processing requirements will necessitate continual maintenance and modification.

A data base is not as susceptible to change. Unless the business of the company changes, its entities and the data that represent those entities will be relatively stable. For example, consider a student records system. As long as the university is in the business of education, there always will be students, classes, faculty, grades, and so forth. These are the university's primary entities about which information must be maintained. Over time, different relationships may be established among these entities, and different processing and reporting requirements may occur. However, the need to maintain information on these subjects will be constant.

Careful consideration of both processing and data-base requirements is necessary during systems design. Although data-base requirements tend to be less volatile, or prone to change, than those for processing, attention must be paid to designing flexibility and maintainability into both. Changes are inevitable. The key to success in systems design is simplicity. Unnecessary complexity and elegance cloud understandability and should be avoided.

System characteristics. The logical and physical characteristics of the system, as described in the specification document, also should be understood at the outset of systems design. Transitional considerations include:

- The designer must understand the extent to which data and processing functions are to be automated. The systems designer is working on the automated, or computerized, portion of the system. However, the characteristics of both the manual and the automated portions of the system must be understood. Considerations include the volume and frequency of each function to be processed, the size of files and the number of records to be processed, and the time frame in which processing must occur.

- Systems design should take into consideration the expectations of users. This consideration involves knowing and understanding the functions of existing systems, competitive alternatives, or other computer information systems that are operating within the same organization.

- Delineations should be drawn between batch and on-line processing requirements. Where on-line procedures have been recommended, the systems designer should understand that this choice reflects a need for timeliness of service. In effect, timeliness is the differentiating factor between on-line and batch processing. This understanding should lead to meaningful design specifications about response times required for the retrieval and delivery of information to users.

- If interactive processing is to be included, the systems designer must understand both the needs and the level of sophistication of the persons who will be using the terminals. Also to be understood are the needs for file security, integrity, backup, and recovery procedures.

- Business timing cycles must be identified and understood for batch processing functions. Volumes of work and the criticality of the jobs should be understood. The updating that takes place on a cycled basis represents, in the long run, the protection provided for the data resources of a system. Also, cycled update functions usually are the points at which files are copied and protected and security measures are applied.

- The specified needs of the new system must be matched, carefully and minutely, against the capabilities of the installation in which the system will be implemented. The designer must be satisfied that facilities exist to support every known operational contingency associated with the new system.

Note that there is a quantitative emphasis within all of these points of understanding. This is one of the differences that emerge in the transition from analysis to design. Analysis collects quantitative information without doing much about it. During the transition to design, it is important to use quantitative data to build a clear understanding of the new system's scope and magnitude.

Evaluate Implications and Challenge Specifications

Understanding comes first. In establishing an understanding, the systems designer should be as objective as possible. That is, there should be no attempt to evaluate or challenge new system design specifications during the first step. Rather, emphasis should be on accepting specifications as they are and on building an understanding of what they mean and why they were included. Questioning and challenging, a subjective process, should not be applied until an objective understanding is reached.

The need for initial objectivity marks an important delineation between the steps in the process of systems design. Ultimately, of course, the primary responsibility is to make sure that the system being built, in fact, meets the needs of users. It is also the job of a designer to challenge all specifications—and to see that the system can be enhanced eventually without major or total redesign. However, questioning and challenging specifications without understanding them first can be destructive. But, challenges based on an objective understanding will be constructive and productive.

In building the needed understanding and applying constructive challenges, the systems designer should look for:

- Questions may arise about functions or capabilities of the system. These questions may lie in the processing area, the files, the availability of inputs, or the specifics of outputs. The total system should be reviewed and questioned in terms of what has to be done to implement it—with an eye toward missing links.

- Processing bottlenecks should be sought and identified. Such obstacles may be encountered with uncanny frequency. Remember, analysis collects quantitative data but does not become unduly concerned with processing and data volumes. At this point, questioning should identify peak volumes, average volumes, volumes in off seasons, and also the timing of these variations. For example, if a system is to support a retailing application, it is worthless to the user unless the system can keep up with the Christmas rush. The place to challenge and deal with such potential bottlenecks is during the initial evaluation of specifications.

- Potential problems that might not have been appropriate to consider in systems analysis should be identified and evaluated at this point. For example, the data or file requirements for the new system must be evaluated for compatibility with existing and future plans for the installation as a whole. Also, physical requirements for processing and data storage must be evaluated in terms of growth being experienced in other applications.

- If users now foresee enhancements to the system that were not apparent to them during earlier phases of the project, the evaluation step in systems design should determine whether to address them now and whether there is enough flexibility to accommodate future demands. For example, in light of increased user sophistication, the water billing system may be expanded in the future to accept electronic funds transfer for payments and to support usage projection models based on history. Both the application software and the physical specifications should be evaluated to see if such enhancements will be feasible.

- During this step, as is true for every activity and phase of systems development, a review should be conducted to identify unsound or questionable practices. As the designer reviews the new system specification, procedures and outputs should be challenged in evaluating *auditability*, that is, ability to trace transactions, including their effects and sources. Special attention should be given to system controls that insure the integrity and reliability of the systems data and processing. For example, in the water billing system, the design must address the need to reconcile payment transactions entered into the computer with bank deposits. Security and controlled access to data are especially sensitive where privacy is an issue. Access to class grades or family financial data in a student records system must be limited to authorized individuals only. Other practices to be observed and maintained would be the separation of duties, access controls, authorization controls, and so on. For example, in the water billing system, it would be unthinkable to have the person who receives payments from customers be responsible for issuing bills. Similarly, in a student records system, information provided to advisors is broader and more personal than that normally delivered to faculty members within a department in which a student is enrolled. In this instance, the need to know governs the providing of information.

As suggestions, changes, or enhancements to the system are encountered during this process, they should be analyzed, documented, and evaluated. Those modifications that are accepted should be either considered for immediate incorporation into the system design or postponed and added to a list of post-implementation maintenance projects or enhancements. For postponed features, the system specifications and, ultimately, the design should be reviewed to verify that the modifications can be accommodated later. However, there may be problems that require immediate attention. Any such drastic problems or oversights should be dealt with before detailed design takes place.

In general, this step also should include an evaluation of the extent to which user requests, as embodied in the new system specification, actually can be met. Faced with the realities of the technical and software environment, it may become apparent that users have asked for more than can be provided economically or practically, given the constraints of budgets and of the availability of resources within the CIS environment. One example may be difficulties in meeting stated schedules because of resource commitments to other projects. Other factors may include changes in the economy, in the marketplace, in corporate goals and objectives, or in management needs. Further, the user commitment may not be as strong at this juncture as it was six or nine months earlier when the project was initiated. Any such problems or constraints should be identified and dealt with before the design process moves into further detail.

Propose Systems Design Alternatives

During this step, systems design proceeds from one or a few generalized black boxes down to the level of smaller black boxes needed as a basis for detailed design. In systems design, remember, activities stop short of *white-box* detailing, or the filling in of a black box with processing details. However, the partitioning of the system continues to a level that facilitates the white-box detailing that will follow.

The study of alternatives appropriate for implementation of any given system will vary both with the system itself and with the makeup of the staff within the CIS organization. As shown in Figure 2-9, proposing design alternatives addresses six major areas: hardware, networking, system software, data base, human/machine interface, and application software. These topics are covered in depth in Chapters 3, 4, and 5.

Hardware alternatives.   Most organizations that support data processing systems rent or buy their own computers and operate the hardware themselves. Therefore, most systems design efforts will deal with in-house computers. However, the systems designer should be aware that there are other alternatives and should consider the possibility that some of these alternatives can benefit specific applications.

As one example, many companies that have their own computers also use outside time-sharing services for special purposes or for specific economies. For example, many time-sharing services have special analysis or engineering programs that involve extensive, expensive development. Thus, companies may elect to take data developed in-house and process them through a time-sharing data base for competitive comparisons or for financial projection purposes. This capability may figure into either a portion or all of the hardware support for any given application.

The same type of reasoning applies to applications that can be processed by service bureaus. For example, many service bureau organizations are so proficient at payroll processing that companies with their own computers find that processing can be handled more reasonably by such outside organizations. This has been particularly true since banks have initiated payroll processing services. Also, many companies that handle their own data processing applications routinely send tapes to service bureaus that have specialized programs for generating general ledger or other accounting reports.

Another option represents, in effect, a hybrid between in-house computing capacities and external services. This option is the facilities management agreement. The actual computers used may be housed in the company for which they are applied. In other cases, two or more companies share an installation that may be housed partly in both facilities or in the offices of one of the companies. The special characteristic of this alternative is that an outside organization assumes responsibility, under contract, to staff and run the computer facility. Services may or may not include systems design and development.

If any one of the three identified alternatives—time-sharing, service bureaus, or facilities management—is elected, outside organizations generally will be responsible for detailed design of the hardware elements of the system. However, if the processing is to take place on in-house facilities, several hardware design concerns should be addressed.

- A determination should be made on whether and to what degree resources will be centralized or distributed.

- An analysis should be done to identify requirements for accommodating the new system in the existing CIS environment.

- Available processing and other service capacities within the CIS environment should be related to the needs of the new system.

- Equipment configuration designs represent trade-offs between computer processor and peripheral device options.

- An evaluation should be done to determine whether existing office automation capabilities can be applied effectively to the new system. Potential areas involve word processing, audio input, audio output, and image processing that encompasses computer graphics, optical character reading input, and the digitizing of images for computer storage and processing.

Human/machine interface alternatives. This is the crucial area in which the automated portions of a system meet its human users. Without thorough consideration of human factors, even the best-designed technical system may be useless or unused. Factors to be considered include:

- Ease of use, or user-friendliness, is a major factor.

- The media to be used for input and output functions must be evaluated. These media can vary from punched cards to video terminals, telephones, or wand readers.

- The places and effects upon the system of human/machine boundaries should be determined.

- The positions and effects upon the system of on-line and batch boundaries should be identified.

- Modularity of input and output functions must be planned so that expanded or improved options can be considered at a later time.

- The patterns and forms of on-line conversations within interactive systems should be covered. These techniques include menus, the ability to bypass menus, and the provision of "help" routines.

- Human access methods to data should be included, along with accompanying concerns for security and privacy.

Networking alternatives.   The term *networking* encompasses all considerations associated with the distribution of processing or the transmission of data. Clearly, networking studies are appropriate only for situations in which distributed data processing will be used. On this basis, networking could be viewed as a subset of hardware considerations. However, because of the vast differences in technologies, a separate category of design consideration commonly is established.

Analysis of networking requirements typically includes the following considerations:

- Data may be transferred in either a batch or interactive mode. With batch transfer, data are collected and transferred in groups between computers. This transmission often occurs at night, is usually two-way, and may include data files, program files, and electronic mail. Interactive transfer of data occurs from a remote location to another location in much the same way as on-line processing takes place through a directly connected terminal.

- The media to be used in transmission of data vary from local area networking schemes, to cable television technology, to remote networks employing dial-up or leased telephone lines or satellites.

- The communication connection, or circuit, can be established as a *peer-to-peer* (no host) or a *master-to-slave* (distributed from a host) configuration, thus determining the path that data will follow during transfer.

- *Protocols,* or transmission standards, required by alternate communication links provide means to insure that data that are communicated can be understood where received.

- Data encryption techniques may be considered for privacy or security of data. Encryption can involve the use of signal scramblers to render data useless if they should be intercepted during transmission.

Systems software alternatives.   Systems software forms an integral part of the operating environment in which the programs for the new system will be processed. Thus, it is necessary to review existing, planned, and other available systems software products with an eye toward determining the options available for integrating the new system into the environment in which it will function. Systems software components that probably will be available in most computer installations and should be considered include:

- Operating systems and the utility programs they incorporate
- Data management software
- Database management systems
- Report writing programs
- Application generators
- Advanced programming languages.

Decisions about systems software for any given application may involve trade-offs to match the requirements or constraints of existing systems. That is, even though an installation may have the software to support a desired feature in a new application, the needs of existing applications may make it undesirable or unprofitable to do so.

Data base alternatives. As stated earlier, the underlying corporate data model is one of the primary resources of an organization. Data are probably second only to people in importance to the organization. Because data are at the heart of a CIS, the designer must address the needs for completeness in content and flexibility in structure. More specifically, design concerns must include:

- Data structures chosen must be reviewed in the light of processing throughput requirements. The trade-offs in compromising such structures can involve gaining performance and introducing redundancy at the cost of flexibility. (The process of optimizing data structures is discussed in depth in Chapter 5.)
- File access and organization methods must be matched to processing requirements. Storage media for data must be reviewed to provide cost effective storage of data.
- The alternatives in file design must address processing requirements, response times required, volume of activity of the application, volatility of data, backup and recovery plans, and storage capacities.

- The introduction of a database management system must be weighed against supporting data requirements using more traditional data management approaches.

Application software alternatives. A major, continually growing issue that should be dealt with at every level of systems development, particularly during design activities, centers around whether to use application packages or to design and develop custom application programs. This decision involves alternatives and trade-offs as well. For example, it is becoming increasingly feasible to buy standard application packages and then modify them as necessary for the custom requirements of a specific user.

Application packages fall into two broad categories—industry applications and generic applications.

Special-industry applications have proliferated as quickly as needs have been identified. Today, there are thousands of application packages, available from hundreds of suppliers. In some cases, regulatory conditions serve to make standardized application programs a hands-down choice. For example, consider the case of common-carrier trucking companies that operate interstate. These organizations must report mileage to government agencies. Since this is a standard requirement for large numbers of companies, many highly efficient application packages have been developed and marketed. Today, it would be a questionable practice for a trucking company to develop an application program from scratch. Similar standardization has taken place in many industries such as insurance, manufacturing, and banking.

The same basic principles have been brought to bear in a number of generic application areas. For example, it would have to be an unusual situation, indeed, for a company to develop its own payroll program from scratch in today's marketplace. There are so many payroll packages available, with so many options, that it simply doesn't pay any more not to consider packaged options.

If a new system involves analysis of financial data, it now is almost axiomatic to consider use of "electronic spreadsheet" packages. These are specialized programs that accept accounting data and present them in standardized report formats. The programs have the capability to accept factors provided by managers and to project financial results for an organization. For example, factors such as growth or sales increases can be entered into the spreadsheet. Financial data for the current year then may be used to project accounting results for future years.

Another, similar area in which packaged programs play an important role is in computer graphics. A wide range of programs is available that makes it possible to convert financial or operating data to graphs and charts. Programs also make it possible to prepare illustrations or animated presentations through the use of graphics programs. For business-oriented systems, particularly systems delivering results to top management, graphic techniques can illustrate necessary points quickly and dramatically, saving time and placing emphasis where it belongs. This is an excellent illustration of a situation in which it simply would not pay to begin developing application software from scratch.

Available report writing packages, some of which are provided with existing operating systems, incorporate output or screen formatting capabilities, file or data element selection and sorting functions, and also interactive inquiry capabilities. Where such tools are available, they can reduce significantly the amount of programming needed for relatively mundane jobs. Report writing programs also are valuable in situations requiring preparation of special reports that may be used only occasionally.

In general, identification and review of application software alternatives should include these considerations:

- Compatibility with existing or purchased database software should be considered, if appropriate.

- Application program designs, whether they are to be purchased or developed internally, must meet stated requirements in the areas of security, privacy, and auditability.

- There should be some investigation of the possibility of including any existing word processing or office automation software into new applications. Such integration is becoming increasingly common. Where feasible, extensive savings can be realized in time and money for development of system inputs and outputs.

Even if an organization determines that it is necessary to develop custom application software, a study of available packages is still beneficial. At the very least, there are benefits to be realized from looking at designs, control structures, and security/privacy measures incorporated into packaged programs. If the software package has been successful, it probably has gone through a series of modifications and enhancements that now make it a mature product. A review of this alternative provides an opportunity to draw from amount of time, learning, and detail that the software house has put into the design effort. This effort usually is greater than an in-house group will be able to devote feasibly to the project.

If custom development is seen as a necessity, the issues reviewed at the systems design level should include:

- Strategies for file maintenance subsystems differ between batch and on-line systems. The single transaction that is initiated by a terminal user in an interactive system requires a different software structure from the update of a file by a batched set of transactions.

- Error detection and correction differ in timing and method between on-line and batch systems.

- If batch processing is to be used, modules should be designed so that conversion to on-line methods can be considered at a later date if decision criteria change. That is, it is possible to design batch systems through designation of a series of modules that can function in an on-line environment with minimal revision. The primary goal is to build the system with modules that perform single business functions, that are coupled to other modules only by data parameters, and that are independent of the physical environment in which they are performing. Some modules in the batch system will address management of processing, and some will handle input and output functions. These modules will change in a conversion to on-line processing, but the business-function-oriented modules usually can be carried over without modification.

- Especially for on-line systems, attention should be given to strategies for creating backup files, for the procedures to be followed if services are interrupted, for recovery when service is restored, and for restart in the event of a major interruption.

Select Desired Alternatives

By now, if any serious problems have been encountered, the specifications for the new system should have been reconsidered and adjustments made. All of the required changes have been incorporated into the specifications for the new system. Usually, two or more design alternatives have been considered seriously for implementation. The project team develops each alternative to a preliminary point at which technical feasibility is shown, the cost/benefit is clear, and the solution is understandable to the user and management steering committees. Then, an alternative is selected for greater detailing to achieve the objectives in the analysis and design phase. As shown in Figure 2-9, the design process has a feedback loop that provides for the consideration of alternatives. Even after selecting an alternative for further work, refinement of the design is likely to be necessary until all involved are comfortable with the solution and the cost/benefit analysis. The final decisions can be incorporated into the new system design specification. This document then serves as the basis for detailed design activities.

Under some circumstances, it may be desirable to stagger, or phase, development procedures. As the design evolves, schedules and budgets should be checked carefully. Situations often arise in which portions of a system must be implemented independently and expeditiously to meet either regulatory deadlines or management priorities. If this is the case, the development effort may be divided into a series of phases, or stages. Nothing changes except an awareness of priorities. Each phase or version represents a functionally complete, working subset of the entire, intended system. Design and subsequent implementation and testing are done. The completed version then is installed in the user's environment and made operational. The next phase supplements the installed one, rather than replaces it, as the system evolves toward its goal. The approach provides a means to meet user needs quickly and control the use of limited resources in a productive manner.

Figure 2-10 is a structure chart that presents an overview of the total systems design process. In completing the steps described above, the designer has finished the work of systems design that is positioned at the top level of the structure chart representing the design process. The new system design specification document produced during this

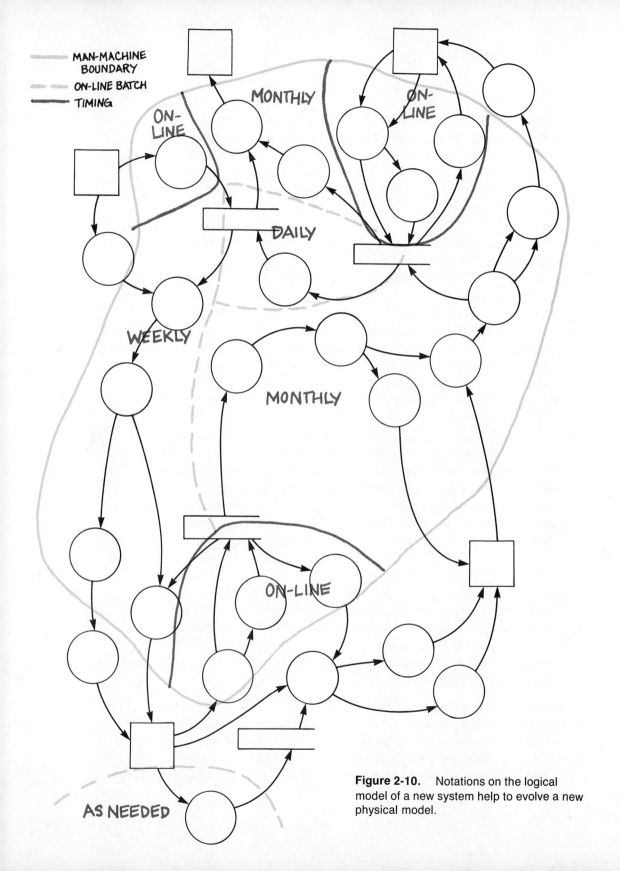

MAN-MACHINE BOUNDARY

ON-LINE BATCH

TIMING

MONTHLY

ON-LINE

ON-LINE

DAILY

WEEKLY

MONTHLY

ON-LINE

AS NEEDED

Figure 2-10. Notations on the logical model of a new system help to evolve a new physical model.

set of procedures becomes the link to detailed design. This document, in turn, becomes the basis for subsequent design activities, covered in subsequent chapters.

Summary

Systems analysis produces a model for a new system expressed in terms of functional flows and data transformations. A new system remains a black box at the time when systems design is launched. At this point, there is no definition of the computer processing represented by the black box.

Systems design produces a set of technical specifications identifying which computer programs interact with input, storage, and output files, in which sequence, and at what times to perform processing.

Systems design has the effect of preparing the way for detailed design by subdividing large systems into a series of smaller units that can be dealt with at lower, more technical levels. A primary tool for implementing and expressing the results of systems design is the systems flowchart. The flowcharting language is designed to communicate only down to the black box level in describing systems.

Systems design starts with data flow diagrams as major inputs. Systems designers create flowcharts that serve to document their decisions. Systems flowcharts identify job steps. A job step is a processing function that is controlled by a single computer program interacting with one or more input, storage, and/or output files.

Criteria that may be applied at this point are job run times, volume requirements, equipment limitations, availability of utility software, availability of library routines, experience with existing systems, existence of system boundaries, and special considerations related to on-line applications.

For on-line systems, delineation of functions often occurs with the preparation of structure charts rather than flowcharts. The structure chart documents, in detail, the separate procedures that comprise the system.

Flowchart symbols also may be used for conceptual representations of computer equipment configurations.

The first process step in systems design is to build an understanding of the needs to be met. The principle of differentiation may be applied to understand similarities and differences among user organizations that may apply to the project. Differentiation also can occur for internal functions.

Data base design establishes the groupings of data that represent organizational entities. Data base design also devises the methods of relating data to establish processing and reporting procedures that serve current and future needs.

In systems design, evaluating quantitative measurements is one of the major differentiations between analysis and design.

In initial systems design, care must be taken to develop an objective understanding of given specifications without attempting to challenge them. The systems designer eventually must challenge all aspects of the proposed system and must look for unanswered questions, processing bottlenecks, potential problems, likelihood of future enhancements, and detection of questionable practices.

Alternatives are identified and reviewed for hardware, human/ machine interface, networking, systems software, data base, and application software. Hardware alternatives include time-sharing service, service bureaus, and facilities management agreements. Application software alternatives include the availability of packaged programs that may be customized.

These alternatives are classified as belonging either to the technical environment or to the software environment. Based on this classification and on a review of the alternatives, selection of alternatives is documented in the new system design specification document.

Key Terms

1. black box
2. systems flowchart
3. job
4. job step
5. job stream
6. run time
7. background
8. differentiation
9. corporate culture
10. data flow diagram (DFD)
11. data store
12. auditability
13. white box
14. networking
15. peer-to-peer
16. master-to-slave
17. protocol

Review/Discussion Questions

1. How does the concept of a black box represent the proposed system at the beginning of systems design?
2. How is the systems flowchart used during the systems design phase?
3. What is meant by a job step and how do job steps relate to the systems flowchart?
4. What criteria may be applied in identifying job steps?
5. Why are structure charts typically used in conjunction with systems flowcharts in designing on-line systems?
6. When is it appropriate to challenge system specifications presented to systems design?
7. What alternatives are identified and reviewed during systems design?
8. What are some specific areas for consideration of hardware alternatives?
9. How does file or data base design relate to the goals of the user organization?

10. Why is it appropriate to consider incorporation of packaged software in a new system?

11. Into what two categories are alternatives classified and how does this classification assist the systems designer?

12. What is the end product of systems design?

Project Assignments

The Appendix presents three case studies that can be used in support of project assignments. All three scenarios provide functional system specifications that would result from systems analysis. Included are data flow diagrams, data dictionaries, and process narratives that would result from systems analysis. Continuing the work begun in Chapter 1, assignments that relate to the content of this chapter will be found in *PART 1: Analysis and General Design Review.*

SYSTEMS DESIGN SKILLS:

THE TECHNICAL ENVIRONMENT OF SYSTEMS DESIGN 3

LEARNING OBJECTIVES

On completing reading and other learning assignments for this chapter, you should be able to:

☐ Explain the evaluation sequence and methods applied in matching new system requirements to existing technical capabilities.

☐ Describe design alternatives for data capture, input, processing throughput, output, and storage.

☐ Describe the working relationship between the system designer and the technical support staff.

☐ Tell what factors must be considered in understanding technical goals and constraints within a given organization or CIS environment.

☐ Explain how systems may be centralized or decentralized and the design considerations inherent in the degree of distribution.

☐ Give hardware and system software considerations associated with different levels of distribution.

☐ Describe concerns in designing data communication capabilities for individual applications.

MATCHING REQUIREMENTS
AND THE TECHNICAL ENVIRONMENT

During the analysis portions of a systems development project, emphasis is almost entirely upon the business objectives and specific requirements of the application being developed. As the project moves from analysis to design, the realities of the technical environment have to be faced. These factors may force design accommodations to the existing hardware and software within the organization. This technical environment may permit—or even require—the introduction of new technologies—including hardware such as microcomputers, laser printers, or graphics terminals, or software such as office automation, database management, or data communication systems. The capabilities within any given computer installation may represent either new opportunities or unforeseen constraints in the development of designs to implement user specifications.

During systems analysis, quantitative data are gathered about transaction volumes, file sizes, response times, expected growth, and processing volumes for the new system. Analysts may apply rigor to accumulating these data. However, until actual design begins, little is done about matching these requirements with the realities of existing or planned technical capabilities.

Once design has begun, the project team must face the fact that the new system exists in a larger environment. The application under development is one of many. The new application will be faced with the hard facts of sharing facilities, trade-offs between advanced technologies and existing capabilities, and so on.

In other words, as systems design is begun, evaluations must be made to determine where and how a new system can fit into its overall processing environment. The normal data processing cycle can provide a structure for matching the requirements of individual systems with the data and throughput capacities of the environment in which it will exist. The evaluation sequence can run the following course:

- Data capture
- Input
- Processing throughput
- Output
- Storage.

Data Capture

There was a time when systems designers assumed, routinely, that source data would exist on source documents and that some transcription step would be necessary for data capture. This assumption no longer holds true. Today, data capture can be regarded as a frontier for innovation in the design of computer information systems. For example, increasingly, supermarkets are installing scanners at checkout counters that can read the universal product code on an item's packaging. Key entry at the cash register/terminal by the checker is reduced greatly but not eliminated completely. This innovation speeds up the checking process and reduces errors.

Consider another example—a hospital supply firm that offers a special service to important customers. Using a tone-generating telephone, a customer can place a call to the supply firm and enter an order directly into the computer without relying on human intervention at the supply house. The customer responds to computer-generated audio prompts and "keys" in customer number and, for each item to be ordered, item identifier and quantity. The telephone keypad is used as the data capture mechanism.

There is little that a designer involved with an individual systems development project may be able to do to impact decisions about the make or model of computer installed, especially a central mainframe. Also, there is little that a person involved with an individual application can do about the version or capabilities of the operating system software or data communication networks on which the application will be processed. However, selection of data capture and input devices can be expected to fall within the decision responsibilities of an individual project team.

Modern data capturing equipment is small enough and inexpensive enough so that almost any business-related application can support individual selection of equipment in this area. In the era of microprocessors, this opportunity means that the systems designer can have a profound impact upon system capabilities through selection of locally installed devices used for data capture and possibly also for direct input to and receipt of output from a system.

Input

Individualized input capabilities form a logical extension of the tailoring possible through selection of data capture methods or devices. There was a time when input meant transmitting media to a central point for reading into a computer system on high-speed devices. Today, a better rule is that users should control and operate the input function as close to the point where transactions take place as possible, unless there is a compelling reason to do otherwise.

Local control of the data capture function is a clear-cut trend. Statistics gathered by the U. S. Department of Labor, for example, show data entry operators as the only job category within the data processing field for which diminishing demand is anticipated. It makes sense, given today's technologies and continuing improvements expected in the future, to incorporate the capture and input functions within transactions rather than to require separate, centralized data entry functions that are costly, time-consuming, and error-prone.

One compelling reason for placing the input function within the user group is that responsibility accompanies performance. That is, users who handle their own input also are responsible for verifying the accuracy and reliability of the data they enter. By controlling the quality of data entry, users also help to assure the quality of results they will receive. If all other things were equal, quality assurance alone would be enough to justify decentralization of input.

Also, given the low costs and efficiencies of microprocessor-based input devices, substantial savings usually can accrue from decentralization of input. Even though actual keystroking may cost less if the work is performed in volume in a central facility, there must be some savings involved in incorporating data capture and input into the actual transaction. The example of the hospital supply firm illustrates the potential savings. Similarly, when placing a charge call at certain public telephones, the caller is asked to enter his or her calling credit card number. Intervention by a long distance operator is not required. When data entry is part of the transaction, there is no cost at all for the labor involved in data capture and input. Thus, the trade-off lies in a simple measurement of the cost of inexpensive equipment against the repeated, continuing need for separate, possibly redundant keystroking and verification at the central location.

Again, as with data capture, input represents an area of potential creativity and meaningful breakthrough; the systems designer probably can expect to have little impact on the way things are done at a central computer facility.

Processing Throughput

The systems designer, working with technical specialists, must develop an understanding of the current and planned computer installation and be responsible for accommodating the new system to it. The application under development will become, in effect, a customer of the processing facility. As a representative of the customer organization that will use these services, the systems designer is responsible for making the best deal possible for his or her own application. To do this, the systems designer must be able to present, factually and intelligently, the processing requirements of the individual application to members of the technical staff who will be responsible for developing support services.

Since the system under development is likely to be a replacement, it is natural to compare its processing throughput requirements with the current ones. The objective will be to determine the impact on the computer resources and operating environment. Specific types of information that should be reviewed with the technical and operations personnel would include:

- *Activity volumes.* The number of transactions for both on-line and batch processing and the frequency of processing, especially of batch jobs, will affect resource requirements directly. Volumes should include highs, lows, and averages. An increase in the volume usually will accompany the new system.

- *Pattern of activity.* Volumes alone do not tell the entire story. The pattern of activity will identify peak and lull times. This measure is important for determining the maximum times needed for resource support. In addition, it indicates lull times during which unneeded resources may be shared. For all subsystems, the activity patterns should specify seasonal trends (such as holiday sales) and calendar trends (such as heavy end-of-fiscal-year reporting). For on-line systems, the pattern of transactions throughout the day should be specified. For example, consider a national distribution firm with a centralized on-line order entry system at its East Coast headquarters. That activity is likely to be greatest within a time period from 11 A.M. to 4 P.M. EST, when locations in all time zones across the country are open for business. For a financial institution, end-of-year interest earned and paid statements require heavy batch processing and printing.

- *Processing time per activity.* Certainly, the longer the time necessary to complete a processing activity, the fewer such activities can be completed within a fixed time and with processing resource. This fact applies to both on-line transactions and batch job streams. Various components must be included in determining the processing time of an activity. These timings include CPU processing, data transfer to and from storage devices, printing or other output, and data transfer across communication lines. Terminal operators' response and fatigue factors also should be addressed.

- *Mix of processing activities.* Because a system's processing will comprise a mix of different activities, the use of resources will vary, depending on that mix.

- *On-line response time.* Although related to all of the above, performance of an on-line subsystem—and terminal response times in particular—are important enough to be highlighted. The timeliness of computer feedback at a terminal, in response to an operator striking the <ENTER> key, will impact, in turn, the performance and productivity of the user. The level of acceptable and desired response times will vary. A range usually is specified. Trade-offs weigh heavily here in the design of the technical configuration, including CPU memory sizes, disk access speeds, and data communication transfer rates.

The above factors are major considerations in evaluating designs for processing throughput capacities. Although interwoven tightly, these factors can be combined to determine a minimum and maximum impact on the installation.

Output

New output options also are emerging. Traditionally, the predominant output method was to create either an inquiry file or voluminous documents at a central location and provide results as needed through on-line inquiry or physical delivery.

Today, although centrally printed reports or standard terminal displays still may be at the heart of system outputs, there are other options. In a three-campus student records system, for example, a remote facility might maintain files for its own class schedules, combining this information with course descriptions from a central source. Individual departments or faculty members may have their own microcomputers connected to remote minicomputers. The central computer would remain a universal resource, but each faculty or staff member also might have the ability to tailor the delivery of outputs to specific needs. That is, users of such a system would not be limited to a standard menu of inquiries or to fixed content in reports. Rather, the capability exists to tailor outputs to individual needs.

In many instances, the same microcomputers or intelligent terminals that make it possible to control inputs remotely also can format and generate tailored outputs. For example, an investment of just a few hundred dollars will add a printer to virtually any microcomputer. For this relatively small sum, a user is able to document individual inquiries or to be selective about the data to be incorporated in a report. It usually is preferable to allow creation of selective reports remotely instead of requiring users to wade through large volumes of paper that are distributed universally.

Other options include the ability to accept and manipulate data at local terminals, to use graphic outputs, to incorporate data into a word processing document, to synthesize data and produce audio outputs, or even to install devices that create documents in a variety of typefaces.

To summarize, the systems designer no longer should feel constrained to the limitations of report or screen layout forms. Rather, the context of user operations can be matched closely with available, affordable output operations that already exist and are proliferating rapidly.

Storage

Many alternatives also exist today in the area of data storage. Data are among an organization's most valuable resources. To be useful to the organization, data must be stored in structures and formats that are planned, predictable, and readily available. Data resources must be safeguarded carefully to assure continuing availability and integrity.

Further, there must be compatibility of data resources throughout an organization.

Therefore, in the data storage area, the challenge lies in, first, identifying existing resources that can be of value to a new application and, second, making sure that the new data elements developed in support of the new application are compatible with the organization's overall data base. Remember, data structures emulate the organization they support. Constraint and discipline are essential in planning for and maintaining data resources.

UNDERSTANDING TECHNICAL GOALS AND OBJECTIVES

As discussed previously, each systems development project has certain business goals and objectives that are the underlying motivation for the project. These factors often involve increasing corporate revenues through expanding market share or marketplace, avoiding or reducing expenses, or providing enhanced customer service and satisfaction. These business objectives translate into systems specifications that address functional capabilities and data usage. In addition, though, the project may have technical goals and objectives that are concerned with the use of technology by the organization. These must be understood by the project team, especially those who are responsible for the design of the technical aspects of the system, including hardware, system software, and data communications. First, however, it is necessary to understand the organization's attitude toward technology.

In any given organization, management will have formed attitudes and adopted computer technology policies that may be unique to the organization. As one example, there are companies with computer technology policies shaped around a drive toward newness. These companies are first to acquire whatever is newest and most advanced in any given year. Conversely, there are companies in which management strongly resists any new trend. These companies want to be sure that a new technology is proven and solid before they commit to it. Of course, encountered attitudes comprise all gradations between these extremes.

Thus, in designing a system, one of the ingredients should be an awareness of where the company stands in its attitudes and approaches to computer technology. The policies relating to the technical computer environment can become either driving forces or constraints on each individual application that is developed. There is little that an individual project team or designer can do—or perhaps should do—to alter these attitudes. All that a systems designer may be able to do is to watch for indicators of these attitudes and adopt designs to fit them.

The technical environment within a computer installation may inhibit some potential for a new computer installation. For example, a large aerospace company considered a proposal for installing word processing terminals or standalone devices throughout the company. The purpose was to achieve compatibility with a computerized publishing system used to produce proposals, catalogs, and manuals. It was decided that replacing electric typewriters in all affected departments would involve purchase of some 4,000 pieces of equipment. Faced with this expense, the company elected to have the typing done on existing typewriters and then read text into the new publishing system through optical character recognition (OCR) equipment. Clearly, this company did not believe in rushing to install the newest and latest equipment just to be at a technological forefront. As this example shows, almost any corporate stance or policy in the area of computer technology is the result of a trade-off of decisions.

There are also companies in which gradualism is a preferred stance. Management seems to prefer, figuratively, to dip a toe in the water before jumping in. To introduce a new technology required by the application, it may be necessary, in such an environment, to plan for a phased conversion in which new devices or methods are installed incrementally and proven before a full commitment is made. Especially in projects in which computer resources are distributed into areas not previously automated, the typical approach is to install pilot sites to evaluate the system before beginning a universal installation.

For example, insurance companies that automate their agencies take this approach. A computer system with limited functional capabilities, usually marketing and sales support tools, is installed in pilot agencies for a trial period. This approach allows evaluation of the technology, its operation and usefulness, and the training that is necessary. Depending upon the success of these pilots, a decision on commitment to the system is made. Managing costs during this trial period is important. If such a situation exists, a strong argument could be made for version implementation, or phased introduction, of the new application system.

Having an understanding of the organizational attitude toward the use of computer technology facilitates the identification and understanding of the technical goals and objectives for the project. These technical goals and objectives complement the business and application objectives for the system. While allowing for the constraints imposed by corporate policy and the existing technical environment, these objectives usually are stated as opportunities. Technical objectives include opportunities to:

- *Overcome current technological constraints.* The existing technological environment may limit the ways in which business is conducted. For example, single-user minicomputers do not allow simultaneous sharing of data and program resources. A technical objective of a project may be to introduce a *local area network (LAN)*. This technology provides access to a common data base and program library by allowing CPUs to share peripheral devices such as disk drives and printers. As another example, if the minicomputer currently supporting the Central City water billing system can provide only for batch processing, the business goal of providing more timely customer service may support the technical goal of adding on-line communication capabilities. Upgrading to new generations of computer hardware or versions of software also may become technical objectives of the project.

- *Introduce a database management system.* A database management system (DBMS) offers advantages over traditional file access methods for both application development and technical support. Applications based on DBMS may be modified or extended easily, since the underlying database is quite flexible. These benefits are discussed in greater detail in later chapters. In addition, DBMS software provides upgraded technical support capabilities over traditional approaches. For example, programmer productivity aids often are contained in DBMS software. These productivity aids include data dictionary, file definition, query, and report generator capabilities.

- *Increased use of computer technology.* A technical goal may be to introduce or expand the use of technology within areas of an organization. Clearly, this objective must be coupled with sound business reasons. These efforts to broaden computer support usually are directed at areas with little or no automation.

- *Create networks of distributed intelligence.* Remote terminals or distributed central processing units can provide valuable computer resources to people at decentralized locations. Such locations may include branch banks, manufacturing sites, insurance agency offices, or catalog sales centers. Resources include processing functions, data, and communication capabilities between locations. The objective is to extend automated support while retaining centralized control through network management. There are significant business advantages to installing a distributed system. From a technical view, these networks are sophisticated endeavors with new and exciting challenges. Networks also extend the influence of the CIS department throughout the organization.

- *Introduce office automation and advanced technologies.* Increasingly, organizations are incorporating all forms of information—data, text, audio, and graphic—into a system's design. The automated merging of a word processing document with data extracted from a data base is a common example. Such capabilities provide an effective alternative to typing correspondence such as customer service letters. Also, audio technology and electronic mail are receiving increased interest. Suppose, for instance, that a manufacturing or distribution company elects to make inquiry capabilities available to its dealers or distributors. The best way to implement these capabilities may be through audio response systems or through electronic mail services that are already available. With audio response, any tone-generating telephone can become, in effect, a computer terminal. With electronic mail, inquiries and responses can be processed through existing facilities and services rather than by having to set up an entirely new network. With such technologies, designs that formerly were considered impossible now can be implemented effectively and at relatively low cost.

It should be stressed that the technical goals and objectives for the system are not sufficient reasons, in themselves, for the project to exist. The underlying business reasons are the primary objectives and provide the motivation for the project.

Technical Perspective

Just as important as an understanding of organizational policies and biases in the area of computer technology is the need for each systems designer to develop a perspective about his or her own feelings and positions on technological issues. There are people who are enamored with technology for its own sake and eager to apply the newest features. Conversely, there are also people who tend to wait for reassurance before they apply new methods.

In assessing a technical computer environment, the thing to remember is that technology should be a tool, not a driving force. The job of a systems designer is to differentiate and understand thoroughly the needs of the organization and of individual users. The challenge lies in finding the best way to implement a system that will do what the users need. Technology is a means for achieving this result, not an end in itself.

Given a set of automated tools for a system, the systems designer should avoid asking what human tasks can be replaced. Rather, technical alternatives always should be evaluated from the standpoint of users. Technology should not be employed to perform those tasks that can be done better by humans and *vice-versa*. In short, decisions should be based on the alternative that gives the greatest support to the people who will be involved in using a system and applying its results.

A particular technology never should be forced into a design, especially if users feel it is inappropriate or unnatural to the surroundings in which it will be used. At the same time, technical resources should be applied for the most convenient use by and the greatest support of the people who will depend upon a system.

Given these considerations, design decisions fall into comparatively few, general categories, including:

- Centralization or distribution of CIS resources
- Hardware/system software alternatives
- Data communications.

CENTRALIZATION OR DISTRIBUTION

With almost any application under development today, technical implementation must consider the decentralization, or distribution, of system capabilities. System capabilities are centralized when a single computer system processes all applications and contains all data. Batch-oriented applications that are run on the system use physical media such as cards and printed reports.

Data from remote locations can be processed through *teleprocessing,* or data communications over telephone lines, which expands the reach of the central system. Whether processing is done off-line (batch) or on-line, data are input via a remote device, transmitted to the central processor, acted upon by one or more application programs, transmitted back to the remote location, and displayed or printed.

Distributed systems occur when system components are separated geographically and application processing is done on more than one processor. Further, the application programs and supporting data are not necessarily the same at all locations. The distributed systems may be connected in a variety of configurations including *master-to-slave* or *peer-to-peer* data communication relationships.

In determining whether and how much to decentralize any given application, the main considerations relate to the organization itself. Corporations, for example, tend to be either centralized or decentralized in their management philosophies. The more centralized the outlook of a given management group, the more likely it is that a centralized system will be implemented. Conversely, the more vigorously a company has subdivided itself into operating divisions or geographic entities, the more likely it is that information systems will be decentralized.

Another aspect of corporate philosophy that impacts whether or how much to distribute information processing is the management style of a given organization's leaders. Some executives have the ability to deal with large volumes of data personally. A companion trait often noted in this type of executive is that he or she feels a need to take hold of all of the reins of power personally. The style that emanates from people of this type favors centralization of all aspects of management, including the processing capabilities and data resources that go with information systems.

By contrast, other executives measure their success by the extent to which they can delegate decision making and responsibilities to subordinates who, in turn, are encouraged to delegate their authority further. Such executives are more likely to provide the tools that subordinates can use to implement the powers that have been given to them. Accordingly, in such organizations, budgets, marketing responsibilities, and information system capabilities tend to be delegated. Organizations managed by executives with this type of personal style are likely to have multiple, distributed computer installations reporting to divisional or regional executives.

As time passes, the systems designer needs to be aware of changes in management philosophies. Such changes, if they occur, will not be dramatic. Thus, the degree of centralization or distribution may vary from one project or system to another.

Deciding What Resources to Distribute

When information system resources are distributed, decentralization takes place on a selective basis. That is, distribution of capabilities or resources can be a matter of extent or degree: It is not necessary to make an all-or-nothing decision. Degrees of distribution can be applied in three separate areas for any given application or system:

- Data
- Processing
- Control.

Data.   Distribution of data resources can be carried out in a wide range of options of varying degree or extent. The design decisions made here will have broad significance. Since data represent the primary corporate resource, processing and control designs will follow data. Questions of centralization or distribution of the data resource focus on the location of data, level of detail of data at each location, currentness of data, and the handling of redundancies.

Centralization of data exists when computerized files or data bases are housed and maintained in a single location under the custody of a single operating unit. Centralization exists even though on-line inquiry capabilities to those data resources may be worldwide. Conversely, distribution of the data resource is measured by the physical location of files or data bases on computer systems that are apart from a single, central facility.

To illustrate, the implementation of a three-campus student records system might call for a distribution of data. Data elements common to all three campuses would be centralized. Data elements individual to specific campuses would be distributed. Thus, each campus would have a separate set of files dealing with its schedule of classes, location and time of classes, and faculty assignments for those classes. However, since the courses offered are accredited and uniform throughout the system, these course catalog files could be centralized.

An example of how decisions concerning level of detail affect distribution of the data resource can be seen in the financial record-keeping aspects of the university system. Files could be designed, for example, so that each campus would keep detailed records of its own payrolls, purchases, and revenues. Summaries of these financial detail files could be prepared periodically and used to update a master, budgetary data base at the central facility. The data in the central system, then, would reflect the statewide status of the university budget. Should exceptions develop, inquiries could be initiated for references to the detail files on the individual campuses.

Alternatively, consider the files that support a supermarket checkout system. File references are made in split seconds as items are passed quickly over laser readers. As many as 10, perhaps 20, checkstands may be in operation at any given time in a single store.

Clearly, the supermarket example represents sufficient volume to support decentralization of merchandise pricing files. Sales histories are built up as merchandise is processed through checkstands. This information is accumulated in great detail in individual stores. However, the store manager really doesn't need the detailed records. Individual stores probably can operate effectively on the basis of summaries of sales by product, department, and time of day. Relatively small satellite systems in individual stores probably do not have the capacity to accumulate and process this level of transaction detail records. Thus, these files probably must be transmitted periodically to a large central computer that has the capacity to process and archive such vast amounts of data rapidly.

The currentness of data files depends upon the frequency of data communication between centralized and distributed files. For data that are maintained in the central office and needed also in the distributed locations, either the entire updated data file or just changes can be *downloaded*, or transferred from the host, on a periodic basis. The frequency of this transmission will depend on the volatility of the data and how up-to-date the data need to be at the remote location. Often, these batch transmissions are done nightly or weekly. The scope of data sent depends upon the percentage changed. In a student records system, data updates to courses with prerequisite changes probably would be sent annually. By contrast, an entire file of closing stock prices could be distributed nightly to offices of a brokerage firm. Communication of data from distributed locations follows similar considerations. In some instances, such as an order processing system, transmission may occur throughout the day. Local order files are maintained, and orders that cannot be filled locally could generate an order to the central file site immediately.

Data communication costs are reduced if data are located at each site, or where needed for processing. For common, or shared, data, redundancy of data would be introduced with this approach. An extreme approach, of course, would be to replicate all files at all locations, provided processing and reference volumes supported these expenditures. This level, or degree, of distribution of data resources would call for specific procedures to maintain, distribute, and protect those resources.

Maintenance of the data likely would occur at a single site. Procedures are necessary to provide periodic distribution of data to recreate data at all sites. As is the case with any distributed-site data, backup and recovery procedures must be developed for each site. In this case, such procedures are simplified because data simply can be transmitted from the maintaining site. However, depending on the volume of data and capacity of the communication network, this transfer can be an expensive and time-consuming process.

If files are maintained at a local campus and are not replicated at the central facility, the design of the system must assure that the decentralized files are protected adequately, reproduced periodically, saved at remote locations, and safeguarded generally against major disaster. The same assurance steps, of course, are necessary for central files that are shared with terminals in multiple locations.

Processing.   The distribution or centralization of processing follows functional lines and the handling of data. Separate decisions are made for such areas as data capture, input, mainframe processing, storage, and output. Decisions conform to corporate policies or positions while attempting to meet user needs or desires.

As a general principle, for example, it is considered desirable today, to incorporate data capture and, if possible, input functions within user source transactions. Following this principle, an increasing number of systems assign data capture and input responsibilities to the organization's customers themselves, making customers part of the processing cycle. An excellent example of this practice can be seen in the automatic teller machines installed by many banks. The customer making a deposit or withdrawal actually is posting the bank's files as part of the transaction. Many hospitals or medical practices now use systems under which patients enter their own medical histories, prompted by a series of questions. Similarly, a student records system might call for students to use terminals and enter their own requests for courses during registration.

Decentralized data entry of this type, however, does not necessarily call for decentralization of processing or record storage. Processing functions can be subdivided in almost infinite combinations. For example, an automatic teller machine may function completely under control of a central computer that develops all account balances and simply transmits funds-disbursement codes and transaction ticket information to remote stations. Alternatively, the processing can take place within the teller machine. The decentralized processor then computes balances and transmits information back to the central computer for file updating. Thus, a bank with 100 teller machines in place conceivably could process transactions on all of the machines simultaneously, queuing file update or access transactions at a central computer.

Part of such centralization or distribution decisions centers around questions of equipment configuration. For example, assuring continuous service to a network through a central site would require two or more central mainframes. If the active mainframe experienced an interruption of service, the processing load would be switched automatically to a standby unit. The standby unit might have been used for program development or batch processing tasks, which now are delayed. To continue providing network service, other tasks may have to be rescheduled. However, if processing capabilities are incorporated at the distributed processors, it may be possible to continue processing even though service to the mainframe is interrupted.

Hypothetically, for example, a banking system could assume that all cardholders are entitled to withdraw a certain sum, on a daily basis, if mainframe service is interrupted. Suppose, for example, that this limit is set at $100 daily. The local teller machine could hold transaction data in its memory or store them in secondary file devices, entering them into the main system when service was restored. On completing individual transactions, the teller machine could even encode dated limitations on the account by noting withdrawals on the magnetic stripe on the reverse side of the customer's teller machine access card.

The point is that there are many options for centralizing or distributing computer processing. Particularly in a world that is populated increasingly by microprocessors, there are no hard-and-fast rules—just a vast number of available solutions.

Control.   Control becomes an increasing concern as the extent of distributed intelligence within a system increases. As a basic principle, it is easier and safer to apply controls at a single location and have a single point for all transactions. Additional measures are needed as the number of entry points into or processing points within a system increases.

In general, controls are either operational or systemic. *Operational controls* deal with access, authorization, or verification. *Systemic controls* deal with hardware configurations and software updating to assure that functional compatibility and data comparability are maintained throughout the system.

Among operating controls, access deals with entry into and use of a system. For example, in a student registration system, access to the computer system could be gained through entry of a valid log-on code and student identification number. Authorization, then, would be applied by controls within the system itself. These controls are used to restrict access to functions and data. Further, authorization codes are invoked within applications to control processing. For student registration, authorization also would take place when the system checked its financial records to verify that there were no outstanding bills for the student attempting to register. Verification, typically, is achieved by communication from the system to the user. The intent is to verify to the user that processing was completed successfully. In a student registration system, verification could take place with a display and/or a printout listing the classes for which registrations have been accepted. This verification method would satisfy the user that desired results have been achieved. In the same way, an automatic teller machine typically prints a transaction summary each time a depositor uses any device within the system.

Systemic controls are designed to make sure that the system is geared to implement policy or design decisions. With a distributed system, data updates may need to take place at different locations within the network. Consequently, controls must be built-in to address the possibilities of failure in communication to or in processing at another location. Usually, verification data are returned for display or printing at the initiating processor to indicate completion of processing or transmission. Procedures must be designed to restore data if partial updates occurred, to retry communications, or to accumulate data for later transmission or processing. Without such controls, the system's integrity is damaged. In a student registration system, it would be necessary to update accounts receivable files in the system with each transaction. If, for example, a central file showed that the student had paid a balance in full while a satellite file showed that a balance was still due, an authorized student might not be able to register. Thus, if there are multiple files, multiple updates may be necessary. The same need for updating would apply to modifications in application programs. That is, each time a procedural change was made, all copies of software affected by the change would have to be updated concurrently.

In addition, system controls must be incorporated to prevent concurrent updates of a particular data file or record. For data in a centralized system, the second update transaction can be locked out from processing. However, such controls become much more complex if the data being updated are redundant on several processors.

DESIGN CONSIDERATIONS

Separate sets of issues or concerns must be considered for centralized and for distributed systems.

Issues

Standards or requirements should be specified in a number of critical areas, including:

- The level, or extent, of service to be provided to specific users or to a business as a whole must be specified. Part of this consideration lies in spelling out the cost, or consequences, of service interruptions. For example, one decision might be that an interruption of service of one to two hours could be tolerated in a student registration system. However, an airline that transacts millions of dollars of business per hour could not tolerate a situation in which its agents were unable to sell tickets during busy periods of the day. Hardware and software decisions, then, must reflect the urgency attached to any given application.

- In any system, be it centralized or distributed, both system software and application software will be dynamic. As a system becomes increasingly distributed, concerns can mount over the need to install current versions of software so that the same processing rules apply throughout a system at any given time.

- The same concern exists for data files. For example, a customer's record should be identical for supporting transactions at any point within a system. Similarly, pricing or discount data and rules should be uniform for all customers. Data files should be updated to support all points in the system uniformly and on a timely basis.

- Hardware and software problems, including diagnosis and solution of those problems, can vary widely for centralized and distributed systems. The more distributed a system becomes, the more necessary is the ability to log on remotely to distributed systems to isolate or identify problems and to deal with problems once they have been identified.

- The same principle, in general, applies to recovery procedures that are applied following service interruption or disaster. The more widely a system is distributed, the more complex it will be to reconstruct it.

Centralized Systems

If an organization operates with a single, centralized computer facility, it elects, in effect, to put all of its eggs in one basket. On the positive side, considerable computing power can be assembled through concentration. Conversely, however, this type of concentration increases dangers from mechanical, electronic, or accidental disruption of service.

An advantage of centralization lies in the relative ease with which resources may be controlled. All data files and all equipment are in one place. Thus, it is easier to monitor the updating of files and the supervision of data libraries. Further, it becomes easier to coordinate with equipment vendors to be sure that preventive maintenance measures are applied as scheduled and that any equipment failures are handled immediately. In a large, centralized facility, it is easier to store and plug in spare, or backup, devices to maintain service. If installations are large enough, and if most of the equipment in a given installation is provided by a single manufacturer, it may be possible to have full-time vendor maintenance personnel stationed at an installation.

At the same time, centralized facilities are highly visible and considerably more vulnerable than a system distributed over multiple sites. Because of the financial value of the assets and the vulnerability that comes with centralization, special security measures must be applied to a centralized facility. Most computer rooms have access systems for physical security. Provision must be made for remote storage of backup files that would be used for recovery if originals were destroyed. Further, arrangements must be made for backup processing capabilities that could be used on an interim basis if an emergency situation arose. When teleprocessing capabilities are present on the system, telephone lines are used for data communication, and the entire system is vulnerable to unauthorized access.

For the systems designer, there is little that can be done to impact the decision about whether a company's overall facility should be centralized or decentralized. Rather, the responsibility lies in evaluating the existing facility and providing the most practical measures available for assuring users continuing, reliable service. The needs of a given application for file protection and recovery plans will vary in differing technical environments. Even though the individual designer may be able to do nothing about the technical environment, an understanding of that environment can have a strong impact upon the quality of an individual system.

Distributed Systems

In reviewing the impact of distributed systems upon any given application, one of the first considerations should be the scale of distribution, including the size and capacity of individual processing nodes within the overall system. Distributed systems typically result from attempts to off-load processing from the central system. Especially with teleprocessing applications on the central computer, increased demands and activity volumes create a bottleneck on throughput. Performance of on-line systems may be degraded noticeably. The introduction of distributed processors often provides improved user services. With increased purchases of software systems by users, additional computer processors are acquired for running these applications. Finally, distributed systems are developed to bring resources closer to the user and to provide user control over processing of their data. Trade-offs exist in costs. The individual computers may cost less than expanding the mainframe, but the need for an extensive data communication network may offset that advantage.

Concerns also center around the maintenance and protection of data resources. Today, storage devices available for mainframe and minicomputer systems are highly reliable. On the other hand, these capabilities do not exist generally, at this time, for microcomputers. To illustrate the risk, devices that read and write data on diskettes are far less reliable, at least at this time, than hard disk equipment. Evaluation of the methods and relative reliability for recording, storing, and manipulating files at remote sites is essential in planning for the ongoing continuity and reliability of a new application. If data developed and maintained at remote sites are essential to a system, some provision should be made for creating off-line backup files or for transmitting the data to the central facility on a regular basis and incorporating them into master files.

Although distributed data processing gets around the inherent problem of having all of an organization's computer eggs in one large basket, there are also potential problems with the operation of multiple, smaller installations. Certainly, the operations personnel at remote locations are likely to be less experienced and will probably have less training in dealing with exception or emergency situations. Unavoidably, there will be a lower level of maintenance for a site that has a small minicomputer or microcomputer than is available for a large central facility. Spare equipment or spare parts may be in shorter supply at individual distributed sites. In a distributed network, it may be necessary to ship small units of equipment to central points for maintenance and to maintain a central store of backup units. In such networks, each equipment failure does mean some degree of service interruption at the remote point.

Maintaining, diagnosing, debugging, and updating software can be a lot more complex if an overall processing system is spread out over multiple sites. Even though processing is decentralized, all elements of the network still must be planned for, supported by, and managed by a centralized CIS staff.

Because distributed sites often will not have technical support at the location, the system support must be provided through remote log-on capabilities. These capabilities permit access to a distant system, using the communications network, so that it may be unnecessary to dispatch personnel to the site in case of trouble. Problems can be recreated, diagnosed, and corrected from a central support location. Both hardware and software can be addressed. Software modifications usually can be downloaded over the network. Hardware problems, though, will require sending an engineer to the site to do the actual repairs.

Although there is some inherent safety in dispersing processing facilities, there are also some exposures to risk. Instead of one massive backup or recovery plan, for example, a company may need multiple plans, depending upon the extent of distribution. The primary concern is ensuring that the plans are executed properly at each site. Even though the exposure at any individual facility is smaller than for a central location, destruction of resources is a possibility at each location—a contingency that must be planned for in advance.

There are also some security and privacy consequences involved in distributed processing that should be considered in designing each individual application. One frequently used precaution, for application programs, is to establish a rule that only object code will be located at remote sites. Greater security assurance results if the source code is maintained, protected, and accounted for carefully at a single location.

HARDWARE/SYSTEM SOFTWARE ALTERNATIVES

In general, computer processing equipment and related system software can range from large mainframes to minicomputers and microcomputers, and dependent terminals. Processing design challenges shift with the size and capabilities of equipment and related software.

With mainframes, for example, emphasis is on high volume productivity. The designer has to anticipate and avoid situations that can cause program execution to abort entirely. Rather, the idea is to plan for and deal with exceptions before they occur, making provision within programs for noting exceptions for later resolution without interrupting the main flow of processing. As units of processing equipment get smaller, systems become increasingly dependent upon human intervention. The smaller and less sophisticated a hardware device becomes, the greater the provision that has to be made for manual procedures covering recognition and resolution of exception situations or problems. On a mainframe, it is possible to have literally millions of transactions processed without involving people in resolution of exceptions. Exceptions may not even surface until the end of a business day or possibly the end of a week. On a microcomputer, by contrast, each exception may need a human response.

For each type of configuration, the system must have an appropriate design. Individual applications may use a full range of equipment, from mainframes to microcomputers, and may, therefore, require programs and procedures to deal with all of these.

Mainframe Considerations

Processing to be done on a large mainframe assumes extremely short cycle times, large throughput capabilities, and the availability of mass storage to support processing. Also assumed is system software with extensive utilities and multi-tasking capabilities. A large mainframe system can be assumed to have both batch and on-line capabilities. There usually are large batch applications to justify the kinds of processing power delivered by such equipment. However, many large processors exist just to deliver on-line applications.

The system software supporting a large mainframe will be structured to implement some sort of priority scheme. A mainframe installation typically serves many departments of an organization and may, at any given time, be processing dozens of different applications.

Usually, a modern mainframe installation will be isolated from users through physical security measures. Hardware devices will be in enclosed rooms accessible only through locked doors. Individual equipment units may be in locked enclosures of their own. For example, many installations keep the central processor separate from file peripherals and printers. If there is a centralized data entry facility, it is segregated from all of the other equipment components.

A processing center built around a large mainframe can be regarded as an information factory. The designer must recognize that users and operators associated with any given application will have little impact upon the processing or scheduling of jobs in such an environment. Because production delays are costly, the heavy volume makes it vital that programs anticipate and incorporate recovery capabilities for every imaginable contingency. In some critical applications, duplicate or redundant CPUs may be used to insure nonstop processing with automatic cutover, or switching from one host to another, if one processor fails.

Minicomputer Considerations

Most minicomputers are designed specifically for use in an interactive environment. Super minis, or minicomputers used as mainframes, may have up to 128 terminals, eight megabytes of memory, and more than 10 billion bytes of on-line storage. Many of the minicomputers now being offered are larger and more powerful than the largest mainframe of a decade ago. Therefore, many of the large minicomputer systems have, essentially, all of the processing capabilities of mainframes. That is, even though minicomputers may be installed largely for interactive use, they probably will be capable of handling batch jobs as background work with considerable efficiency.

The minicomputer fits a wide range of business environments, from the first-time user who needs a system with expansion room to the large, sophisticated user who employs minicomputers in an extensive communication network. Today's minicomputer can serve quite adequately as the sole host for centralized processing systems in small- to medium-sized corporations. In larger firms, department solutions often are offloaded to a minicomputer located within the user's area. Personnel, purchasing, and corporate legal counsel often are examples of areas where microcomputers are found doing CIS applications. Many minicomputers also are designed for the fast "number-crunching" needs of engineering or scientific research firms.

Microcomputer Considerations

Microcomputers are being incorporated into medium- and large-scale information systems with increasing frequency. These devices have moved rapidly from the arena of video games and hobby uses into the realm of serious business tools. For the systems designer, a decision to incorporate microcomputers in an application can present some special problems. For one thing, a microcomputer is a fully integrated computer system on its own. Thus, a microcomputer needs its own operating system and its own application software, regardless of what other provisions have been made for application or system software within the overall application. A microcomputer also has its own secondary storage that probably is physically incompatible with the storage devices or schemes used by other computers within the same overall system.

These conditions create some special challenges:

- Each microcomputer must be regarded as a standalone processing system, with plans and designs structured accordingly.

- Within a distributed processing environment, the microcomputer must be configured for compatibility with central capabilities, and other distributed processors, through data communications as a remote terminal or by direct attachment.

- The microcomputer offers additional, standalone processing capabilities that can support branch or remote offices—possibly through applications independent of the one under development.

- Microcomputers are highly communications-oriented. They can be linked into data communication networks using either leased lines or dial-up connections.

- The expanding presence of microcomputers is well-known. First-time users and experienced professionals all are among the purchasers. Even though microcomputers are considered to be in the infancy of their business potential, there are already vast program and application libraries available. Among the more popular are spreadsheet, graphics, and word processing. It is entirely possible that existing software can be used to support input or local processing functions performed on microcomputers, reducing overall software development costs for a new application.

This proliferation of microcomputers is one of the major concerns for organizations, and for the systems designer in particular. Because microcomputers are attractively priced and, in some organizations, are considered status symbols, many systems already may have been acquired. Thus, as a new system is developed, incongruities can arise quickly. Software under development likely does not run on all machines, and, consequently, either new equipment must be purchased to replace nonstandard processors, or dual sets of software must be maintained. Data communication capabilities also will vary, possibly precluding attachment to a network. These conditions represent roadblocks to the project and major challenges to the designer. A management policy decision is required that states what configurations will be supported. Supporting multiple systems is extremely expensive.

Terminal Considerations

An additional area of concern during design portions of a systems development project deals with the physical work areas, or *work stations,* at which people and computer systems interact. Of necessity, the logical-level work done in systems analysis does not deal in depth with work station alternatives. The systems designer, faced with the realities of throughputs and controls, must be aware of conditions within the working environment and their practical effect on the business application. Some of the critical human-machine interface considerations that should be taken into account in the design of computer applications are described below.

Distributed intelligence. Individual work stations can contain a range of capabilities from complete standalone computers to "dumb" terminals that have no processing capability at all and are totally dependent upon the processor to which they are connected. Obviously, the intelligent terminals cost more. Part of system design lies in determining whether such distributed intelligence costs are worth the price or whether it is more cost effective to use host-dependent terminals. On one hand, a "smart" work station makes it possible for some processing of transactions to take place independently of the work load at a central facility. The use of a microcomputer as a terminal represents, in effect, the ultimate intelligent terminal. Thus, at least a considerable amount of transaction processing can take place without interrupting and adding to the work load of central processors. Because the terminal's resource is dedicated, performance is optimal. Conversely, nonintelligent terminals put a demand on the processor's resources and cause interruptions of host activity every time an action is taken by the terminal user. This type of work station is less expensive to purchase, but there is the trade-off of eventual decreases in the processor's performance. However, depending on the nature of the application and the extent of terminal usage, savings can be extensive. Possibly an extreme example of the savings potential can be seen in systems that use in-place telephones as terminals. If such capabilities are adequate, it may be possible to implement extensive on-line services without installing any terminals at all.

Portable terminals. At this writing, portable microcomputers are gaining rapid acceptance. Two types of devices hold some application potential. One type is battery operated with some sort of recording medium, such as a microcassette. This storage capability makes it possible for individuals to record data or text at home, on airplanes, in cars, and so on. Many of these devices have built-in modems (see below) that make it possible to transmit recorded data from any telephone.

Another alternative is the full microcomputer built into a carrying case. Such units can be set up anywhere electric power is available. Addition of a modem makes it possible for these units to function as intelligent terminals.

With devices of this type, as with others, the designer should keep the business function in mind rather than the technology. For example, it might be attractive to establish procedures on which salespersons carry portable terminals into customer offices, recording orders right on the spot and transmitting the data directly into a central computer. However, in reality, the comparison should be made between the cost of a terminal for each sales person and the alternative of calling in orders on the same telephone that would be used by the terminal anyway.

Special terminal features. Color monitors are available for most microcomputers or video terminals. Obviously, color monitors cost more than simple monochrome units. The systems designer should be aware of the potential of color and should be able to assign a value to this potential that, in turn, can be compared with the cost involved.

The same principle is true for large displays or displays with special graphic or symbolic capabilities. For example, if a work station is to be used by a group of people rather than by an individual, it may be desirable to install a larger video monitor that is more legible from a distance. Again, corresponding costs may be determined readily. The challenge to the designer lies in establishing a value to be compared against these costs.

Microcomputers or terminals are being designed with a number of special *ergonometric,* or human engineering, features that may be attractive to executives and operators. For example, there are movable or tiltable screens that purport to ease eyestrain or neck-muscle strain. With concern for the usage of desk space, a user, for example, may select a feature that makes it possible simply to push the terminal screen out of the way when it is not in use.

A similar feature is the detachable keyboard that can be positioned separately from the accompanying processor and screen. Detachable keyboards may be seen in ads showing people with keyboards on their laps while conveniently sitting back and looking at attractive screen displays. Such features may serve to break down inhibitions for executive users. However, in production situations, the designer should be aware that keyboarding productivity probably will fall off markedly if such work practices become standard.

Still another work station option can be selection of keyboards that include function keys, as distinct from those that have little more than conventional typewriter keyboards. Typical functions that can be programmed into individual keys might be text insertion, file creation, file updating, data transmission, and so on. The alternate is to have a <CONTROL> key that is depressed while other keys on the keyboard are used to implement system functions. Labeled function keys can be attractive to occasional or inexperienced users. On the other hand, proficient touch typists may be comfortable using standard keys that they can find without looking. Either alternative, using dedicatated or standard keys, serves the same purpose.

A useful feature is multiple window displays on terminals. Usually, the display is dedicated to the data used as input or output of a single application program. However, instances exist in which a user wants to view data from two separate tasks simultaneously. Multi-tasking operating systems allow concurrent execution of separate tasks. Multiple window display of software at terminals allows the screen to be subdivided into sections, or windows. Because the terminal's screen is fixed in size, only a portion of the display from an application can be shown in a window at one time. The subset of the display data associated with each task, then, is displayed in its assigned window. This feature allows the user to initiate, interact with, and display data from several applications simultaneously on a single work station.

For example, suppose that the financial aid officer at a university is preparing a letter to an applicant. He or she is using word processing to create the text of the letter. To retrieve and quote financial data, a second window can be opened and an inquiry task initiated in it to display that data. While displaying the needed information, the user can go back to the first window and continue with the letter, incorporating information as necessary. If grade history needs to be browsed in order to determine an impression or pattern, a third window and task can be opened (or the second one closed and reopened) with this inquiry task.

Other features include graphic-type inputs such as light pens, touch-sensitive screens, or graphics tablets. As with the other features, each of these has its own price tag and place in the configuration of information system work stations.

Detailed consideration of specific work station features is beyond the scope of this book. However, the systems designer should be aware of the ever-increasing offerings that will be available at any time. Given this awareness, the challenge still remains to solve business problems while avoiding giving in to a fascination with technology for its own sake.

DATA COMMUNICATIONS

A data communication link or network is nothing more than another way to join the equipment components that form a system configuration. Looking at data communications in this way can impart a valuable perspective to the design of applications that will use distributed processing capabilities.

A computer processing system is an integrated entity. Any computer processing system, be it in a single room or in multiple sites around the world, consists of interconnected devices that can be used in coordination to produce a desired result. If all of the components of a system happen to be at the same site, the components probably are connected by cables that operate at system speed. Such devices are said to be *hard-wired*. If the individual components are dispersed geographically, the connections are through data communication links that usually operate at slower speeds and that, in addition, impose requirements for controls over completeness and integrity of data transmissions.

Given this relatively straightforward distinction, a decision to investigate data communication capabilities opens a Pandora's box of alternate options and configurations. An information systems designer need not be expert in the technical details of these alternatives and options. However, the designer must understand the basic concepts and trade-offs and know when to draw on whatever specialized technical expertise is available to the project team.

When elements of a processing system are connected through communication links, the choice of communication method to be applied depends mainly on a few critical factors. These factors include volume of data to be transmitted, frequency of transmission, and the importance of privacy or secrecy of data. For all systems, controls must be applied to assure that all transmitted data are received and accounted for. Selection criteria lie in the magnitudes and relative importance of the three factors—volume, frequency, and privacy.

Processing Options

In thinking of a communications-supported computer system as a single entity that just happens to be spread out geographically, much of the mystique often introduced into this field can be eliminated. With this outlook, the same kinds of processing options and services can be obtained from a distributed network as are available in a single installation. That is, processing options tend to center around batch or interactive capabilities. Remote, communications-supported sites can accumulate transactions and send those transactions to a central facility in batches. Remote sites also can do local batch processing. The same is true for interactive service. On-line users can interact with a central computer; or they can use a local minicomputer or microcomputer for interactive processing.

Given that systems will need to communicate in either batch or interactive mode, data communication capabilities add another option of concern to the system designer. For networks with a centralized, controlling computer, transmission to a processing facility within a data communication network can be initiated either through *polling* or *demand* transmission.

Figure 3-1 presents a diagram of a typical polling system. In a system that uses polling, the central computer contacts each transmission point in a network sequentially and allows each node time to transmit or receive information. If the central computer has data to transmit, special characters are sent to the remote site to ask whether it is ready to receive data. If a positive response is returned, the central node begins transmitting. In the case of a negative acknowledgment, the central computer waits before sending another control signal. A similar sequence then follows for data transmission from the secondary node. This process is repeated until all points are contacted, and the entire cycle repeats. As an alternative, each outlying facility is allowed access to the network according to a predefined schedule. The central node assigns each secondary node a time slot during which it can access the communication line.

In a demand transmission system, the outlying user facility simply begins sending data as they become available. Provision must be made within the network and at the central facility to accept whatever comes in and buffer transmissions in some memory device.

Transmission Methods

Data communication is the movement of encoded information from one point to another by means of electrical transmission systems. Computers communicate with on/off electrical signals, or binary digital pulses, as shown in Figure 3-2A. Most long-distance communication between computer devices occurs over telephone transmission lines. The problem is that most telephone systems are designed to carry continuously variable, or analog, signals, as shown in Figure 3-2B, rather than discrete, digital pulses.

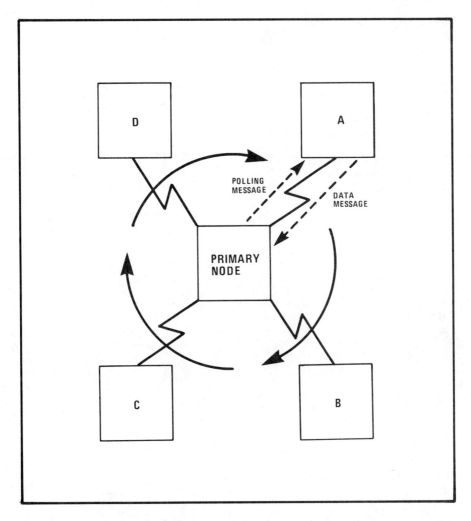

Figure 3-1. A polling system that provides a type of data communication for networks with a centralized, controlling computer.

When data are transmitted between the central processor of a computer and an attached, hard-wired terminal, the signals form a series of bits in a digital mode. When remote communication across analog telephone lines is needed, it is necessary to take the digital pulses and impress them onto an analog signal that varies by changes in the value of the information pulses. This translation step must take place before data transmission can take place.

There must be a device at each end of a communication link that translates computer digital code into analog code for communications

A.

1 0 1 1 1 1 0 1 1 0 1 0 1 1 0 1

DIGITAL SIGNAL

B.

ANALOG SIGNAL

C.

MODEM

MODEM

Figure 3-2. Waveforms of signals used in data communications.

and then, on receipt, translates the analog code back into the digital code again. Devices that handle these functions are called *modems.* This term is an acronym for *modulator/demodulator,* which describes the functions performed. On transmission, the modem *modulates* the digital signal, creating binary analog code. On receipt, the device *demodulates* the signal, translating the code back into digital format. If digital telephone service is available, a modem is not necessary. The action of a modem is diagrammed in Figure 3-2C.

There are two common methods for modulating signals, *amplitude modulation (AM)* and *frequency modulation (FM),* as illustrated in Figure 3-3. Amplitude is a measure of volume, while frequency measures timing. To indicate the presence of a bit of data using amplitude modulation, the signal is amplified, or made louder. With frequency modulation, ''on'' bits are indicated by causing the signal to oscillate faster, and ''off'' bits cause it to oscillate slower.

Volumes, or rates, of data transmission are measured in terms of *bits per second (bps).* Communication line speeds also are measured by *baud* rate, which refers to the number of times per second that a line signal can change its status of a communication carrier. The term baud is an abbreviation of the name Baudot, the Frenchman who developed the coding scheme for machine telegraphy. Baud rates are not the same as bps. Speed expressed as a baud rate is lower than the equivalent expression in bits per second, or bps.

Transmission rates are related to the capacities of the transmission media used. For example, traditional teletypewriter lines have transmission rates of 1,200 bps. These are single lines connecting two points. A normal, voice-grade telephone connection like those normally used in *point-to-point* dialing systems has a minimum capacity of 2,400 bps and a maximum capacity, with special conditioning, of 9,600 bps. Coaxial cables like those used for television systems can be equipped for transmission rates of up to three million bps. Similar capacities are available through microwave and satellite transmission systems, both of which use radio transmission principles. In addition, high-speed transmission is now coming into the marketplace through the use of fiber optics and laser methods, both of which use coherent light as a transmission medium. These light-oriented carriers have the advantage of being relatively independent of weather or environmental interference.

AMPLITUDE MODULATION (AM)

FREQUENCY MODULATION (FM)

Figure 3-3. Two forms of modulation used in signal transmission.

Service Options

When data communication is called for, a major aid to incorporating this capability into a computer processing system lies in the fact that vast communication capabilities already exist and, depending upon cost justification and needs, are readily available. That is, before computers ever entered the scene, a highly developed telephone network already existed. Through the years, measures were adopted that joined the two technologies, making it possible for each to extend the other. Telephone networks became, in effect, large systems of interconnected computers as computer technology was applied to such applications as call switching and transmission monitoring. As telephone networks became more electronic in their content, they also became more supportive of data communication requirements of computer systems. This evolution accounts for the wide range of options currently available, some of which are described briefly below.

Dial-up service.   In *dial-up service* used for data communications, small, inexpensive modems can be connected to any telephone instrument, anywhere, to serve as transmission or receiving points for computer communications. Thus, in its simplest form, a temporary, point-to-point connection for data communications can be implemented from any telephone, in any home or office. At this level, the possible combinations of transmission and receiving locations are virtually infinite. Modems that make this type of service possible are available at prices down to a few hundred dollars each.

Wide area telephone service, or *WATS,* is an option under dial-up services. WATS service can be contracted for with a telephone utility by a user for either outgoing or incoming calls at any given location, or for a combination of incoming and outgoing service. Contracts can be full time or part time. Tariffs for WATS service range from unlimited use for a monthly fee down to agreements covering a minimum of 10 hours per month, with additional time charged incrementally. Data communication capabilities are implemented in the same way as for dial-up service. The only significant difference lies in economies of scale. For large users, WATS service costs less per call than dialing individual numbers. However, the same dialing methods are used and calls are routed through the regular telephone utility network.

Dial-up, or switched, lines have the advantage that the destination points are not fixed, and, consequently, changes in the physical locations of nodes are accomplished easily. Costs are based on usage, so low volumes of transmission can be affordable. If data volumes are high, these cost savings are offset by the slower transmission speeds and the resulting longer connect time. Because lines are not dedicated, busy signals may be encountered, forcing retries at establishing the data link. In addition, since these are standard, voice-grade telephone lines, *noise*, or interference, can occur. Noise will degrade the performance of a communication channel, alter the signal, and add connect time for correction or require lower transmission speeds.

Leased-line service.   *Leased-line service* establishes fixed connections between identified points. In effect, the users contract for sole use of permanent interconnections. Leased lines can connect any number of points. In some situations, just two points are involved. There are also leased-line networks interconnecting literally thousands of points. For example, the news agencies that serve newspapers and broadcast stations can have thousands of ''drops'' served by single transmissions.

Leased-line service is more expensive than WATS service. However, the quality of the connections is usually better on leased lines than on dial-up service. Leased connections are best when high volumes of traffic originate from and terminate at known points. Filtering elements usually are added to leased lines to support high data transmission rates. This process is called *conditioning.*

Leased lines offer several advantages. The higher rates possible for transmission support higher volumes of data. Costs are fixed and the line is always available. However, the destinations are fixed also, which enforces some restrictions on the network. If volumes are low, the fixed costs result in a high cost per unit of data.

Special data transmission services.   A number of companies now offer data transmission services as alternatives to telephone utilities. These organizations use some of their own microwave links, lease capacity on satellite systems, and also tie into telephone networks as appropriate. Generally, for large volumes of data, these services are more economical than telephone tariffs. Further, some of these companies have specialized data switching capabilities, making their services more compatible with data communication needs than use of voice-grade lines might be.

Networks

Networks are multiple-user interconnections for data communication purposes. Virtually all dial-up and WATS traffic is carried over network systems operated by communications utilities or other specialized companies. A growing trend has developed in the establishment of *private networks* by large corporations or governmental organizations. Private networks serve multiple points within individual user organizations. Many companies that have gone into distributed data processing have implemented these capabilities through private networks.

Within any given network, there will be two or more computers serving as data communication processors. These processors, in turn, will be connected, through modems and transmission links, to each other and to terminals, microcomputers, or a variety of input and output devices. In a multiple-computer network, the processors that serve as control points, monitoring and logging transmission traffic, are known as *nodes*. An alternate term used to describe this function is *concentrator*. A node or concentrator is a hub that services terminals or other user devices. Within the processor, files are maintained to log data entries, data transmissions, and verifications of receipt of transmitted messages.

The transmission and receipt of data communication traffic are governed by sets of rules known as *protocols*. Implementation of protocols has been described as electronic handshaking. The protocols establish the bit, character, or message synchronization between communicating devices. Synchronization signals exist to facilitate the recognition and correction of errors, and to determine how devices can access a network. Protocol standards serve to govern the formation of strings of bits into characters and characters into fields and records. In transmissions, entire messages are assembled under control of these protocols. Each character is validated on transmission and receipt, giving rise to the allusion to handshaking.

Protocols also serve a security function, since they establish the access paths and passwords that will be honored. Where different types of user devices are linked into a network, protocols establish the code formats and patterns that create the needed compatibility. If devices operate on different protocols, the devices cannot be used together.

Data communication networks are set up, generally, under two types of configurations, or *topologies*. Topologies are physical patterns of interconnections between nodes. A network can consist of a series of nodes functioning as peers or can be controlled centrally with a host computer that switches traffic to using nodes or concentrators.

The peer organization method uses a decentralized organization scheme. One common configuration, illustrated in Figure 3-4, is the *ring structure*. Under this approach, any number of processors can be connected in a continuous path. Messages, or *packets* of data, are routed continuously around the circular configuration of a ring structure. Each node accepts and assimilates the data packets intended for its users, passing along those that belong to others.

Under a centrally controlled organization, all data go through a central point that performs all housekeeping functions associated both with data transmission and information control. A *star network* structure, the most common point-to-point network, is diagrammed in Figure 3-5. The central switching equipment logs all receipt and retransmission of messages. Housekeeping software initiates monitoring and control messages to all user points. These messages are used as verification that all transmissions were processed and received as directed. Data can be sent from one secondary node to another by routing through the central controller. Under a point-to-point network, it is no longer necessary to circulate packets of data, passing them through all points to arrive at the desired point. Rather, a message goes directly from origin to central controller to destination.

Systems Design Concerns

In designing individual applications that use data communication capabilities, concerns center around:

- Speed
- Accuracy
- Access
- Compatibility
- Cost.

These factors are not considered in standalone fashion as are many other considerations associated with systems design. Rather, they are highly interrelated. For example, speed of transmission is related directly to the cost of service; the greater the speed, the higher the

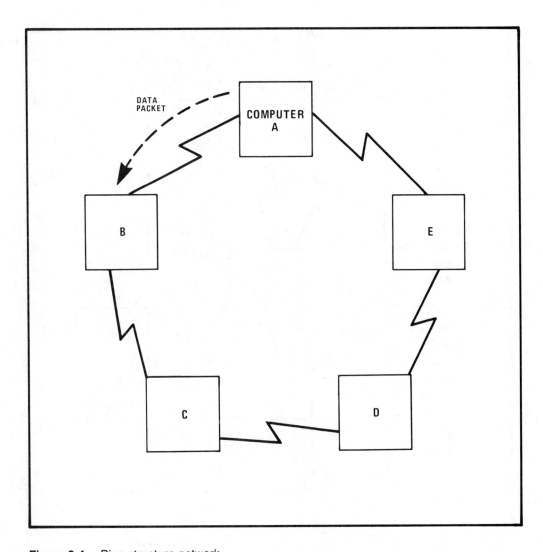

Figure 3-4. Ring structure network.

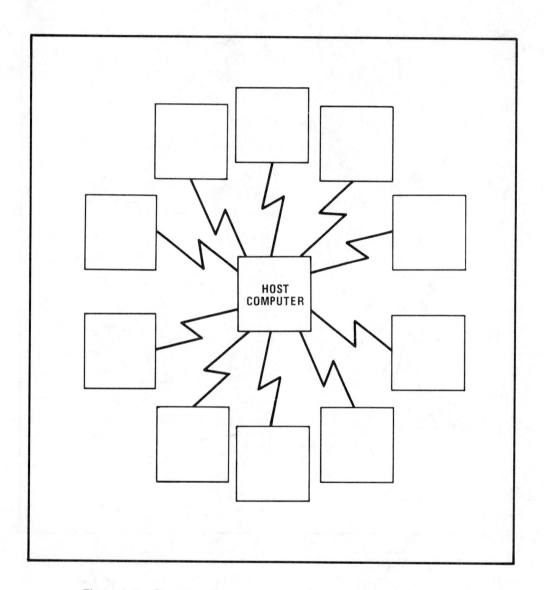

Figure 3-5.   Star network.

cost. Therefore, trade-offs center around identifying the costs of different communication capacities, then measuring these costs against known volumes. These figures, in turn, become criteria for determining whether data transactions will be transmitted as they are created or whether volume jobs will be held for batch transmission during idle overnight hours.

Accuracy also relates to cost—and to speed as well. High-speed service requires dedicated communication lines. The main enemy of accuracy in data transmission is electronic interference, or noise. Dedicated, or leased, lines can be conditioned to be less noisy than dial-up connections.

The question of access is of special concern to the systems designer. Access can be controlled through hardware and software measures. In a leased-line system, for example, access is controlled by the dedicated connections. In a dial-up system, extensive user passwords are necessary to control access. If privacy or confidentiality of data is an important factor, it may be desirable to install special devices that scramble signals on transmission, then unscramble them on receipt of messages. This *encryption* guards against either inadvertent or purposeful connection into lines by unauthorized persons.

Compatibility of data becomes a concern and a systems design factor largely because data communication makes possible the transmission of data between computers with different data formats or binary structures. The binary formats of multiple makes or models of computers also can be rendered compatible by controls, known as *protocol emulation* software, that can be built into data communication systems. From the designer's standpoint, technical compatibility should be thought of as a checklist item. Any time communication capabilities are considered, code formats or protocols should be checked for compatibility. Conversion requirements should be identified and implemented.

An awareness of the technical environment within which a system will function should be thought of as a prerequisite to further design efforts. That is, the environment should be studied and understood as a basis for selecting and applying a strategy for application software design. Alternate strategies and selection criteria form the content of the chapter that follows.

Summary

The capabilities within any given computer installation may represent either new opportunities or unforeseen constraints in the development of designs to implement user specifications.

As systems design is begun, evaluations must be made of existing technologies and capabilities to determine where and how a new system fits into its overall processing environment. The sequence of evaluation can follow the data processing cycle, including: data capture, input, processing throughput, output, and storage.

In applying microprocessors to data capture, the systems designer can have a profound impact upon system capabilities.

As a general rule, users should control and operate the input function as close to the point where transactions take place as possible, unless there is a compelling reason to do otherwise.

The systems designer, working with technical specialists, must develop an understanding of the central installation and be responsible for accommodating the new system to it. Service will suffer if estimates of either service demand or service capacities are inaccurate.

The capability exists either to tailor outputs at a central facility or to bring records from a central facility to a remote location where tailoring can take place. The systems designer should not feel constrained by the limitations of universally applied report or screen layout forms.

To be useful to the organization, data must be stored in structures and formats that are planned, predictable, and readily available. Systems designers must be concerned about identifying existing resources for data storage and assuring compatibility of the new application with those resources.

Information that should be reviewed with technical and operations personnel involves comparing new and existing system volumes and familiarizing the operations center personnel with the new application.

The technical environment that must be understood and accommodated before further design work can be undertaken includes management attitudes toward technological innovation as well as specific technical capabilities and trends within the CIS operation.

Key areas of concern to the systems designer in planning the integration of the new application with the existing technical environment include centralization or distribution of CIS resources, hardware/system software alternatives, and data communications.

Degrees of distribution can be applied in the areas of data, processing, or control. Design considerations relating to distribution include the level of service to be provided, uniform application of software versions throughout the system, compatibility and uniformity of data files, difficulty of servicing hardware or software problems, and provisions for recovery.

Depending upon the degree of distribution, separate alternatives and considerations exist for the utilization of mainframes, minicomputers, or microcomputers.

Work station considerations include distributed intelligence, portable terminals, and special terminal features.

Data communications should be thought of as an alternate method of linking computer hardware units. Communication may be achieved by either polling or demand systems. Systems design concerns for data communications include speed, accuracy, access, compatibility, and cost.

Key Terms

1. local area network (LAN)
2. teleprocessing
3. master-to-slave
4. peer-to-peer
5. download
6. operational control
7. systemic control
8. work station
9. ergonometric
10. hard-wire
11. poll
12. demand
13. modem
14. modulate
15. demodulate
16. amplitude modulation (AM)
17. frequency modulation (FM)
18. bits per second (bps)
19. baud
20. point-to-point
21. dial-up service
22. wide area telephone service (WATS)
23. noise
24. leased-line service
25. private network
26. condition
27. node
28. concentrator
29. protocol
30. topology
31. ring structure
32. packet
33. star network
34. encryption
35. protocol emulation

Review/Discussion Questions

1. What sequence of evaluation may be used in matching system requirements with technical capabilities?
2. What is the role of the project team in making decisions on data capture and input methods?
3. In general, where should the input function be placed and what is the rationale for this guideline?

4. Why is it important for the system designer to work closely with central facility technical staff?

5. What trends should be anticipated in tailoring outputs for users?

6. Why might storage not be considered an area for rapid technical innovation?

7. What factors must be considered in understanding technical goals and constraints within a given organization or CIS environment?

8. Into what three basic categories may systems design decisions be grouped?

9. CIS resources may be distributed in what three general areas?

10. What issues are involved in deciding upon the degree of centralization or decentralization appropriate for a given CIS?

11. What alternatives exist for work stations?

12. What is a basic definition of data communications?

13. What are five concerns in designing data communication capabilities for individual applications?

Project Assignments

The Appendix presents three case studies that can be used in support of project assignments. All three scenarios provide functional system specifications that would result from systems analysis. Included are data flow diagrams, data dictionaries, and process narratives that would result from systems analysis. Assignments that relate to the content of this chapter will be found in *PART 2: Transition into Systems Design*.

4 APPLICATION DESIGN STRATEGIES

LEARNING OBJECTIVES

On completing reading and other learning assignments for this chapter, you should be able to:

☐ Tell why it is important to have a design strategy in developing application programs.

☐ Describe how application packages may be evaluated.

☐ Give application-oriented and technical requirements for modification of program packages.

☐ Tell what program development tools are available for use in developing application programs.

☐ Describe the implications of a decision to do custom programming.

☐ Explain what program development considerations are associated with custom programming for on-line systems.

☐ Explain what program development considerations are associated with custom programming for batch systems.

☐ Describe trade-off issues involved in setting a strategy for application design.

STRATEGIC OPTIONS

Application software development practices should fulfill a plan that is thought of in advance. It is easy to assume, from reading programming manuals and similar documents, that all programs are developed originally and from scratch, every time. Most of the literature, and most texts, understandably, place emphasis upon original design and complete development of programs. Students need this practice. In the real world of systems development, however, alternatives become attractive—particularly for executives faced with the reality of modern programming costs and delays involved in the preparation of original programs. Packaged programs save money. However, cost saving is only one of their attractions. Many packages also make it possible for users to implement programs months ahead of the schedules that would result if custom program development were undertaken.

Thus, it is realistic, today, to consider all available options first. Then, an application software strategy is developed as a guideline for the people who actually will create or adapt the programs to be used. Bear in mind that the setting of software strategies today is far from a binary situation. That is, there is no longer an either/or condition between buying programs that exist or writing your own. Rather, there is a whole range of options that can be used either selectively or in combination. Thus, the programs for any given application may derive from as many as four, five, or six separate sources or methods.

To illustrate, the list below provides just some of the options currently available. This list is extensive enough to illustrate the wide range of sources. From such sources, the designer can formulate a strategy to provide a solution that meets the user's specifications. Options to be described in the remainder of this chapter include:

- Application packages
- Custom programming
- Library routines and utilities
- Generator programs (including data base generators and query software, and fourth-generation, or nonprocedural, languages)
- Skeleton (modular) programming.

The majority of this chapter will address custom programming, since it is also at the heart of the development of other options.

APPLICATION PACKAGES

Depending on the computer to be used, you may find a supermarket-size selection of application packages to choose from. For example, in the range of microcomputers or minicomputers, software packages can be bought right off the shelf from a number of stores that have been set up specifically to show and demonstrate these products. Many packages exist for the mainframe computers as well. With this kind of support, it is possible to buy a computer today and to have an application running tomorrow or the next day.

The range of application packages available is almost as great as the numbers that have come to market. There are packages that set up record-keeping systems for entire companies within specific industries. Mileage reporting for freight carriers, cited in Chapter 2, is one example. Also, it is possible to buy packages that provide all of the record keeping and accounting needs for medical or dental offices. There are also packages that public accountants can use to run their own practices. Many generalized packages have been developed for the legal profession, for manufacturing companies, for distribution companies, for wholesalers, for retailers, for banks, and for insurance companies.

There are also assortments of programs that are almost universally applicable—being usable across a broad range of companies. Payroll is cited as one example in Chapter 2. Other typical examples include general ledger reporting packages to produce financial statements, inventory reporting and replenishment systems, invoicing systems, accounts receivable systems, accounts payable systems, and income tax computation programs.

These programs are available from a wide variety of sources. Computer hardware manufacturers usually have an inventory of software packages that they offer and support. Also, they maintain a catalog of third-party software vendors who offer software for the manufacturer's hardware. Such resource lists cover both generic and specific industry applications. Finally, trade publications and professional association journals contain advertising for software products. Computer users groups usually conduct exchanges in which members can share programs, sometimes at reasonable fees, sometimes without cost other than membership dues. Most software development companies also produce catalogs that are usually available for the asking.

Computer stores, especially ones that sell microcomputers, offer off-the-shelf software.

It is during the analysis and general design phase that a decision should be made on whether application packages are to be used and, if so, which packages should be selected. Significant advantages can result, but the trade-offs of using packaged software must be evaluated for each project.

The most obvious potential advantage of application packages is that, if they fit, they can save substantial time and money. A software package that requires little modification can be installed in less time than custom development of a system from scratch. Users can realize a financial benefit from a more timely answer to their business needs. In addition, a CIS department with already stretched resources probably can afford more readily the expense of package selection and installation. This approach can provide better service to the user department and to the company in general.

There also is a degree of assurance—since other users have applied and found success with the package—that the programs are workable and of relatively high quality. Such programs may present an opportunity for a high-quality solution to a problem that is complex from a business view or is technically difficult to solve.

The main disadvantage is that, if the package does not represent an exact fit with identified needs, either the purchased package or the procedures of the user organization may have to be modified. To use the package, the company may find that it has to trade off desirable system features or change some of its procedures. In some instances, the user may have to take the system as it is or not at all.

With packaged software, the interface with other parts of the system or related systems may not be straightforward. If the mismatch is great, portions of programs may have to be revised or rewritten to tailor them to the specific user needs.

Another potential disadvantage of purchased software is that maintenance may be a problem. Maintenance can be performed internally or can be provided by the software vendor. Since the programs were not written within the organization, different standards may have to be applied in designing, coding, or documenting the programs. Further, program designs may be too inflexible to change readily. Even if the package is of generally high quality, time and money will have to be committed to training one or more CIS staff members to maintain it. Typically, a maintenance contract may be purchased that covers error correction and continuing enhancements to the package. This needed resource, though, is outside the direct control of the CIS department and, consequently, timeliness of response may be a concern.

Packaged programs cannot be purchased and installed just anywhere they seem to fit. Rather, before a decision is made to use packaged application software, careful study and evaluation are necessary.

Evaluating Application Packages

In considering the potential and appropriateness of application packages, the new system specification provides a good basis for evaluation. In the new system specification, requirements are described in terms of data and functions—at a highly logical level. Physical requirements usually have been addressed only to the user's view of interacting with the system.

Thus, on the basis of processing and data retention requirements alone, it is possible to set and weight priorities for the desired features of any given application software program. These weighted values can be checked against the specifications for a variety of available application packages. Though this type of evaluation may be too simplistic for final decisions, it is possible simply to combine weighted values assigned to each package, coming up with a total and buying the one with the highest rating. Thus, it is critically important to be sure that the package is capable of handling the data and functions specified. In practice, there are many qualitative factors that also would be considered. But this weighting approach can be a basis for identifying candidates.

Evaluation should be based also upon a number of other factors, including:

- The reputation of the software house or vendor can be checked and evaluated. Companies that have been in business for some time and have developed programs known to be reliable are usually better bets than individuals or persons who are dealing in application software as a "sideline." References should be sought. Such references should include organizations currently using the software package in which you are interested. References should be checked carefully and skeptically. Usually, suppliers will provide the names of customers believed to be completely satisfied. It may take detailed questioning, but an effort should be made to uncover weak spots or potential problems with any application package. A good technique to use in checking out a specific piece of software is to find out which version currently is being marketed. Also find the dates of earlier versions. See how frequently and extensively programs are updated. In general, the greater the stability for a program, the more reliable it is likely to be.

- The extent to which the vendor supports a software package is crucial. Evaluate the usefulness of the manual provided with the package. Find out if schools or training sessions are available. Also, determine whether the vendor offers a standard contract that provides assurance that the bugs found in the program will be corrected and that enhancements or future versions will be provided on a regular basis. The means of distributing changes should be evaluated for convenience, accuracy, ease of installation, and minimal interruptions to the organization.

- Attention certainly should be given to verifying that the software will run on existing, planned, or acceptable hardware. This same concern also applies to existing or planned system software. If a package runs in an environment—hardware or software—that is not installed presently in the organization, an evaluation must be made to determine the acceptability of introducing new vendors. The same evaluation factors presented here would apply in such a decision. Compatibility of the purchased software, and any associated technical requirements, is a major factor for evaluation.

- The software vendor's policy and willingness to customize or enhance a package will affect how well it matches the specific user's needs. Cost and timeliness of such changes must be weighed against benefits. In addition, maintenance of the package to meet ongoing changes in business needs must be considered. This aspect of maintenance is especially important in businesses that are volatile or subject to regulatory changes that require technical business interpretation. If a log of changes to the existing system has been kept, such a log can be a source of information about what changes may be needed. These concerns, then, should be discussed with the vendor to secure agreement on an acceptable maintenance policy.

- Quantitative measures must be evaluated to determine the package's ability to provide the necessary capacity, throughput, and performance. The purchased software also should match application requirements in terms of business and CIS processing cycles.

- Attention also should be given to: installation requirements, including any difficulties that may occur in bringing the purchased system into use; the completeness and quality of documentation; and the flexibility of design, or architecture, of the programs themselves.

Using the new system specification that contains both logical and physical needs, it is possible to compare individual data elements, key processing functions, and even specific process narratives to those of a software package. Those comparisons and the above considerations will assist the evaluation of whether an available package fits both the system and the organization. If the fit is not exact, the evaluation either will disclose what needs to be done to adapt the package or will reach the conclusion that custom development is necessary.

Package Acquisition Options

In the evaluation of application packages it is necessary to find out just exactly what is being bought. This evaluation is not made easily based only on the brochures read or even on the sales letters received.

There are often wide variations, for example, in the features incorporated with or included in the sale of a package. Determine, for example, whether a purchase delivers source code, object code, or both. Determine also what support materials should be received: systems manuals, user procedure guides, training documentation, or actual training classes. Installation assistance and customization of programs to specific user requirements also can be either included or available as options.

For any or all of these options, be aware that arrangements are negotiable. Depending upon the maturity of the package and the situation of the vendor, there can be wide ranges in what features, including support services, are offered without cost and what features or services result in additional charges. Charges involved, particularly for large systems, can be far from trivial. Before signing any contracts, make sure all of these bases have been touched.

Other considerations involve whether a package is offered for sale or whether it must be leased, with continuing payments for some period into the future. Also, prices are set according to different patterns. For example, some companies offer application packages allowing unlimited use so that once the package is purchased, it can be used as desired. In other situations, incremental charges are applied according to the number of processors on which the program will run, usually with substantial discounts as the number of processors increases.

Modification of Application Packages

With application packages, the reality is that there usually is no exact match between organization and user needs and available software packages. This becomes increasingly true as the scope and size of the system increase. If the software vendor will provide customization or will permit local modification by the CIS staff, a software package can be tailored to provide a complete solution. Modifications are needed to meet criteria of either the business application or the technical environment within a CIS facility.

Application-related requirements for modification of program packages can include:

- Needs or practices of the business itself

- Compatibility with data structures that already exist for a given application or as interfaces to other systems

- Conformance to processing schedules or requirements established for a given application.

Technical reasons for modifying a package can include:

- The need to conform to existing input, output, or data access methods within a computer operations center

- Conversion of packages from one programming language to another

- Modifications to make a package compatible with a given operating system used in a specific environment, particularly in the area of file access techniques.

- Modifications involving compatibility with changes in make or model of computer equipment.

Such modifications are made often. Thus, it is both logical and desirable to set up detailed procedures for the review and evaluation of application packages, and for the determination of which modifications may have to be applied. The effort required to modify an application package may include the following steps:

- The coding of the original application package, as well as coding for the modified version, should be checked in the same manner usually applied in walkthroughs of programs written in the shop. Typically, a line-by-line review of program code is conducted to identify business, function, and technical changes necessary.

- Conversion programs may be needed to translate files or programs to match the technical environment of a given installation. The same type of inspection and validation should be applied to any such programs.

- Individual pieces, or modules, of application programs may have to be rewritten. If so, programming should be handled under the standards that normally apply in the installation.

- Some minor redesign or change of system structure may be necessary.

Combinations of these modification efforts also may be involved. Thus, one of the evaluations that should be made lies in determining the extent of the required modifications and the costs involved. These costs, in turn, should be added to the cost of acquiring or renting the basic program. The combined cost, then, should be compared with the expense, the availability of staff, and the timeliness associated with custom program development. Cost is a major factor in determining whether an existing package should be modified or a program should be developed from scratch. However, cost is only one factor. Total benefits should be considered.

CUSTOM PROGRAMMING

Despite the extensive growth in the use of application software packages, the amount of custom programming done for business-related packages will continue to be considerable. Although software firms or consultants are retained to develop unique or custom software, most of this effort is done by an organization's own CIS staff. In general, there are three major reasons why CIS organizations develop their own programs:

- Although there is considerable breadth to the offerings of application packages, software is not available to satisfy all applications in all industries. Further, what is available may not be acceptable to an organization after an evaluation of the above factors. In such cases, custom development is the only remaining option.

- In-house development may be preferred for compatibility with other existing systems and business practices. A given company may have adopted a philosophy or unique set of standards for all of its application software. If so, management may feel that continuing compatibility is important enough to override any convenience or economy available through packages.

- Some systems people are simply more comfortable developing their own application software. Others may be skeptical about programs they did not invent or develop themselves. In general, this line of reasoning can be seen as a do-it-yourself syndrome. If this preference is strong enough, and if the manager who holds the preference has enough influence within his or her organization, in-house programming will prevail.

Separate concerns and considerations are associated with the development of custom programs for on-line and for batch use.

On-Line Considerations

On-line systems may be more expensive to develop and operate than batch applications. Therefore, one of the starting points in custom programming of on-line systems should be to build an understanding of why the option was specified. Criteria that may impact a decision in favor of an on-line system include a need for immediacy in response, a necessity for instantaneous file updating to provide current information continually, customer service or competition, and safety or protection. These factors are highly interdependent.

Justification for the immediacy of response of an on-line system usually is based on cash considerations and dependent upon timely retrieval of critical information. For example, airlines earn revenue by selling tickets. They can sell more tickets, with greater assurance of service, through on-line reservation systems. Also, banks retain and service their business through on-line inquiries about customer balances.

File updating and the potential for improved operating effectiveness also can be important requirements. In these situations, the transaction being conducted this minute may depend upon something that took place a few minutes ago, or even a few seconds ago. When the last seat on an airplane is sold, the person who calls one second later is put on a waiting list. This is a big difference in transaction outcome: The latter transaction actually depended upon the completion of the first only a moment earlier. If two parties to a joint bank account step up to different teller windows and attempt to clear out the balance of that account, the bank needs to be able to determine the exact status at the moment of each transaction. In fact, many banks went to on-line transaction processing because existing systems were unable to stop dishonest people from writing multiple checks to withdraw many times the amount of money they had in accounts. In the case of banks, on-line service actually can reduce operating costs—as well as protect the institutions through current file updating.

Applications such as on-line checkout at supermarkets contribute to improved customer service and provide competitive advantages.

Several factors are involved. Customer orders are processed faster by computers than by people punching cash register keys. In addition, the store realizes a considerable savings by eliminating the need to label each item placed on a shelf. Further, in a fiercely competitive market like food retailing, operators maintain the ability to reduce or change prices through simple keyboard entry alteration of a single set of computer-based price tables, rather than having to relabel merchandise. Operations also become more efficient because computers record exact sales by product, rather than by the classification system to which markets were limited with cash register entries. Besides accomplishing data input more rapidly, these systems facilitate the capture of detailed, useful data. This level of classification means that reordering and inventory restocking can reflect consumer buying trends more closely.

As an extreme example, on-line systems can provide safety and protection for users. This advantage, certainly, is the main rationale behind systems that enable law enforcement officers to check vehicle registration numbers through on-line systems while the officers are in pursuit or before they approach cars that have been pulled over.

The point is that, if an on-line system has been authorized, there will be some extenuating or overriding business reason for committing the special efforts and resources that are necessary. The systems designer should be aware of this background and develop an understanding of the rationale used.

Specific areas of concern associated with the custom programming of on-line systems include:

- Human factors

- File maintenance

- Continuity of service

- Security and access controls.

Human factors. With on-line systems, the concepts of *human engineering* and *user-friendly* software are extremely important, since any on-line system will involve high levels of human interaction. To maximize the effectiveness of the human-machine dialog, care must be taken to facilitate the input of data and its retrieval. Concern must be given to the design of:

- Initiating and terminating processing
- On-line assistance or "help" functions
- Confirmation of data input
- Output options.

The first consideration, therefore, lies in determining how a human operator communicates needs or commands to the computer. Today, human-machine interaction is controlled chiefly through menus of processing options. The computer terminal may be assumed to be available to the human user. Given this condition as a necessary starting point, it follows that the first step in initiating interaction must be taken by the human. However, the designer controls the processing options presented and the reactions that will be taken by the system. The structure of the human-machine dialog, then, is one of the pivotal design decisions associated with on-line systems.

Initially, of course, the human must announce himself or herself to the system. This action is referred to as *logging in,* or *signing on,* to a system. This initial step involves access control and security considerations covered later in this discussion. In overall design, the first critical decision lies in determining how and where a dialog begins.

The options lie in determining whether the system gives the human a choice, or range of choices, or whether the human is required to announce his or her requirements. On one hand, the system can present a menu of available options, requiring a choice by the person. For example, the customer service menu for the water billing system is found in Figure 4-1. On the other hand, the system simply can announce that it is ready and let the person request services through a computer instruction or command. For example, Figure 4-2 contains the command that could be entered to retrieve the account history for account 01-520-656 in the water billing system.

The best choice to make in this situation depends upon both the needs and the activity patterns of the user. For example, an executive

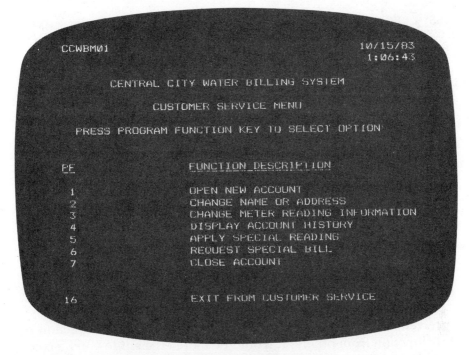

Figure 4-1. Water billing system customer service menu.

Figure 4-2. Retrieving information by a command.

user who will operate a terminal only intermittently probably would do better with a menu. The menu serves as a convenient reminder because occasional users might not become familiar enough with the system to remember all of the available choices from one session to another. Also, an intermittent user cannot be expected to memorize even a relatively small vocabulary of commands. Thus, if an intermittent user would have to study a series of command options each time the terminal is to be used, that person probably would find other ways to get the needed information. On the other hand, an operator who is going to use the system on a daily basis quickly will master a command vocabulary of reasonable length. Such a person's time may be wasted by tumbling through multiple levels of menus to find the desired option. For example, if the payments for the water billing system were entered on-line by a clerk twice a day as the mail was received, a menu would not be necessary that required selection of this function from others. It would be more efficient to enter a command plus the necessary data.

In short, the approach used in designing the software for an on-line system must be tailored to the needs and projected learning curves of prospective users. One way to establish these factors and associated criteria is to run tests. Such tests could involve either devising a small program using menus for various purposes or, if available, running some actual tests with microcomputer equipment and software to determine projected learning curves and user preferences.

Part of the design concern here lies in determining the extent to which menus have to be customized to applications and users. Different types of users may need different menu options. Sets of options can be offered through multiple menu levels, or hierarchies. That is, if a user asks for one option on a first-level menu, he or she may be directed immediately to the application. In other instances, an uncertain user may have to go through two or three levels of menu selection before reaching the desired application. For example, Figure 4-1 may be a second-level menu, whereas Figure 4-3 could be the first-level menu for the on-line portion of the water billing system.

Similarly, design care must be given to terminating processing. Usually, some key or combination of keys permits escape from the

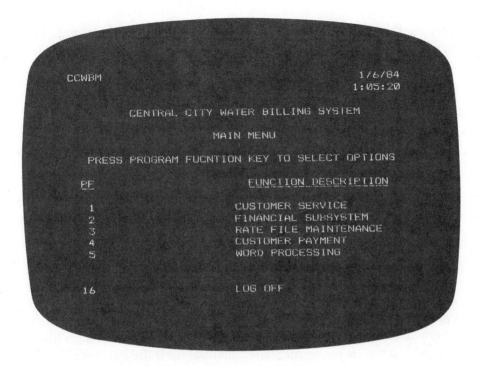

Figure 4-3. Water billing system main menu.

current function and return to the menu at the previous level. For example, while executing a water rate update within the water billing system, an escape key interrupts processing, leaves the data record under consideration unchanged, and returns the user to the menu. A similar exit or escape menu option should be provided upon completion of a task. In some instances, exiting a task may require presenting the user with options to close off processing. Ultimately, the user can return through the menu levels to the initial level.

Associated with—or as a supplement to—menus as a means of interactive communication, "help" functions often are added to application software. These functions are mechanisms that users can request in seeking assistance. When the user depresses a key to go into "help" mode, the system generally starts a dialog. This dialog is aimed at helping the user to identify his or her problem to the system. The system then responds with directions for accomplishing the task at hand. In effect, the "help" feature may be thought of as an inter-active instruction manual. Choosing the "help" option stops the processing that is underway. After the necessary assistance is ob-tained, the operator resumes the suspended task from the point of interruption.

With assistance subprograms, as with menus, the designer must make a choice of options available for user responses. One approach is to ask the user to position the cursor to indicate selection. Another is to ask the user to enter a *mnemonic,* or letter code. Another option is the use of programmable function keys.

Whatever approach is taken, the decision to use the "help" feature—and the specific assistance to be offered—should be based on extensive analysis. Given the design of an on-line system, the designer should create a scenario covering expected problems that a user may encounter. These problems then should be analyzed to determine if interactive assistance is necessary, if the user should be left on his or her own to return to the published manual, or if the designer should provide an option for escaping from the program and starting all over at the beginning.

An example of where a "help" function can be useful is in "forms completion" tasks in which the operator, in effect, fills in blank fields on a form displayed on the screen. For those fields that require a code input, the "help" function can be used to display the valid options and the interpretation of each. The operator can remember the code, return to the suspended task, and enter the code. Alternatively, the "help" display may allow positioning of the cursor at the appropriate code location and permit recording by striking an <ENTER> key. With fields that have several valid options, this approach provides an efficient means for entry. In addition, note that "help" can be built into a field prompt. For example, the notation "SEX (M=MALE, F=FEMALE):" may appear on a student records system admission screen.

Another human requirement that should be considered is whether users may need the ability to interrupt and restart their interactive con-versations. For example, suppose, in the water billing system, a person in the city finance department who is entering payments is interrupted for a status inquiry. There should be a way to take care of the immediate inquiry without losing the work that has been done. The operator should be able to apply a special command that saves the application being processed at its current status, permits the inquiry to be processed, then permits the operator to return to the suspended task.

Another valuable option is user review of keyboarded data and displays before actual transaction entries are processed. This provides the user with an opportunity to recheck keyboard entries and to correct typographical errors before erroneous processing can take place. For example, in a student records system, it might be envisioned that students themselves will use terminals and interact with the computer to process registrations.

One option would be to set up conversational entry patterns for each class requested. The student would enter class numbers. The system then could display the title of the selected course and ask the student to verify course identification before the request was processed. If the student had made an error in the initial entry, or wanted to reconsider, a change could be made at this point without affecting file status.

Such provisions often are made in invoicing, payroll, or accounting programs. In such programs, that deal with critical monetary balances, the operator is asked specifically to verify the accuracy of the information before the accounting entries are processed. If the operator indicates an error, the computer moves the cursor to each entry in the previous transaction, giving the user a chance to accept or alter the entry.

The point of these descriptions is to stress that the custom development of on-line programs requires tailored services that anticipate and meet needs of specific users. Thus, the design of on-line programs should start with—and stress—analysis of user needs and the human interface.

Human factors planning also has its output dimensions. Today, for example, many video terminals can support graphics output for users. Thus, it is entirely possible to build a customized system that uses standard output software to deliver graphs and charts instead of screens full of numbers. For many users, particularly among executives or professional planners, graphic outputs represent a more natural, more easily interpreted system response than do statistical reports.

Another factor to be considered is whether a terminal user may need *hard copy,* or permanent records, of outputs that may be presented initially on terminal screens. If this contingency exists, an evaluation should be done to determine how frequently, and with what urgency, hard copies may be needed. Examples of hard-copy requirements could occur at the executive level: An executive preparing for a meeting may want copies that can be distributed as well as viewed on the screen. Another possibility is that displayed data may be needed in the form of documents for external use. For example, the credit manager may call up a display of a customer account history during a telephone conversation, then want a copy to send to the customer.

Internal distribution of documented or displayed outputs also can be incorporated as part of the human factors of an on-line system. An executive, for example, may wish to route copies of a report to others within a company. Routing can be done easily if the company has an electronic mail system that can be interfaced with the data processing system under development. Two approaches are readily available: First, information recipients may be advised, through a displayed message, that certain reports have been routed to them. The recipients then can recall and review the information on terminal screens and decide whether hard copies must be created. Another option would be to create hard copies at the closest available locations to the desired destinations and then have the hard copies routed manually.

Audio response is computer output by digital voice storage or voice synthesis. This increasingly prevalent medium should be considered as an output option for interactive systems. Audio response has the potential of extending the range of the system to every telephone station in a company. Further, the availability of audio response would open the potential for delivering information directly to the homes of executives or even to selected customers.

For example, an order entry system might include a telephone number that customers could call to find out if the items they want are in stock. This application of computerized response would save considerable time for a company's own order-taking personnel. Also, such capabilities are being implemented widely by credit reference services. Credit inquiry stations in retail businesses are equipped with devices that read magnetic code on plastic credit cards. The CIS at the credit reference service, then, can give credit limits over the phone. As an alternative, the amount of the purchase can be keyed in at the point-of-sale terminal and authorization obtained over the phone.

File maintenance. File maintenance requirements take on special dimensions in on-line systems. Several factors must be taken into consideration in the design, including:

- Placement of boundaries for on-line and off-line file maintenance
- Configuration of on-line files
- Logging of transactions and records for backup and recovery
- Audit trails
- Handling of erroneous transactions.

The basic characteristic of an on-line file maintenance system is that the processing centers around entering a single transaction. The processing that the transaction initiates is to change one record or a group of related records. Thus, with a single transaction, it may be possible to access a record from a file, apply the transaction to update its data, and rewrite the complete record back to the file. By contrast, in batch file maintenance, a transaction file is processed against a file to be updated, changing some or all of the records in that file.

A special concern lies in establishing boundaries between on-line and off-line file maintenance activities. Three approaches may be taken to maintaining a file:

- Entirely batch updating of off-line files
- On-line data entry with batch updating of off-line files
- Entirely on-line data entry and real-time updating of on-line files.

Note that a file available for on-line inquiry may be maintained in either an on-line or an off-line mode. Usually, some combination of these approaches is used in any given system. Each of the files within the system may be maintained differently, or a file may be the object of multiple file maintenance subsystems, each usually with responsibility for different portions of the file. For example, recall the several approaches used in maintaining the files of the water billing system.

The first approach is discussed later, in greater detail. The second is a hybrid that has the advantages of up-front error detection and correction capabilities over the data entry techniques typically used with batch processing. The simplicities of batch updating in the areas of backup and recovery, audit trails, and control are retained. However, the availability of the most timely data is dependent upon the timing of the batch update. The last approach provides the necessary benefits to businesses for which current data are crucial to ongoing activities. However, special design considerations are necessary to insure data integrity and security with on-line maintenance. All on-line systems though, require some off-line attention to files in terms of maintenance and protection of data assets.

The extent of on-line and off-line maintenance functions depends primarily upon the nature of the system and the conditions under which processing services are provided. For example, a bank might support on-line services for teller transactions from 9 A.M. to perhaps 4 P.M. or even 6 P.M. However, this still leaves most of the hours in a 24-hour day available for off-line procedures that can include master file updating, backup and protection, copying of files, and exception reporting. However, in a law-enforcement system or in an airline reservation system, access to some files must be available around the clock.

A related concern deals with the configuration of on-line files. In some applications, entire master files are made accessible for on-line processing. Where sensitivity of data is an issue, or where batch processing efficiencies are indicated, access may be allowed to update only portions of records or designated records within a file. In some instances, special on-line files are created that are subsets of master files. Such a file might be maintained interactively to support other activities, such as inquiry or control, but represents an "unofficial" master file. Transactions are created by the on-line processing and are merged with other entered transactions for batch processing against the real master file. After this update, special on-line files may be rebuilt entirely, or modified records may be replaced with current data.

Files for an on-line banking application, for example, may have minimal account identification and balance information. Master records, such as name and address of depositor and detailed transactions for the current month, need not be available for such applications as check cashing, balance inquiries, or transactions on automatic teller machines. Instead, the on-line file would contain skeleton information covering only such data as account number, current balance (as of the close of the previous business day), and possibly a digitized signature for verification. The current day's transactions then update the available balance and are collected for inclusion in the nightly batch processing.

At some point, either the on-line file or a separate log would be used to update master records. Some provision would be made to reestablish the on-line file to reflect new balances. If the system provides enough time for off-line processing overnight, this updating is no problem. However, if the system provides user service around the clock, some intermittent method for file updating may be necessary. For example, groups or batches of records can be updated during time windows that run from a few seconds to perhaps a minute. During this brief interval, the records being updated would be locked out.

Designs for file maintenance in on-line systems should take into account quantitative factors that include the length of the service day, the amount of data required to support on-line service (as distinct from the total amount of master file data required to support the system), and performance/response needs. From the standpoint of response time, for example, a 15- or 30-second wait at a teller machine would be less critical than the same response time for a law enforcement officer giving chase to a suspect. Thus, usage patterns and volumes must be studied in depth with recognition of the fact that files must be maintained, and adequate backup and logging must take place. The file maintenance strategy for any installation should accomplish these necessary functions with the least penalty in terms of user service and processing interactions.

In addition to maintaining the on-line reference or master file, provision must be made for logging all transactions and for creating file copies as part of maintenance procedures. These files are backup for the system and are used for data recovery. At stake ultimately is the protection of the system's data resources against loss or destruction either from natural causes, unauthorized changes, or from system failures. A standard backup procedure with on-line file maintenance is to create a serial log of input transaction data and either a pre-update or post-update copy, or both, of the record being modified. Each entry has the date and time appended to it. Single or multiple physical files may be used. Copies of these logs normally are removed from the installation site and sent to remote storage on a regular basis, usually daily. Along with the log files, the backup copies of the on-line reference or master files are dispatched for remote protection as well. These copies are made as frequently as feasible, depending upon the length of processing time required for the copy and the available resources.

Restoration of the data may be required for an entire file or for selected records in the file. In either case, a date and time must be selected as the point to which the file must be restored. To recover the entire file, a backup copy, the most recent taken before the targeted date and time, is used to recreate the file. The log file of post-update record copies then is used to bring modified records up to targeted date, based upon the time stamp on the copied record. If only pre-updated images are available, logged transactions must be processed to update records using a batch process that applies the transactions in a simulated on-line mode.

To recover selected records, the cause of failure needs to be understood. If the cause can be isolated, the above approach can be used by limiting the records recovered to those needed through some means of record identification and selection.

To recover from a software failure in a transaction, an affected record first must be restored to its correct pre-update status. Then, the transaction is reapplied using corrected software and is followed by any subsequently logged transactions to bring the record up to date.

The logging of transactions for an on-line system also provides the basis for a requisite *audit trail,* or means of tracing the changes to a record back through its transactions and to their sources. In the restoring of an on-line record, financial statistics and audit trails may be affected and may have to be adjusted after the update has taken place.

The capability provided by an audit trail is necessary to answer questions about the current status of a record and its origins. For example, a customer may call the water billing department of Central

City and ask why he received a shut-off warning for a past-due bill. The customer claims he has the cancelled check for the amount in question. In such cases, the audit trail provides the necessary path of information for resolving the problem. In addition, audit trails are used to verify that input data can be traced through the system and that master files actually were updated. Auditors may be able to use the same log files that were created as system backup. However, additional or separate logging may be necessary to assure the integrity of audit trails and any necessary additional data such as authorization codes or links to outside paper sources such as correspondence.

In general, the degree of attention to audit trails centers around the impact of a given data item on the financial status and liability of the company. Auditors are guided by the factor of *significance,* such as *materiality* of amounts, in establishing monitoring and review procedures. Also, audit activities are carried on within a financial framework. Thus, if a given on-line system had only minimal impact upon the financial assets of an organization, it would be of less audit interest than one involving a high portion of revenue or liability. The more important a system is financially, the higher will be the degree of attention given to its audit trails.

To illustrate, suppose a bank were transacting only 3 percent of its business through automatic teller machines. Audit materiality generally centers around an estimated 5 percent of assets. Auditors would want to assure themselves that controls are in place over automatic teller machine transactions. However, this application might get less attention than reviews of trust accounts in which bank employees can use their own discretion in the handling of many millions of dollars of funds.

Still another, related concern deals with detecting, processing, and correcting erroneous transactions in on-line systems. All data processing systems must include controls that verify inputs before processing into outputs. When errors are contained within input transactions, the transactions cannot be processed. To protect the quality and reliability of the system, something must be done about such transactions. All errors committed by people or transaction records containing erroneous or unprocessable data should be intercepted and a means provided for correcting or, if possible, discarding, the transaction.

In terms of systems design, on-line procedures present both opportunities and potential problems in the handling of errors. Opportunities come from the fact that, because of the nature of on-line systems, errors are detected sooner than they are in batch procedures. Errors involving incorrect entries by operators usually can be corrected immediately, and, consequently, processing is not delayed as in batch updating. Such errors usually are announced to the operator by returning the screen display with the data fields having detected errors highlighted or blinking. In addition, a message explaining the errors may be displayed. The operator, then, either can delete the transaction entirely or can key in corrected data as needed and resubmit the transaction immediately. The ''help'' function, or on-line user manual, can be of assistance in such error correction.

In some instances, such as on-line data entry, it may be desirable for an operator to ask the system to accept the erroneous transaction by flagging the error field and holding it in some temporary file until the operator can determine the problem and correct it. Then, a ''tickler,'' or reminder, must be provided to the operator that the flagged transactions are pending. However, provision also must be made for the recording, handling, and correction of errors that stem from other causes, such as software failures or the omission of batch update processing that must occur as part of processing the transaction.

As a design principle, remember that all systems experience the introduction of errors, and all systems require ways of dealing with errors. In an on-line environment, a special opportunity is presented to detect and correct errors at the time of entry and, typically, at a point close to the source of the transaction.

Continuity of service.   At some time, any on-line system is apt to experience service interruptions. Therefore, systems design should concern itself with procedures for dealing with unexpected, unscheduled *down time.*

The exposure and vulnerability of a business must be considered; the consequences of loss of service should be evaluated and estimated. In some cases, an organization will be out of business, literally, during the time when an on-line system is unavailable. For example, when an airline reservation system is not functioning, no reservations can be confirmed or taken and customers may go elsewhere for their travel plans and arrangements.

Service can be interrupted by failure of either hardware or software. In the case of a software failure, it may be possible to disable a particular on-line transaction while keeping the remainder of the system available. After correction of the problem and restoration of any affected data, the transaction may be enabled. Extensive testing of software, as discussed later, can help reduce the likelihood of such an occurrence. Hardware problems may involve some combination of central processors, work stations, storage devices, or data communication network components. In some instances, problems can be isolated and taken out of service for repair or replacement. Thus, ongoing service may be maintained, at least for a subset of users. However, failure of critical components may bring down the entire on-line system.

If the situation is serious enough and the probability of down time is great enough, equipment redundancy may be supportable. Many on-line systems, for example, have two or more central processors and alternate availability of communication connections. Also, redundancies can be supported by the mix of processing in a given CIS area. For example, systems that normally do batch processing or are dedicated to CIS development may be switchable in the event that an on-line processor goes down. Such design decisions center around trade-offs between the costs of interrupted service and those for redundant equipment.

Typically—at least where business systems are concerned—alternate manual procedures are established for the usually infrequent incidents of service interruption. To deal with such situations, there must be some manual method of recording transactions. There also must be an alternate reference that can be used. In banks, for example, master files typically are copied to paper or microfilm and used as reference whenever the computer is down. Either the documents themselves or alternate records, posted off-line, are created to record transactions. When service is restored, all transaction records are input to update the files as though on-line service had been continual.

Security and access controls.   High levels of concern for security and privacy invariably surround on-line services. In this context, security deals with value inherent in assets. For example, by getting into on-line files, people may be able to implant bogus records or alter files to derive income unlawfully—or even create life-threatening situations.

Privacy is a slightly different issue. People or organizations represented in computer records have a right to restrict access to and distribution of these data. For example, student records maintained on administrative systems at universities are treated as private. Access is strictly on a need-to-know basis and, usually, only with student permission.

Design concerns over security and privacy stem from the fact that access to data may be available broadly and from remote points. To deal with these concerns, a series of special security and privacy protection measures have been evolved through the years. In designing any on-line system, the needed levels of security and privacy for all accessible records must be established. Then, procedures must be designed that assure the meeting of these standards.

At the very least, all on-line systems should require that users sign on to a system before they have access to data. Sign-off procedures also should be established and enforced strictly. To begin using a system, a user must, in some cases, turn a terminal on by using a key. Then, before accessing data, a user must enter some identification code and a password that is known only to that individual. The system must then validate both the identification and the individual password before permitting access to the system. Other access controls can be implemented by limiting use of a designated system to only specific terminals. Authorized individuals may be restricted to using terminals in secure areas only. On many systems, terminals are logged off and access is denied to the system automatically if there has been no activity for a set amount of time.

Control over access to a particular application system can be granted through limiting acceptable user identifications and passwords. The features of the system to which a user has access may be restricted further by the assignment of a "user class" to an individual by a systems administrator, or, again, by identifications and passwords. Additional codes or security information may be requested to specify and validate the user.

Besides restricting processing functions, security can be introduced at the file level. Even records and data items within records may be restricted for access and use. For example, in a student records system, students may have access to their individual records only. Within those records, students would have the ability to view all stored data but would be able to change only data describing future coursework. On the other hand, an admissions officer could have access that allows the viewing of all data for all students but that permitted only limited ability to make changes. Also, advisors could have access to a student's coursework, including grades, for counseling but would not be able to view financial information.

The issues of security and access control are related closely to privacy, especially aspects of privacy covered by government legislation. System intruders have received considerable publicity. The protection of corporate resources from such unauthorized access and potential damage is especially crucial in an on-line environment in which resources are not under "lock and key." Continual changing of security codes and requiring several levels of access codes are necessary measures.

Batch Processing Considerations

With computer information systems, time is truly equal to money. Response times can be virtually immediate. But each reduction in turnaround time of service carries an increasingly heavy price tag. For this reason, even in an age of highly sophisticated on-line capabilities, much data, in many organizations, are processed on a batch basis. Even if a system provides a measure of on-line service, substantial portions of that system still may be processed in batches.

In general, then, a characteristic of batch processing is that there is a lower level of urgency and, accordingly, less tension associated with the time factor. Processing functions that are batched often involve long running times to complete execution and prepare outputs. However, time still can be a critical factor in batch processing. All output deliveries should be scheduled. Schedules should be met. These schedules can be daily, weekly, monthly, or even annual. Whenever the output is due, the system should be structured for timely delivery. Further, batch systems may be designed to provide their services on demand. If this is the case, turnaround times usually will be scheduled for relatively quick delivery of results.

Considerations associated with application design for batch systems include:

- Human factors
- Cohesion and coupling
- Job-step boundaries and structures
- File maintenance
- Security and access controls.

Human factors.   Separate considerations apply to the input and output aspects of batch systems.

For input design, operational productivity and reliability of results are the prime considerations. To achieve these goals, many systems are using on-line methods to capture input, even though actual processing will be done on a batch basis. The on-line input can be handled either on the mainframe system or, perhaps, on separate devices such as microcomputers.

In any case, readability and usability of input source documents should be reviewed carefully. In particular, problems can arise when only a relatively few fields from a complex form will be used for input. Situations become even more error-prone when carbon copies and/or copies with multiple handwritten notations over machine entries are involved in input. The general rule to follow is that transaction documents used for system input should be as readable and clean in appearance as possible.

Forms should be designed so that they can be completed easily, correctly, and conveniently. The primary concern in any form or document design must be ease of use. For example, the original source document should be used itself for input rather than a coding sheet that requires someone to transfer data manually for use by a data entry operator. To improve the likelihood that a data entry operator will capture data correctly the first time, the order of the data recorded should match the order and placement on the source document. This correspondence allows for normal left-to-right and top-to-bottom reading of source documents by the operator. The computer, then, can reformat the data for processing.

Except for possible high-volume production data entry operations in which efficiency is a major issue, end users can provide a valuable data entry function. Users are better able to resolve errors and omissions because they are closer to the source of the transaction and are knowledgeable in the application. In areas of high-volume input, alternatives to key entry of data by humans are being used where feasible. Such techniques reduce errors due to human fatigue and boredom and provide increases in production. For example, in the water billing system, the computer-printed stub that is returned with the payment could be read by a scanner. For all full payments, the account number and any other pertinent information could be captured in this way. This technique would eliminate human errors in rekeying the computer-prepared data. However, the feasibility of purchase of the necessary scanning equipment must be weighed against the input volume and production needs and costs.

Increasingly, opportunities are opening to integrate inputs of batch data through word processing or microcomputer systems. At this writing, these office automation developments are relatively new. Integration of office automation methods into major computer information systems appears to be in its infancy, but growing. Both word processors and microcomputers are, of course, programmable. Thus, it becomes possible to incorporate editing, error detection, and screening functions into the capturing of input. Further, reliability and accuracy tend to be assured because operators in user departments are familiar, in many cases, with the multipurpose devices used as part of everyday jobs.

Concerns in the output area center around *readability, accuracy,* and *timeliness* of documents delivered to users. These factors need no elaboration. Requirements in these areas should have been defined as part of systems analysis. In design, it is important to make sure that standards have been set and will be met.

A final consideration in human factors for batch systems deals with the appropriateness and usefulness of outputs. There are many horror stories in the CIS field about regularly produced reports that, somehow, got skipped and were never missed. Surveys have been conducted indicating that, in a mature installation, it is possible that up to 50 percent of all outputs are no longer used by the persons who receive them. Thus, even during the design phases of a new system with outputs that have been requested specifically, some mechanisms should be established to monitor the appropriateness and usefulness of all outputs. Urgencies for specific outputs may have changed even since the initiation of a given systems development project. At the very least, challenging the use to which outputs will be put enhances an understanding of the business problem being addressed.

Cohesion and coupling.   *Cohesion* is the degree to which a process has a singular business purpose. *Coupling* is a related concept that refers to the interdependencies between two higher-level processes.

In the design of processing tasks, maximizing the degree of cohesion and a corresponding minimizing of the level of coupling should be the goal. For batch systems, the objective is to organize processing into highly cohesive jobs, each providing an essential business function. The only coupling between jobs may be access to the same data base.

Within a job, then, each job step should be highly cohesive. The coupling between steps should be by data only, such as the sharing of an intermediate file. The goal is to design job steps so that each has no need to know what specific processing precedes it or follows it. All that is known is the data that is input and output. These same concepts later will be applied to the modules that make up a good job step. Any control information from a job step that would affect the subsequent processing in the job stream should be handled by a procedure or job control language that is written to manage and execute the job stream. Passing such information from job step to job step to modify the second step's processing seriously increases coupling.

Care should be taken to avoid combining processing tasks in a single job simply because the tasks occur in the same time frame. For example, suppose an installation had three jobs that were run each evening: employee attendance reporting, job cost accounting, and inventory status. Good coupling exists between the employee time records processing and the job cost information processing, since labor costs will become part of job costs. However, there is not necessarily any relationship between these two applications and inventory updating. Thus, in designing batch processing, a job stream containing the two applications that share data could be created.

Job step boundaries and structures.   The practical usefulness of a batch application depends greatly upon how effectively jobs and *job-step boundaries* are defined. Job-step boundaries, in effect, provide increments, or segments, for processing within any given job stream. Within a job or job-step boundary, a program is relatively self-sufficient. That is, the program has its own inputs, processing routines, and outputs.

The independence, or free-standing nature, of job steps is critically important when it comes to reruns of relatively large jobs. In the realities of operations within a computer processing center, there are occasions when a job that is running is interrupted or aborted. Reasons can include anything from demands for priority jobs to power interruptions, operator errors, or even mechanical problems. Such occurrences require that computer centers allocate time for rerunning jobs.

In general, the smaller the processing increments within any job stream, the lower will be the penalty for job reruns. When a rerun is required, work begins after the last job step with intact output work files or system files and continues from that point. For a particular situation, a rerun may require recreating lost files or restoring partially updated files using backup copies. These additional efforts to get the job restarted plus the time required for actually rerunning job steps represents the total time lost. Generally, a large job with several job steps is less vulnerable if an unplanned interruption occurs. Each job step, then, represents a finality in a portion of the job's processing. For example, a job that requires 12 hours to run and contains eight steps likely will take less effort to rerun than a job that has three steps. There is, however, a cost associated with having several job steps.

Generally, an intermediate, or working, file is needed whenever a job crosses a job-step boundary. That is, the working file becomes the means of passing data from one job step of a job stream to another. As described above, intermediate files serve as buffers to minimize the effects of service interruptions and the need for reruns. However, other considerations also determine when and how effectively working files should be used.

Though working files provide protection for processing already done, there are costs and penalties associated with the creation of working files. Working files occupy storage areas and also require processing time for creation and reading. Therefore, the efficiency and integrity of the job as a whole should be considered in determining where and when to establish job steps and the required working files. Trade-offs, in addition to protecting work already done, include consideration of the complexity of each job step and of the job stream as a whole. Also to be considered is the impact on the total job run time of creating and using each working file. Writing data to a work file takes time, as does reading it in a subsequent step. With the introduction of job steps and the necessary intermediate files, design consideration should be given to validating any data that cross from one job step to another. Assuming that no programming problems exist, data received from a preceding job step should be in the proper format and have valid content. Even so, a standard practice is to validate file content by such means as accumulating statistics during processing and checking them against input control totals. These totals may be record counts, hash totals of all account numbers, or financial totals such as payment dollars.

There are other factors that influence configuring a job into job steps. The hardware capacities or some constraint of the configuration may force splitting processing into smaller job steps for which there are adequate resources. Security standards and audit controls must be considered. For example, in the water billing system, payments are batched and processed nightly. Audit controls will require that, after these transactions are edited, verification takes place to insure that the total dollars in the payment transactions match the bank's payment deposit total for the day. Such a control likely will dictate a job step separate from the update run, if not an entirely separate job. Job steps may be introduced to take advantage of appropriate system processing utilities, such as sorting or file copying, or application packages, such as a report writer or a liquidity analysis program.

File maintenance.   Files are the foundation for any information processing system. However, this foundation must be maintained and protected if the system itself is to build and maintain a sound structure. Thus, file processing and maintenance strategies are important if quality is to be designed into application systems. Separate sets of strategies are appropriate in dealing with files to be stored on tape and magnetic disk.

In maintaining master files stored on tape, transaction files must be sorted into master file order. This sorting is necessary because the sequential update process is accomplished by a collation of the master file with the transaction file. Also, in processing sorted transaction files, cross-edit procedures may be applied to validate a group of transactions that relate to the same master record and to determine that these transactions can be processed.

Routines to update master files with transactions should take advantage of the basic nature of tape media. In file updating functions using magnetic tape, new master files are created as a byproduct of each update run. That is, transaction records come from one tape file and master records from another. The file maintenance program correlates these and creates an entirely new file, which is written to a separate tape. Thus, the source tapes remain intact and a new master is created. The new master file is referred to as the *son file.* The original master file, the *father file,* is saved along with the transaction tape used to update it. In addition, the father to this original master, referred to as the *grandfather file,* is saved as additional backup along with the transaction file that was used with it to create the father file. Thus, the use of tape facilitates addressing security and backup needs and is relatively inexpensive, since tape is the least costly storage medium. Typically, these files are stored in vaults, with the grandfather file being sent to a secure site away from the computer center. Thus, if tapes are destroyed, recreating a current master file is a straightforward matter of running the backup tapes again. In addition, copies usually are made of key business cycle master files, such as year-end, and month-end, and these files are retained on a long-term basis.

In processing master files stored on disk media, special provision must be made for backup files, since new physical master files usually are not created automatically as byproducts of disk updates. A characteristic of direct-access file processing, typical of disk-file applications, is that updating is done in place. That is, for a transaction, the master record with matching key is accessed from the disk; its fields are changed; and the entire record is rewritten to the disk, probably in the same position. No new physical file is created. Consequently, backup of master file must be provided explicitly. The simplest approach is to *dump,* or copy in its entirety, the disk file to tape for inexpensive storage that can be removed from the site. If tape media are not available, the copy can be made to a removable disk pack. Copying may take place either before or after file updating, or both. For large files, the backup approach used with on-line systems can be applied. In any case, transaction files must be retained to provide for re-creation of a particular file from its parent, if necessary.

Since disk files can be accessed at random, it may not be necessary to sort transactions into specific sequences. However, the organization scheme used for the master file should be studied and, if feasible, the most efficient processing sequence should be established for transaction batches. Sorting the transactions may be unnecessary if the transaction volume is small, or if only a single transaction occurs per master record, or if the distribution of transactions is scattered throughout the master file. In these cases, activity is low, and accessing the master file in random order (usually the order in which transactions were recorded) should not result in unacceptable I/O performance. However, if the volume of transactions is high or multiple transactions can occur for a single record, sorting the transactions into the key sequence of the master file will result in efficiency of search and access.

File maintenance operations require editing transaction records to identify and deal with errors. Therefore, to the extent possible, transaction files used in file updating should be as clean, or error-free, as possible. Transactions are input, edited, and, if valid, applied to the master record. In batch processing, error detection and notification to the user depend upon the timing of the update cycle. For example, in a batch order entry system that is run nightly, an error in a product code that is caught by edits tonight would cause a notification to the user tomorrow. The user's correction, then, would not be processed until the following night's cycle. This delay primarily affects how quickly the customer's order will be processed.

Increasingly, batch processing techniques are gaining improved capacity for identifying and eliminating errors in transaction records prior to input. One method is the use of on-line data capturing techniques, as described above. Such techniques as optical input or data capture on the creation of such documents as invoices are inherently more accurate (and usually less costly) than separate keyboarding operations.

When error detection is handled as part of batch master file updating procedures, some alternatives must be considered and an option selected. One option is to include edits within the same job step as the update processing. Other options are to establish separate steps within the job or to create separate jobs just for editing. Typically, whenever editing is performed, the result will be an error exception report that will identify each erroneous transaction and its error conditions.

The choice of option may be a factor of complexity. The edit function may be segregated into a separate job step that only inputs transactions, validates the data, and writes valid records to an intermediate file for use by the update. In this case, each separate step becomes simpler than doing both in a single job step. However, certain edits in posting the transaction record to the master record still require a second step. For example, edits would be needed to reject an "add" transaction for an existing customer or an unmatched account number on a name and address change.

Decisions about which editing technique to apply also depend upon the nature and timeliness requirements for individual transactions. For example, in some applications, all transactions for an entire batch may have to be corrected before processing can proceed. Such an approach would be used to balance water bill payments against bank deposits for the day. Thus, any errors must be dealt with before the processing of a given batch can proceed. In other situations, it may be possible to make adjusting entries and carry on processing with the valid transaction records. Then, separate batches are created for the correction and entry of errors. Trade-offs on which method to use can depend upon requirements for timeliness within a given application and upon processing schedules.

Security and access controls.   Security and access controls within batch systems are largely a matter of custody. That is, the records exist in one place and tend to be physical and visible. Such records cannot be accessed from terminals usable by unauthorized persons. Therefore, security and access controls are mainly a matter of establishing authorization procedures, distribution procedures, and schedules. New batch jobs that must read an existing file must be authorized. Only persons designated to receive outputs should get them. Along with the delivery of documents, a transfer of responsibility also should take place, possibly by requiring that persons receiving sensitive information sign for it and acknowledge their responsibility for protection of these resources.

Procedures also should be set up so that sensitive information is used only selectively and only as long as it is needed. In recent years, shredding machines have become standard features in many data processing facilities. It is sound practice to insist that sensitive documents be returned after the documents have served their purpose and that they be destroyed as soon as they are no longer needed.

LIBRARY ROUTINES, UTILITY PACKAGES, AND SKELETON MODULES

It has been a long-standing practice of computer manufacturers to provide programs known as *utilities* to perform standard functions, such as sorting, collating, and reporting. In addition, most installations have software modules that are usable across different applications. As programs have been developed and refined to meet these needs, many systems or operations managers have set up what are known as *library routines* to preserve these standard program modules for reuse. Included can be data editing routines, programs to establish tables, and so on.

Utility programs are used by the designer to handle typical, mundane tasks. For example, report writing packages can be used to generate both standard reports and unique, one-time reports. After defining the files being used, such tools provide for the entry of criteria by which records can be selected for reporting. Report or screen display formats then can be specified, in a manner much simpler than with usual progamming languages. Data elements to be displayed and their format are specified, as are sorting and level summary totals. Thus, reports can be prepared with relative ease.

So-called *skeleton programming* is an extension of the same principle, adding to capabilities of vendor-supplied utilities and library routines. In this case, a designer has available pre-coded programs that serve as the framework to which specific applications are added. These programs may range from a listing of all the necessary headings in the four divisions of a COBOL program to the actual coding of certain high-level modules, lacking only application-specific names. For example, the design and implementation of a sequential, batch master file update is a standard task. A skeleton program could be derived so that all file descriptions are copied and the main level processing logic is already present. All that is needed are the details of each transaction's processing.

The objective of these tools is to provide the designer with building blocks for the development of a system. Such an approach has the advantages of improving productivity, using proven, tested components, and focusing custom efforts on unique parts of the system.

APPLICATION GENERATORS

An *application generator* is a level of program development above that of a compiler or translator. One distinguishing characteristic of a generator is that it does not generate object code but produces source code in an intermediate, high-level language. For example, there are generators that produce source code in such languages as COBOL or BASIC. This source code then can be used as though it had been written by programmers, modified as necessary, and compiled for execution.

Another characteristic of generators is that their users deal in *parameters*, or specifications, rather than in writing code in executable sequences. Typically, forms or screens are provided in which the user specifies functions to be performed, file sizes, record sizes, and field sizes. Given sufficient specifications, the generator then produces a source program.

The most obvious advantage of a program generator is that programming time is reduced. Unquestionably, the computer can generate code faster than a person can. Overall programming time is reduced because it is possible to go directly from design documentation into coding without actually writing lists of instructions.

Another characteristic of generators is that they usually are designed around functions associated with files or data bases. Most generators provide a capability for defining and building files. Editing and validation criteria can be expressed as parameters within input documentation. Major and minor file keys can be designated.

Often, specialized parameter functions are provided for maintaining, updating, querying, or reporting from files.

Programming tradition holds that, while program generators add convenience and reduce costs for the creation of programs, the programs produced are not as efficient, in terms of processing throughput, as those that are coded by experienced programmers. Thus, in the past, there has been a tendency to use generators primarily for programs or applications that are either run infrequently or involve minimal processing. However, these programming tools are part of an evolving trend. *Fourth-generation languages* are non-procedural and typically require that parameter input be supplied to generate applications. This structure permits higher productivity and more closely resembles natural language. Such languages become especially helpful to end users in developing distributed applications.

TRADE-OFFS

Setting a strategy for application design, as with any other aspect of systems development, involves a number of potential options and trade-offs among those options. In general, the options lie between custom development for specific needs and the so-called "short-cuts" offered by application packages or by special programming tools such as utilities, library routines, or skeleton modules. The issues to be faced include:

- Productivity
- Costs
- Timeliness
- Compromise of business requirements.

Productivity

Emphasis on productivity stems from a continually increasing demand for information systems to support organizations. The greatest demand is for packages, tools, and techniques that provide quality solutions with a minimum investment in time. Proven, tested solutions are attractive in the light of otherwise long development schedules. Specifically, the writing and testing of program code is time-consuming. Thus, anything that significantly reduces, or even eliminates, the writing of code enhances programming productivity.

On the other hand, there is another dimension to productivity that has to be considered—the actual effort and cost required to install a "short cut" and the productivity of the job once it is running. It could be ultimately wasteful to save a small percentage of programming cost at the outset, only to double development time or to multiply other problems once the job is in use.

Further, productivity also extends to maintenance of operational programs. This dimension, too, should be considered as part of the trade-off in establishing an application design strategy. This productivity factor can have an effect lasting several times the length of the development cycle.

Costs

Costs, too, should be examined fully. Saving money at the front end may be an illusory advantage if increased costs of using and modifying programs are encountered later.

In considering application packages particularly, total costs of acquiring the system, putting it on-line, maintaining it, and securing other services from the vendor should all be taken into account. As one example, a systems organization may consider acquiring a package because a part of it fits the needs of a system under development particularly well. The decision may involve buying some elements of a package that will not be used. Further, in buying a package, a company may be undertaking to update and maintain those parts that will never be used. Thus, costs should be projected and evaluated over the full, estimated life of the system.

Timeliness

Sometimes, a system has to be up and running by a given date, no matter what the circumstances. If deadlines are real and must be met, decision criteria should center around achieving these goals with the least wear and tear upon the people and the organizations involved. If timeliness is a paramount factor, concerns over cost may have to be disregarded almost entirely.

When timing is critical, it is also important to look to available resources within the organization. It may, for example, be possible to finish a program in time—if all of the existing staff is thrown at the job. However, if there is any question about meeting the deadline, or if the people involved are already committed to other important projects, packages may wind up being the only solution.

Compromise of Business Requirements

There are different schools of thought about compatibility of computer programs and business functions. One classic outlook is that the automated system must adapt itself always to the needs of a business, regardless of other situations or circumstances. Under this outlook, there is heavy pressure within an organization to design all applications from scratch, tailoring the programs to the stated needs of users.

Another outlook holds that business procedures usually have a greater potential for flexibility than operating managers are willing to admit. Persons who hold this view look at an automated system and its implementation as a totality. It is felt that programs and procedures that do the best job in terms of all factors—cost, productivity, availability, maintainability, and so on—should be adopted. If business processes or procedures must be modified to accommodate the best overall solution, changes may be both justified and desirable.

In the real world, compromise tends to be a way of life. Decisions rarely go all one way or another. Rather, trade-offs are considered and compromises are established to fit individual situations. Users can expect to modify their lives to some extent if they become involved in development of new automated systems. Similarly, technical purists who would like to optimize every system, in every way, are given the opportunity to do so only rarely.

Design strategies, in the end, represent a compromise in favor of what has to be done and the best way to do the job.

Summary

An application software strategy is developed as a guideline for the people who actually will create or adapt the programs to be used. There is a whole range of options that can be used either selectively or in combination. These options include application packages, custom programs, modification (customization) of application packages, library routines and utilities, generator programs, and skeleton (modular) programming.

Application packages are available for specific industries, and there are also assortments of programs that are almost universally applicable, such as payroll.

On the basis of processing and data retention requirements presented in the new system design specification, it is possible to set and weight priorities for the desired features of any given application software program. Evaluation also may be based on vendor reputation, reference checking, and vendor support policy.

Application-related requirements for modification of program packages can include the needs or practices of the business itself, compatibility with data structures that already exist for a given application, and conformance to processing schedules or requirements established for a given application.

Technical reasons for modifying a package can include the need to conform to existing routines within a computer operations center, conversion of packages from one language version to another, modifications to make a package compatible with a given operating system used in a specific environment, and modifications involving compatibility with changes in make or model of computer equipment.

Custom programming may be preferred in some CIS shops for compatibility with existing systems or because management is prejudiced in favor of in-house development.

Separate considerations are involved for on-line and for batch applications. In general, on-line systems are more expensive to develop and operate than batch applications.

User requirements that may affect a decision to employ on-line systems include a need for immediacy in response, a necessity for instantaneous file updating or other inherent efficiencies, customer service or competition, and safety or protection.

Specific areas of concern associated with the custom programming of on-line systems include human factors, file maintenance, continuity of service, and security and access controls.

A characteristic of batch processing is that there is a lower level of urgency than exists with on-line systems. However, schedules still must be maintained and processing deadlines must be met.

Considerations associated with application design for batch systems include human factors, cohesion and coupling, job-step boundaries and structures, use of intermediate or temporary files, file maintenance, and security and access controls.

Trade-offs involved in selecting program development tools include productivity, costs, timeliness, and compromise of business requirements.

Key Terms

1. human engineering
2. user-friendly
3. log in
4. sign on
5. mnemonic
6. hard copy
7. audio response
8. audit trail
9. significance
10. materiality
11. down time
12. readability
13. accuracy
14. timeliness
15. cohesion
16. coupling
17. job-step boundary
18. son file
19. father file
20. grandfather file
21. dump
22. utility
23. library routine
24. skeleton programming
25. application generator
26. parameter
27. fourth-generation language

Review/Discussion Questions

1. What are six options, or program development tools, available for development of application programs?

2. What factors may be used to evaluate application packages?

3. What are two commonly encountered reasons for custom program development?

4. What criteria, or user requirements, might impact a decision to develop an on-line system?

5. What specific areas of concern are associated with custom programming of on-line systems?

6. What basic characteristic of file maintenance is associated with on-line systems?

7. What factors must be considered in developing the strategies for file maintenance for on-line systems?

8. What is the controlling factor in attention of auditors to a given audit trail?

9. How are erroneous transactions detected in on-line systems?

10. What considerations are associated with application design for batch systems?

11. What is meant by cohesion and coupling and how are these concepts applied to batch systems?

12. What is the importance of identifying job-step boundaries in batch processing?

13. What issues are involved in evaluating trade-offs in applications design?

Project Assignments

The Appendix presents three case studies that can be used in support of project assignments. All three scenarios provide functional system specifications that would result from systems analysis. Included are data flow diagrams, data dictionaries, and process narratives that would result from systems analysis. Continuing the work begun in Chapter 3, assignments that relate to the content of this chapter will be found in *PART 2: Transition into Systems Design.*

FILE AND DATABASE DESIGN 5

LEARNING OBJECTIVES

On completing reading and other learning assignments for this chapter, you should be able to:

- ☐ Explain how an organization's files describe entities, attributes, and relationships.
- ☐ Tell how data structures may be derived as a basis for file design decisions.
- ☐ Give the principal methods of file organization and access.
- ☐ Describe the physical characteristics of magnetic tape and magnetic disk devices and the file design considerations associated with each.
- ☐ Tell how files are implemented physically for different file organization and access methods.
- ☐ Give the key concerns for the systems designer in designing files.
- ☐ Describe the general organization and main advantages of database management systems.

FILES AS A CRITICAL ASSET

In building and using computer information systems, files are a major focal point. An information system revolves around the need to capture data and record those data in files, to maintain the files so that the information contained within them is current and accurate, and to extract information from files and make it available to executives, managers, and operations personnel for conducting the affairs of a business. In short, nothing happens in information processing unless files are available and reflect current information. Further, the output of an information system has no meaning unless its files are complete, reliable, and accessible.

It has been said that an organization's collection of files—its data base—models the organization itself. At the very least, files require structures that parallel the structure and functions of the organization. This correspondence should exist because the organization depends upon files for operational, supervisory, and planning support. In turn, the components of files are the data that represent and reflect the key assets of an organization, including its customers, its inventory, its production, its employees, and other assets or entities important to the functioning of the business.

For these files to be useful, the relationships among the organization's assets also must be reflected in the relationships among the data elements, records, and files within the data base. File design, therefore, requires a modeling of the relationships and communication functions that exist within the organization. Establishing this replication between files and the organization that will use them represents a major challenge in the design of computer information systems.

FILE DESIGN DECISIONS

The inputs for file design come from the outputs of the analysis and general design phase in the form of documentation that outlines a logical structure for files. User specifications include identification of data stores, data elements within those stores, and access paths among the data stores to be used by the system under development.

The logical data structure that supports the processing of a new system is derived through use of *normalization* techniques. Normalization, which is described in depth in the companion text, is a definition process that results in a comprehensive set of data stores in which redundancy among component data items has been eliminated. Also, access paths to data components have been built around a series of *primary keys,* or fields within the data records used to define relation-ships among those records. Normalization produces data structures that can be packaged physically as two types of files:

- Attribute
- Correlative.

Attribute Files

Attribute files contain data components that describe an identifiable entity or object of importance to the organization. In a sense, attribute files correspond to master files that will be maintained in support of processing.

Correlative Files

Correlative files establish relationships between attribute files through key values that associate records in one attribute file with those in another attribute file. Correlative files establish access paths among attribute files so that data elements in physically separated files without common keys can be related logically for integrated use within information systems.

Correlative files can be implemented in several different ways. In a traditional file environment, correlative files can be packaged separately as direct-access files that can be searched randomly to find associated keys. Also, the correlations can be maintained within in-ternal tables held in memory during processing. If attribute files are organized as indexed-sequential files (as described later in this chapter), correlations can be implemented as *secondary keys,* fields that are used to define alternate relationships associated with the files. In a database environment, the database management software formats the files to allow the relations among attribute records.

To review and illustrate, Figure 5-1 shows a data access diagram for a portion of the normalized data stores that might be defined for a student records system. This subset of files would be necessary for maintaining student and class information and the relationships necessary for producing reports such as class rosters and grade reports.

Three main entities are represented by the files—STUDENTS, CLASSES, and GRADES. The STUDENTS data structure contains

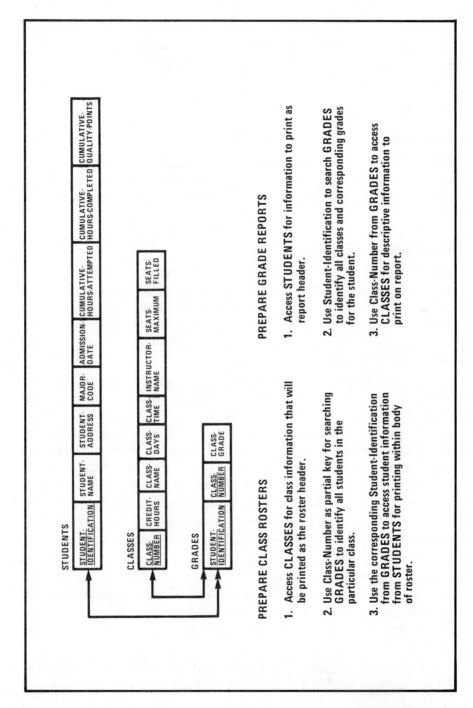

STUDENTS

| STUDENT-IDENTIFICATION | STUDENT-NAME | STUDENT-ADDRESS | MAJOR-CODE | ADMISSION-DATE | CUMULATIVE-HOURS-ATTEMPTED | CUMULATIVE-HOURS-COMPLETED | CUMULATIVE-QUALITY-POINTS |

CLASSES

| CLASS-NUMBER | CREDIT-HOURS | CLASS-NAME | CLASS-DAYS | CLASS-TIME | INSTRUCTOR-NAME | SEATS-MAXIMUM | SEATS-FILLED |

GRADES

| STUDENT-IDENTIFICATION | CLASS-NUMBER | CLASS-GRADE |

PREPARE CLASS ROSTERS

1. Access CLASSES for class information that will be printed as the roster header.

2. Use Class-Number as partial key for searching GRADES to identify all students in the particular class.

3. Use the corresponding Student-Identification from GRADES to access student information from STUDENTS for printing within body of roster.

PREPARE GRADE REPORTS

1. Access STUDENTS for information to print as report header.

2. Use Student-Identification to search GRADES to identify all classes and corresponding grades for the student.

3. Use Class-Number from GRADES to access CLASSES for descriptive information to print on report.

Figure 5-1.   Data organization and access for a portion of a student records system.

information to identify the students and to keep track of academic progress. The CLASSES structure contains identification information, along with data elements that keep track of the availability of openings in each class. The GRADES structure relates students with classes for the purpose of maintaining grades earned in each class. These three structures can be packaged as attribute files that store basic information associated with the entities represented by those structures.

Suppose, for example, that class rosters are to be prepared for each of the classes represented in the CLASSES file. First, access would be to the CLASSES file to extract class identification data. Then, the CLASS-ID field within this record would be used as a partial key for searching the GRADES file to find the STUDENT-ID numbers of students enrolled in the class. These numbers, in turn, provide access to the STUDENTS attribute file for extracting student identification data to be printed on the rosters.

Being able to identify entities important to the organization and to define the attributes associated with those entities is an essential feature for flexibility and maintainability of an information system. The motivation is to structure data so that those data represent the business organization rather than support specific processing urgencies of any given moment. Unless the organization changes its line of business, the entities with which it deals will remain relatively stable. There may be need, over time, to expand or contract the sets of attributes that represent these business elements. In general, however, the elements will remain the same.

Processing needs, however, probably will change. In a business organization, new reporting obligations will have to be included within the system to meet changing business needs and government regulations. Different report formats will be solicited from users of the information system. Additional processing steps will be inserted into work procedures to reflect changing business policies. These and other changes are to be expected, since organizations are dynamic and are evolving new methods of operation continually.

The data structures that support an organization, therefore, also must be adaptable to processing changes. This adaptability is built into the structure of attribute files through the correlations established between those files. Note in Figure 5-1, for example, that all three attribute files within this subset of student information files are inter-related. A relationship has been established between each entity and the other two entities through their primary keys. Thus, any new processing need that requires access to any combination of attributes from any of the files can be serviced without changing the file structure. In other words, the overall data structure can support any particular logical viewpoint or processing need involving these attributes.

Derivation of the logical data structure for an organization or a subset of the organization can be done through analysis of current processing procedures. Data flow diagramming, for instance, documents the data stores needed to support current business procedures. Analysis of these data stores and their input and output data flows makes possible identification of entities and attributes. The process of normalization distills this information into sets of attributes and relationships characterized by nonredundancy of data elements among files and simplicity of access through primary keys.

A second method for deriving an organizational data structure begins by soliciting the information from managers and other persons in the organization. Over the span of several weeks—or months, depending on the complexity of the system under study—entities and attributes are defined. Relationships among entities and attributes are established through anticipation of processing needs and through documentation of current needs as expressed with graphic tools such as data flow diagrams. Normalization is applied to the resulting structure in deriving the final design that will be implemented as files or databases.

It should be remembered that a normalized data structure is an idealized solution to an organizational problem. Although it represents the optimum design for supporting that organization, a given normalized data structure may not be the most efficient design for any one processing need. For example, note in the STUDENTS data structure in Figure 5-1 that no grade point average (GPA) attribute is included. The GPA can be calculated from the values of other attributes included within the structure. To include the GPA within the structure would be redundant from the standpoint of a strictly normalized structure.

Suppose, however, that this structure were to be implemented as a file that would support on-line inquiry and that the grade point average would be displayed. For each access to the file, the GPA would have to be calculated. If, in a high-volume situation, the computer time necessary for this calculation caused a degradation in service, it might be best to include the GPA separately within the data structure.

An opposite situation, however, also might occur. Suppose the GPA were to be retained as a separate data element for the overall average for each semester and also as an accumulated GPA at the end of each semester. In this case, a grade change would require a considerable amount of recalculation during file updating. The point is, there are trade-offs that must be considered between the idealized data structure and its processing efficiencies. Usually, however, these trade-offs impact the attributes associated with particular entities rather than the entities themselves or the relationships among them.

The final data structure for an organization or a subset of that organization can be packaged physically in one of two ways. In a file-oriented environment, the attribute and correlative structures are implemented as traditional sequential, direct, or indexed-sequential files; the designer determines the appropriate organization methods and establishes the formats of data fields within the records. These file access and organization methods are discussed later in this chapter. In a database environment, a database administration (DBA) group works closely with the designers to implement the *data model* with database software.

File design decisions include determinations on whether on-line file maintenance capabilities will be used, whether transactions against the files will be batched, whether inquiry capabilities need to be supported, whether access times are sufficient to support proposed applications, and whether the hardware and systems software available have the capacity to store the files and handle the transaction volumes. A number of trade-offs must be considered in connection with these decisions.

Such considerations arise during systems design and are reflected in the systems flowcharts that describe the physical packaging of systems into jobs, job steps, and on-line procedures. When systems flowcharts first are prepared at the end of the analysis and general design phase, the flowcharts reflect the designer's best initial estimate of the physical structure of programs and files that will support processing. As these design efforts are reviewed and carried to a more detailed level early in the detailed design and implementation phase, some of the assumptions may be modified to allow for hardware, software, and processing constraints.

The following sections of this chapter discuss some of the considerations that impact file design. After that, a brief discussion of design within a database environment is presented.

FILE ACCESS AND ORGANIZATION METHODS

File access is a term describing the ways in which data maintained by a system can be sought, extracted from, and replaced within data files. A file access technique represents the user's outlook on data as a resource. That is, file access represents the way in which data will be used.

File Access Methods

There are three basic access methods:

- Physical sequential (serial)
- Logical sequential
- Direct.

Physical sequential (serial). *Physical sequential access*, or *serial access*, represents, in a sense, a chronological approach. That is, records in a file are read or processed in the same order in which the data were recorded initially. An example of a file that usually is accessed serially is a transaction file of weekly payroll records captured from time cards or time clocks. Another example might be data entries recorded at the point of sale in a retail store. Both types of files are built chronologically, as data are generated.

Logical sequential. *Logical sequential access* follows a keyed sequence. That is, records are read in order, according to a logical identifier, or key. Keys that drive logical sequential access functions are data fields within the records that contain the data. Payroll master files, commonly organized and accessed successively by employee number, represent a typical example of logical sequential access.

Direct access. *Direct access* is also known as *random access*. These names come from the fact that access is directly to the record being sought, without regard to the storage sequence. Given any key value, the corresponding record can be found at random from within a collection of records that comprise a file. Virtually all on-line or interactive systems require random access capabilities.

File access methods are determined during the analysis and general design phase of the systems development life cycle. These methods represent the user's requirements for accessing data and having files available for processing. Thus, file access is a logical consideration in the development of information systems. At this point, little concern has been given to the physical design of files to support the access needs. As development moves into the detailed design stages, an important decision relates to how data files will be placed on storage devices to support these needs.

File Organization Methods

File organization refers to the physical patterns in which data are recorded on storage devices. The organization scheme chosen for packaging the logical data structures into physical files either can support or constrain access capability. For example, of the commonly available methods for organizing files, some provide serial, sequential, and random access while others provide only serial and sequential access. A challenge of design, therefore, is to match file organization with logical file access requirements.

Whereas file access represents the user's processing viewpoint of data, file organization is from the system's viewpoint; a computer system manages data according to their physical arrangement on storage devices. Therefore, file design involves determination of how data will be placed on these devices and how they will be made available by the computer system to provide needed user access. There are three basic file organization methods:

- Sequential
- Direct
- Indexed-sequential.

Sequential. *Sequential file organization* is common to all computer systems. Under this method, records are written onto a storage medium contiguously, one after another, in the chronological sequence in which they are presented to the system. Sequential files support serial access. This organization scheme also can support logical sequential access, as long as the records were placed in the file in a physical order corresponding with the logical order of keys embedded within the records.

Sequential file organization is appropriate for applications in which either serial or sequential access is necessary. Most often, this requirement occurs when the majority of the records in a file will be accessed during each program execution and when it is possible to anticipate the order in which records must be made available. A payroll master file with employee number as key, again, is a common example. Each time the file is processed, most of its records will be accessed and used in processing. Further, it is both convenient and efficient to sequence transaction files so that they correspond with the key sequence of the master file. Thus, it is possible to process transaction records efficiently against master records to update the file. For these reasons, many payroll files use sequential file organization.

Direct. *Direct file organization* is a method in which the physical location, or *address*, of a record in a file is determined by the value of its key field. Typically, a mathematical formula is applied to the key value to determine an address in the file. With availability of this address, the system can access the record directly, without the need to search through the file in physical sequence to locate the desired record. Although the file may be accessed serially, the physical sequence of records on storage media may have no logical meaning since the records are not necessarily ordered according to keys.

A direct file organization is applicable where there is a need for immediate, rapid access to records in the file and where it is impractical or impossible to anticipate the order in which records will be processed. A common example of direct file use is an airline reservation system. In this case, the file is composed of flight information, with records usually keyed to the airline and flight number as the *composite key.* These records are accessed whenever customers inquire about the booking status of a flight or when seat reservations are to be made. Under these circumstances, it is virtually impossible to anticipate the order of inquiry about flights. Further, flight information is time-critical. The customer requires immediate information and cannot wait while the computer system searches serially through a massive file to locate and present information about the desired flight. Thus, airline information files are excellent candidates for direct organization. Within these systems, airline identification and flight numbers are used for calculating addresses of records and for accessing records immediately and in random order.

Indexed-sequential. *Indexed-sequential file organization* is a compromise between sequential and direct organization. To oversimplify to some extent, records are written onto the file in sequential order, by key. In addition, an *index*, or table, is established to associate key values with physical addresses. For sequential processing, the file can be accessed in order, by key. Also, where random access is required, the index can be referenced to identify the physical location of the record with the desired key. This approach represents a compromise, in that supporting both access methods comes at a cost—namely, trading off efficiency of access.

Obviously, applications in which both sequential and random access to records are desired are candidates for indexed-sequential file organization. A typical example is an inventory control system. Such a system is used in keeping track of merchandise on hand and for ordering and restocking depleted goods. The primary file in such a system is an inventory file in which product records are kept. As merchandise is removed from stock, the records are updated to reflect withdrawals. When merchandise is restocked, the corresponding records are modified to reflect the additions. Since withdrawals and replacements do not take place in a predetermined order, it is necessary for this file to support random processing. However, in support of other applications, such as preparation of inventory listings or merchandise reports, it becomes necessary to access the file sequentially, ordered by product number. Thus, an inventory control system demands both random and sequential access capabilities for which an indexed-sequential file is the appropriate mechanism.

As noted previously, decisions on file access methods, typically, are made during the analysis and general design phase of the systems development life cycle. Tentative file organization decisions are made during the general design activity of that phase and are documented in systems flowcharts. Then, during the detailed design activity, these decisions are reevaluated and crystallized within specifications that define the exact methods for packaging files physically.

As with virtually all systems design decisions, file organization trade-offs must be considered. Options relate to storage device and media availability and to file organization characteristics. The remainder of this chapter discusses these options.

STORAGE MEDIA AND DEVICES

Two broad categories of peripheral devices, which include hardware units for the storage function, are used to organize data within files:

- *Sequential access devices* read and write records in order, one after the other. These devices, therefore, require that data records be organized in either physical or logical sequence, corresponding with the access method to be used. Magnetic tape drives are the most popular type of sequential access device.

- *Direct access devices* make it possible to process records either sequentially or at random. The physical constraint of sequential access devices, requiring correspondence between organization and access methods, is removed. Records can be accessed in the physical order in which they are recorded on the file, in logical sequence according to keys within the records, or on a random basis, without access to any preceding or succeeding records. The most common secondary storage device for direct access is the magnetic disk drive.

Design considerations for file access center around trade-offs between sequential and direct access devices. The most common of these devices are tape and disk drives. Accordingly, the presentations that follow concentrate on these two widely accepted groups of devices.

Magnetic Tape Media and Devices

Magnetic tape consists of a ribbon of acetate material, typically one-half-inch wide, coated with a magnetic material. Tapes are wound onto reels of varying lengths, including 300 feet, 600 feet, 1,200 feet, and 2,400 feet. The longer reels are used most commonly. The reels—or cartridges used under some systems—are loaded manually into tape drive mechanisms that keep the recording surfaces moving in close contact with *read/write heads* during operation. The rate of movement, or *transport speed,* of tape ranges from 75 to 200 *inches per second (ips).*

A typical recording pattern for data on magnetic tape is shown in Figure 5-2. This diagram shows that characters, or bytes, of data are recorded along the length of the tape. Character codes can use either six or eight bit positions for data. In addition, one bit in each position is used as a *parity bit,* or *check bit.* The presence or absence of this bit is used to verify recorded codes. Thus, magnetic tape recording patterns consist of seven or nine bit positions, or *tracks,* of data that are recorded in parallel.

From a file design standpoint, one of the most important characteristics of tape is the high *recording density* that is possible. Recording densities available for magnetic tape are measured in terms of *bytes per inch (bpi).* Densities available include 200, 556, 800, 1,600, 3,200, and 6,250 bpi. Consider the storage power implicit in being able to place 6,250 bytes, or characters, per inch on 2,400 feet of tape. Even at lower recording densities, considerable storage capacities are available on magnetic tape, at relatively low cost.

Some devices, particularly microcomputers and terminals, record data on magnetic tape cassettes. Most cassettes use recording patterns and densities similar to those for traditional, half-inch magnetic tape.

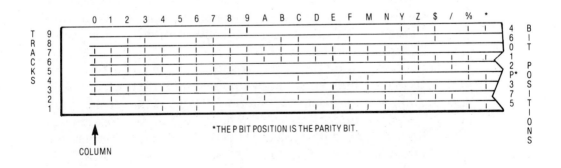

Figure 5-2. Data recording pattern for 9-track magnetic tape.

Magnetic Disk Media and Devices

The typical disk used in high-capacity, direct-access devices for main-frames and some minicomputers is a 14-inch-diameter platter made of a rigid material, such as aluminum, and covered with a magnetic coating. The disk rotates around a spindle that positions the platter beneath read/write heads that operate at small fractions of an inch above the surface. Disks may be packaged in devices that include a single platter or a stack of platters. The disk platters may be either fixed or removable. In other words, disks may be interchanged on some drives. On others, the disks are mounted permanently. Devices with permanently mounted disks usually have read/write heads in fixed positions. Removable-disk devices usually have heads that move back and forth over the surface of the disks. A device with a single, removable platter generally is called a *disk cartridge*. Stacks of disks are referred to as *disk packs*.

Microcomputers often use small disks, or *diskettes*, made of flexible plastic sheeting materials that are magnetically coated. Typical diameters for diskettes are 5.25 and 8 inches.

Data are recorded and read in concentric circles as the disks spin under read/write heads. Recording and reading are in serial patterns, bit by bit. As shown in Figure 5-3, data are written on circular tracks, or recording positions, around the platter. Individual data records are recorded along the tracks and are separately readable and addressable.

Figure 5-4 illustrates one of the recording patterns used for magnetic disks and shows the way in which record addressing schemes are defined. Note that data records are written onto the disk surface along tracks that run concentrically around the disk platter. When a track becomes full of records, the next track to be used is the corresponding track on the next disk surface in the pack. Thus, a file of records may be built beginning with the outermost track of the top surface and continuing onto the outermost tracks of the second, third, and subsequent surfaces, down through the bottom-most surface. When this set of outermost tracks is filled, recording continues from top to bottom on the set of next innermost tracks. As shown in Figure 5-4, the collection of corresponding tracks on each of the disk surfaces is referred to as a *cylinder*. In effect, a cylinder is composed of the set of tracks accessible by all of the read/write mechanisms at any one position.

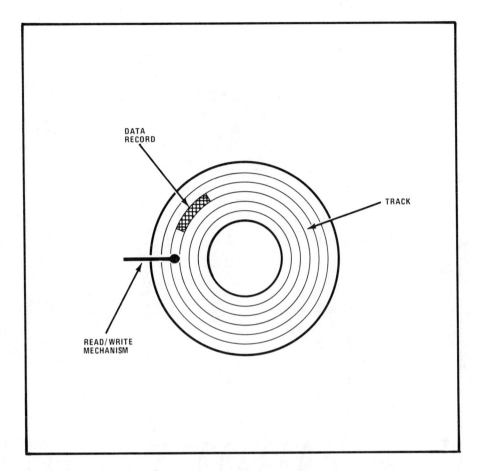

Figure 5-3.   Data recording pattern for magnetic disk.

As mentioned above, records written to disk can be addressed separately. One type of disk addressing uses the record's cylinder number, track number, and record sequence number as the identifier for the record. The cylinder number refers to one of the corresponding sets of tracks that span the disk surfaces; the track number refers to the particular track within that cylinder; and the record number refers to the location of the record on the track relative to the first record. Thus, a record with an address of CYL=1, TRK=3, SEQ=5 would be located as the fifth record in sequence on the third recording surface of the set of outermost tracks on the disk pack.

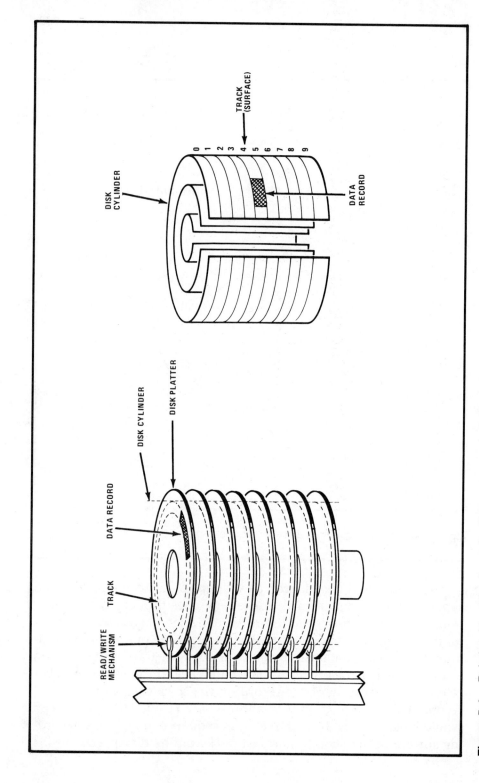

Figure 5-4. Relationship of tracks and cylinders for magnetic disk pack.

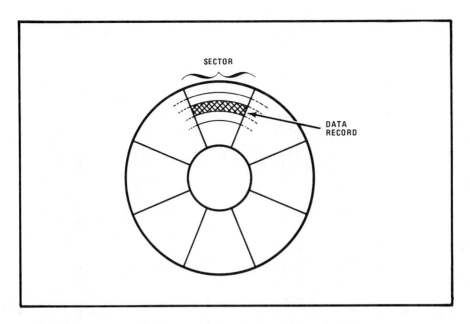

Figure 5-5. Sector method of data recording for magnetic disk.

The cylinder approach represents one method for formatting and handling data on disk packs. Another method is to divide individual disk surfaces into *sectors*, as shown in Figure 5-5. The sector approach is used for most diskette devices and also for some *hard disk* systems. As shown in Figure 5-5, a sector consists of all of the track areas within a pie-shaped portion of a single disk surface. Each sector presents a fixed-length storage area that can be addressed separately. Typical capacities for sectors include 128 bytes, 256 bytes, and 512 bytes. Thus, addressing schemes are based on track and sector positions.

Total disk capacities, then, represent a multiple of the capacity of a sector or track. Capacities of disk units range from as low as 128,000 bytes for some makes and models of diskettes up to more than 2.5 billion bytes for hard disks. In discussions of disk storage capacities, the term *megabyte (MB)*, is used frequently. A megabyte is approximately 1 million bytes of data. To illustrate the range of capabilities typically available to systems designers, Figure 5-6 lists capacities for a number of commonly used models of IBM disk drives.

In addition to capacity, disk drives are commonly evaluated on the basis of data access speeds. The *access time* for any given disk unit is the elapsed time from the execution of a READ command until the desired record is delivered to memory.

IBM DEVICE TYPE	CAPACITY (MB)
3330	200
3340	70
3344	280
3350	317
3370	517
3380	2520

Figure 5-6. Disk storage capacity in millions of bytes (MB) for selected IBM disk devices.

Access times are influenced by a number of factors. One obvious factor is whether the drive has fixed heads or movable access arms. With fixed-head systems, there is a separate read/write head above each track. Thus, access occurs within one rotation of the disk. Access time for movable-arm systems is, of necessity, expressed in averages. There will be a variance in *head positioning* times based upon relationships between the position of the access arm at the time a command is executed and the track location of a desired record. The further the arm has to move, the longer the access will take.

Other factors affecting access time include the rotation speed of the disk itself and *transfer rates,* or data transmission speeds, of the circuits that connect the disk drive to the central processor. Transfer rates usually are given in terms of thousands of bytes per second, or *kilobytes (KB).* The factor of rotation speed in determining access time is known as *rotational delay.* This factor, too, is expressed as an average, since it depends upon the rotational position of the desired record at the time the command is executed. Figure 5-7 lists typical disk performances in terms of head movement, rotational delay, and transfer rate. As indicated in this table, transfer rates are shown in KB intervals. Head positioning and rotational delay factors are indicated in *milliseconds (ms),* or thousandths of a second.

RECORD FORMATS

Application programs process files of *logical records.* A logical record is the combination of data fields presented to a program for processing in a single READ operation. The fields in a logical record represent a set of attributes concerning one of the organizational entities about which information is stored.

A *physical record,* on the other hand, consists of one or more logical records grouped to conform to the storage and processing requirements of secondary storage devices. A physical record is the unit of data that is either written or accessed by the computer system at any one time. There may or may not be direct correspondence between logical and physical records. That is, a physical record area may contain one or more logical records, depending on the file organization preferences of the systems designer. Whereas a single logical record is made available for program processing with each READ operation, one or more physical records may be accessed by the systems software.

File design involves consideration of the physical sizes of records residing on secondary storage devices. These physical records, or *blocks,* can take on various sizes, depending on the characteristics of the devices and on needs for processing efficiency. Thus, one decision that must be made by the systems designer relates to the blocking factor (discussed below) for records in the file.

IBM DEVICE TYPE	HEAD POSITIONING (AVG.MS)	ROTATIONAL DELAY (AVG.MS)	TRANSFER RATE (KB/SECOND)
3330	30	8.3	806
3340	25	10.1	885
3344	25	10.1	885
3350	25	8.3	1198
3370	20	10.1	1859
3380	16	8.3	3000

Figure 5-7. Access speeds for selected IBM disk devices.

Unblocked Records

Unless a computer is instructed specifically to do otherwise, files are formatted with one logical record to each physical record, or block. Such formats are also known as *unblocked*.

An unblocked format layout for a portion of a file recorded on magnetic tape is shown in Figure 5-8. As indicated, a physical record is identified by a contiguous collection of characters blocked off by the appearance of *interblock gaps (IBG)*. These gaps, also referred to as *interrecord gaps (IRG)*, are blank spaces on the tape that serve the dual functions of separating physical records and providing the slack necessary for stopping and starting the tape transport as it reads through the file. When a block of records is accessed, it must be read at a fixed speed. Therefore, the gaps provide space for the transport to accelerate the tape to reading speed and, when the block has been read, to decelerate and stop the tape. These gaps usually measure between 0.6 and 0.75 inches.

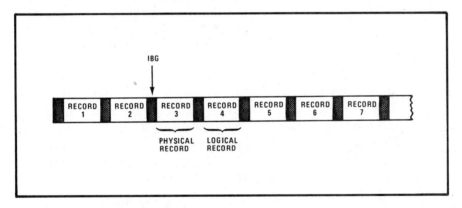

Figure 5-8. Unblocked record format for magnetic tape.

In unblocked format, a single logical record is contained within the block. Thus, when a program issues a READ command, the systems software gets the next physical record and makes it available for processing. In this case, a single logical record also is accessed.

In recording unblocked records on magnetic disk, there is a choice of formats available, depending upon whether keys are to be used in conjuction with the records. Without keys, records are written in

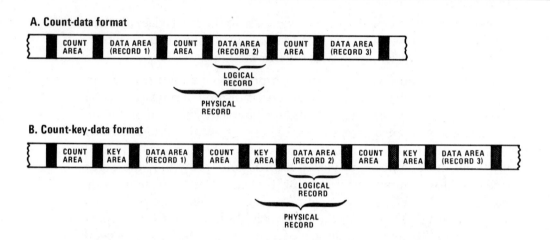

A. Count-data format

B. Count-key-data format

Figure 5-9. Unblocked record formats for magnetic disk.

count-data format. This format is shown schematically in Figure 5-9. As indicated, each logical record occupies one physical record area known as the *data area.* Appended to this area is the *count area,* containing the disk address of this physical/logical record. On free-format disk devices based on the cylinder method of storing data, this count area contains the cylinder number, track number, and record sequence number for the record. In this format, records can be written and accessed only serially—or sequentially if they were organized into key sequence before being written to the file.

To implement a direct-access capability for unblocked records stored on disk, a *count-key-data format* is used. Under this approach, record keys from the data records are repeated within the *key areas,* which appear between the count and data areas. The count-key-data format makes it possible to search for records according to specific key values without having to take the time to read the data content of records in access operations. As indicated in Figure 5-9, a single logical record comprises the data area in this format.

The count-key-data format makes possible serial, sequential, or random access. This format is the basis for files organized as direct or indexed-sequential files.

Blocked Records

Blocking, as indicated above, formats two or more logical records within a single physical record. Basically, blocking is a technique for optimizing computer throughput in the handling of files.

Blocking techniques for magnetic tape are illustrated in Figure 5-10. As shown, the logical records are recorded contiguously, with a specified number of logical records in each physical record. Physical records, in turn, are set off by interrecord gaps.

In processing, the first READ command in a program causes the first physical block of records to be entered into a specially established memory area known as a *buffer*. Then, the system software, operating under logical record descriptions within the program, delivers the first logical record in the buffer to the input area specified in the application program. At each subsequent READ command, the next logical record is moved from the buffer to the program. When the last logical record in a buffered block has been read, a new physical block is sought automatically by the system software.

Buffering techniques are used also in writing blocked files to magnetic tape. Each WRITE operation moves a logical record to the buffer area, where it is placed in sequence following existing records being held for output. When the buffer becomes full, the entire block of records is written to the device and the buffer area becomes available for building the next block of physical records.

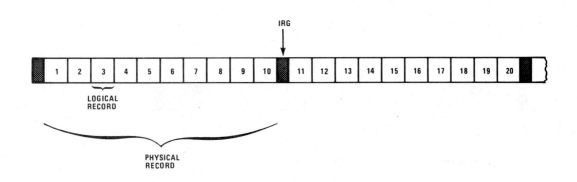

Figure 5-10. Blocked record format for magnetic tape.

In blocking records to be stored on disk, the recording format, as shown in Figure 5-11, is similar to that used for magnetic tape. As with unblocked files, there are format differences depending on whether key areas are included in physical records.

A non-keyed blocked file for disk storage is shown in Figure 5-11A. As indicated, a series of logical records is recorded in the data area. The count area, as in the case of unblocked records, is the disk address for the entire block.

In using a count-key-data format for the recording of blocks of records on disk, as shown in Figure 5-11B, the key area contains the value for the highest key in the entire block. Thus, the first record block shown in Figure 5-11B would have a key area that contained the key value of record number 10. As shown, the physical record encompasses the count area, the key area, and all of the logical records in the block.

As with tape files, an entire block is read into a buffer on execution of the first READ command. The system software then passes along logical records as needed. During WRITE operations, the output buffer is filled with records and then the entire block is written to the file.

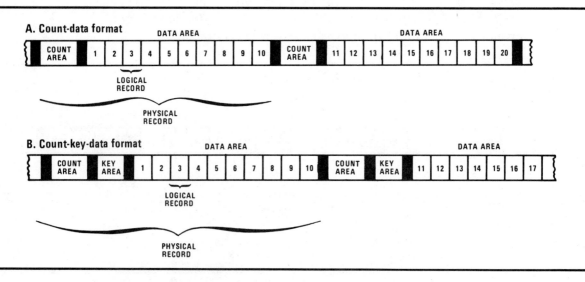

Figure 5-11.   Blocked record formats for magnetic disk.

Fixed- and Variable-Length Records

The examples and descriptions above treat all records alike within each file. That is, all records in a given file are assumed to contain the same number of fields and be of the same overall length. Such a record is known as a *fixed-length record.*

It is also possible to vary the number of fields within records. The result is *variable-length records.* Within variable-length records, record sizes are established by setting up a field that tells the system the length (in bytes) of the overall record.

Obviously, it can be more difficult to establish buffers, memory areas, or file allocations for variable-length records than for fixed-length records. For this and other reasons, the normalization process that establishes the data model on which file design is based identifies and eliminates repeating fields within records. Thus, all records within normalized files are of fixed length.

Record Blocking Decisions

As indicated above, record blocking is designed to optimize computer throughput. Thus, the blocking of records can lead to more efficient processing of any given application program. As a general rule, the larger the block size, the more efficiently the file can be processed. Similarly, the larger the block size, the more efficiently any given file will use its storage media. As indicated, considerable space on file media is occupied by interrecord and interblock gaps. In the case of magnetic tape, the logical records themselves may occupy less space than the gaps that set them off. Therefore, blocking decisions can be important elements in file design.

Note first, however, that whether records are blocked or unblocked for any given file depends upon the application for which the file is being used. Though it is true that large blocks of records tend to decrease the amount of storage space required and also to increase data transfer rates, there may be situations in which large blocks of records cause different kinds of problems. For example, if a file stored on disk is to be used for random access to individual records, it could be defeating to access a block of 50, or perhaps even 100, records through reference to one key. It this case, to find a single record in a large block, all of the records would have to be brought into the buffer. Then, the system software would have to search all records in the block to find the one that is needed, and only then could it make the single record available for processing. Thus, if records are to be accessed randomly, unblocked formats may be better.

Hardware capabilities also may be a factor. For example, on some *fixed-block architecture (FBA)* devices such as the IBM 3370, data are stored in fixed-sized sectors or blocks. Programs have no control over the physical location of data or the formatting of blocks. Storage and location of records are entirely under hardware and software control. Therefore, devices of this type make it unnecessary for the designer or programmer to be concerned with blocking factors.

With a sequential file organization, there are separate decision criteria that apply for blocking of records for storage on magnetic tape and on magnetic disk.

Magnetic tape blocking.   Unblocked tape files have unavoidable inefficiencies. One of these inefficiencies is that data transfer occurs at extremely slow rates because the file device must start and stop in between each logical record—in both reading and writing operations. Another inefficiency involves the utilization of storage area on magnetic tape. As indicated, interrecord and interblock gaps occupy between 0.6 and 0.75 inches each. These blank spaces easily can be larger than the data areas themselves, particularly if high-density devices are used.

For example, assume that an installation uses nine-track tape drives with a recording density of 800 bpi. Assume further that a customer master record contains 240 characters, or bytes. Thus, each master record would occupy the equivalent of 0.3 inches of data area on magnetic tape. This space is less than half the width of an interblock gap. Obviously, then, the fewer the gaps, the more efficiently the file will use tape.

Because of these factors, designers typically strive for the largest practical block sizes to apply to magnetic tape files. In establishing block sizes, or *blocking factors,* some physical constraints must be observed. One obvious constraint lies in the availability of memory buffers to handle input and output operations for file processing. Clearly, the block size cannot be any larger than the available buffer.

In addition, consideration must be given to the possibility that multiple applications will use the same file. In such situations, the blocking factor must be adjusted to the smallest of the block sizes that can be supported for all of the programs that will use any given file. Another program-related constraint might occur if a single program required use of multiple files. The buffer area available for the processing of data from any one file would be limited by the need to share available memory among all the files.

The general principle in dealing with files residing on magnetic tape, then, is that blocks should be as large as feasible—but that all constraints or limitations must be accommodated.

Magnetic disk blocking. Blocking decisions can be both more difficult and more critical—as compared with magnetic tape—when files are to reside on disk devices. An obvious reason is that disk files are organized into discrete storage units that affect block lengths more critically than do tapes. There is much more freedom, for example, in working with a 2,400-foot reel of tape than there is in dealing with tracks or sectors of disk devices. Tracks of a disk have far smaller capacities than reels of tape.

In working with free-format disk devices, the maximum block size normally is limited by track lengths. Thus, block sizes should be selected so that a multiple of the block size is close to the limits of track capacities.

Manufacturers normally supply tables based on formulas for establishing block sizes on the basis of track capacities. Figure 5-12 contains excerpts from a table for the IBM 3350 disk drive. The table is designed to show the number of physical records (blocks) that can be stored on single tracks for varying block sizes. Note that there are two sets of columns giving record lengths. The two columns headed "Without Keys" are used when data are in the count-data format. The "With Keys" columns are used when data are in the count-key-data format.

To illustrate optimum blocking techniques for free-format disk devices, consider a hypothetical situation in which files consist of 100-byte logical records. Assume that the records are unblocked and in the count-data format. Thus, the logical record is equal to the physical record, which is equal to a block. According to the table, 67

| BYTES PER RECORD (BLOCK) | | | | RECORDS (BLOCKS) PER TRACK |
| WITHOUT KEYS | | WITH KEYS | | |
MINIMUM	MAXIMUM	MINIMUM	MAXIMUM	
9443	19069	9361	18987	1
6234	9442	6152	9360	2
4629	6233	4547	6151	3
3666	4628	3584	4546	4
3025	3667	2943	3583	5
2566	3024	2484	2942	6
2222	2565	2140	2483	7
1955	2221	1873	2139	8
1741	1954	1659	1872	9
1566	1740	1484	1658	10
208	216	126	134	48
201	207	119	125	49
193	200	111	118	50
186	192	104	110	51
179	185	97	103	52

Figure 5-12. Track capacities at selected block sizes for the IBM 3350 disk device.

records can be placed on each track. If the file contains 100,000 records, 1,494 tracks would be required to house it. This figure is derived by dividing 100,000, the number of records, by 67, the number of records per track.

Now assume that the same 100-byte logical records, still in the count-data format, are blocked. Assume the blocking factor is 20, resulting in a data length of 2,000 characters per block. The table shows that eight physical records can be placed on each track. Thus, the file containing 100,000 records would require a total of 625 tracks, or less than half the space that would have to be allocated for unblocked records. The requirement for 625 tracks is derived by dividing the total number of records, 100,000, by the number of logical records that can be housed on each track, 160. The number of logical records per track is calculated by multiplying the blocking factor, 20, by the table figure for the number of blocks per track, 8.

For still another example, assume the same 100-byte logical records. Assume the records are unblocked but are formatted with keys and that the key length requires eight bytes. Accordingly, the ''With Keys'' columns of the table must be used as a reference: The key length must be considered in formatting this file. That is, the data length of the record is 100 bytes; the key length then must be added, providing a total record length of 108 bytes. The table in Figure 5-12 shows that 51 records of this length can be stored on a single track. Dividing 51 into 100,000 produces a quotient of 1,961. In other words, unblocked, keyed records with a total length of 108 bytes would require 1,961 tracks. If these same records were blocked with a factor of 20, the overall block size would be 2,008 characters—2,000 characters of data and 8 characters for the key. (Remember, when records are blocked, the key contains only the highest key value for the block.) In this blocked format, the file can be stored in 625 tracks, or less than one-third of the requirement for storage of unblocked records.

In summary, the best method for determining the blocking factor for records to be stored on disk files is to match block sizes to track capacities. That is, the track capacity should be some multiple of the block size.

An extreme example of how space can be wasted can be seen in considering a non-keyed block of 9,443 characters. This block would require a complete track for storage, even though almost half the space on the track would be unused. Thus, in using the file allocation table, the block size should be as close as possible to the figure shown in the appropriate maximum column, rather than to the one shown in the minimum column.

FILE HANDLING TECHNIQUES

Discussions up to this point have covered methods for formatting and recording data on secondary storage devices. The next topic to be covered is a review of physical organization techniques for the creation, processing, and maintenance of files in support of information system applications. The review in this section deals with the physical considerations that are associated with several traditional methods for storing files.

Sequential File Organization

Sequential file organization can be used on both magnetic tape and magnetic disk devices. In this organization scheme, records are written contiguously, one after another, on the storage medium. The records may be ordered chronologically—based on the time at which they were entered into the system—or in logical sequence—based on their keys. In either case, the records will be accessed in the order of their appearance in the file. The keys of the record are not used in building or accessing the file. Consequently, it is up to the user, or programmer, to be aware of the physical organization of the file and to make sure that records are ordered to meet processing demands.

Figure 5-13 illustrates the arrangement of records in a sequentially organized file. Note in this case that the records are ordered logically by key values so that when they are accessed serially they will be presented to the program in this logical sequence.

For both magnetic tape and magnetic disk, records may be blocked or unblocked. Also, sequential disk files maintain records in the count-data format, since record keys are not used in file searches.

When sequential files are processed, records are made available in the order in which they appear on the storage device. Thus, if processing requirements dictate a different order of access, it becomes necessary to create intermediate files through use of sorting routines. If sequential files are used in support of file updating or maintenance,

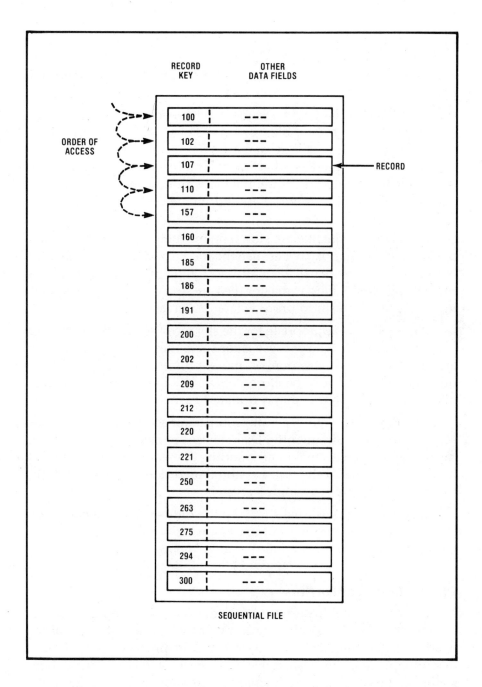

Figure 5-13. Organization pattern of records in a sequential file.

it is necessary to make sure that both the master file and the transaction file are in the same physical/logical sequence. Compatible sequences assure that updating can take place in a single pass through both files.

In addition, updating of a sequential file necessitates the creation of an entirely new master file, regardless of whether tape or disk is used. Usually, the new file is established as the current version, with the source file retained as a backup. However, on some systems, it is possible to "extend" a sequential file with new records appended to it without the need to write an entirely new file.

Sequential files frequently are stored on magnetic tape. Magnetic tape devices are popular partly because the media and the equipment are both relatively inexpensive. In addition, tape processing, involving the writing of new files as master files are updated, provides a built-in backup mechanism to protect data resources.

Disk media are appropriate for the storage of sequential files under certain conditions. One of these is for small files, such as tables, that are used frequently by one or more processing routines. Small numbers of records can be stored efficiently in sequential order on disks and can be accessed and updated easily.

In addition, disk storage is appropriate for sequential files that are sorted or re-sorted frequently. Disk devices present effective sorting capabilities. Tape files that require sorting often are transferred to disk for these functions. If the files are stored on disk in the first place, the transfer operations are avoided entirely.

In some situations, disks may be the only available media. This is particularly true for microcomputers, which typically are provided with diskette drives as standard storage devices.

Direct File Organization

Direct files are used to support random processing. In random processing, the order in which records are processed cannot be predicted in advance. Further, access requests often are processed individually over time. That is, it is impossible or impractical to batch transactions so that all processing can take place at the same time. The nature of this usage eliminates the potential for using sequential organization.

For example, to process each random access request with a tape file would require reading sequentially from the beginning of the file through the records to the point where the required record has been recorded. This access method is inefficient and, essentially, unworkable. An important feature of direct files, however, is that they make possible rapid access to individual records without this type of sequential search.

Despite their orientation toward random access capabilities, direct files also can support sequential processing. Such processing is not particularly efficient or effective, however, because records in direct files may not be recorded in key sequence or in logical order.

Under a direct file organization scheme, the location, or address, of a record in the file often is expressed in terms of its position relative to the first record in the file. When this type of addressing is used, direct file organization is known also as *relative file organization*. Although there is a relationship between the record keys and the positions of the records in the file, there is no requirement that these relative addresses correspond with the logical order of keys.

Direct files function through a capability for equating keys with storage addresses. Two basic methods can be used. One method uses the record key itself as the file address. In the other, more common, method, location of a record in the file is determined through execution of a *hashing algorithm* applied to the key to calculate the file address.

To illustrate use of the technique under which the key is used as the address, consider a small file in which the keys of records have been assigned in sequence from 01 through 20. In this instance, each key can be used as the *relative position*, or address, of the corresponding record. That is, record 01 would appear first in the file, record 02 would appear second, and so on through record 20, which would appear in the twentieth position. If there are missing or unassigned record keys within this sequence, the corresponding positions within the file are blank, or unused. This occurs because the records are not written in a straightforward physical sequence, one after another. Rather, the records are entered into assigned locations in the file that correspond with their key values.

This method works well in a situation in which key values begin at 001, do not increase beyond some fixed upper bound, and use most of the values between the lower and upper bounds.

In most cases, however, these criteria do not apply. Thus, it is necessary to apply a hashing calculation to establish record location.

To illustrate, suppose a file has keys ranging from 1000 to 9999 and that there are large gaps in the key sequence. Suppose, for example, that record key values are assigned in increments of 25: 1000, 1025, 1050, 1075, and so on. If these keys were used as addresses, there would be large amounts of unused space on the disk. Record locations from 0001 through 0999 would be unused. In addition, the 24 key values between records also would be represented by empty spaces in the file.

For such files, it is clearly impractical to use the record key as the file address. In these instances, a hashing algorithm is used to equate the record key to a corresponding storage location. These hashing algorithms, or *randomizing functions*, perform a mathematical calculation on the key to convert it to a relative address. A file organization for this type of direct file is shown in Figure 5-14, which also includes an example of how a hashing algorithm is applied. Note that the algorithm divides the record key by the prime number closest in value to the number of storage positions established for the file. In the example in Figure 5-14, a file with 20 file storage positions has been anticipated. Thus, the prime number used is 19, the closest in value to the number of positions. (A prime number is one that can be divided evenly only by itself or by one.) After the division has been performed, a value of one is added to the remainder. This sum, a number between 1 and 20, becomes the address as expressed by the relative record number. The method illustrated in Figure 5-14 is only one of several hashing techniques used in assigning storage locations within direct files.

Problems can arise when two or more record keys within a direct file hash to the same address. When this happens, a *collision* is said to occur. To illustrate, records 2259 and 9745 in Figure 5-14 both have a calculated address of 18. These records are said to be *synonyms*. One method of handling such a situation is to write one of the synonyms in the next available position, as has been done in Figure 5-14. Thus,

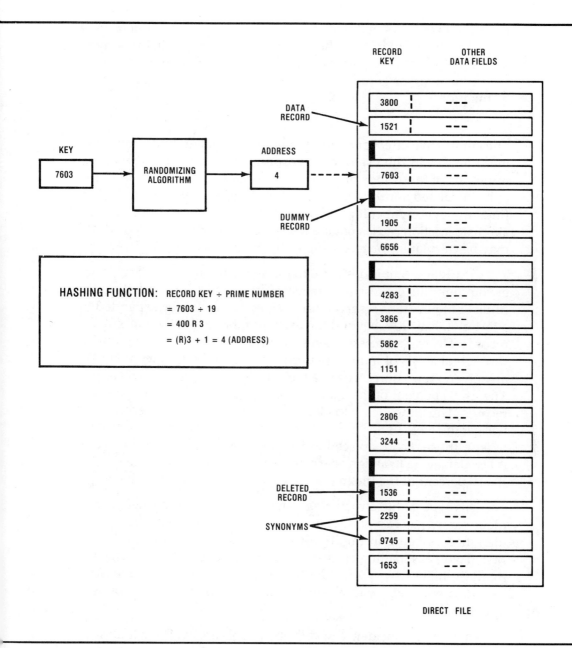

Figure 5-14. Organization pattern of records in a direct file.

record 9745 appears in the nineteenth position, which happens to be the next unused address in the file. The practice of placing synonyms in the next highest position beyond their calculated positions in the file sometimes is called a *consecutive spill,* or *progressive overflow,* technique.

The cumulative effect of alternate placement of synonyms can cause inefficiencies in the use of a file, especially in large files containing hundreds of thousands of records. For each record that is not found at its calculated address, the system must search sequentially until it finds the synonym record. Further, each synonym recorded in an alternate location causes potential displacement of another record that may be calculated for recording in the same location.

Clearly, then, it is desirable to minimize the number of collisions encountered in processing against a direct file. As a first step, the number of positions reserved for storing the file can be expanded by a factor of, say, 20% beyond the exact number needed for all current records. Delays in accessing synonym records also can be minimized through analysis of inquiry patterns anticipated for a given file. For any given group of records in a file, the *80-20 rule* usually will apply. That is, 80 percent of the usage can be expected for 20 percent of the records. Thus, the idea is to identify the 20 percent of the records that will get the most usage and then to make sure that these become *home records,* those records placed exactly at their calculated addresses. It is less damaging to assign alternate locations to records that fall within the 80 percent of the file that will experience 20 percent of the usage.

This type of file structuring usually is achieved through a technique known as *two-pass loading.* Under this approach, the records identified as being in the high-usage portion of the file are loaded first. Two-pass loading assures maximum usage of home addresses by these high-activity records. In the second pass, all of the other records are loaded, with alternate addresses assigned synonyms as appropriate.

To minimize incidence of collisions, tools are available that make it possible to process a series of record keys against several alternate hashing algorithms. These programs, in effect, emulate the setup of a file, describing the distribution of record keys across available storage locations. Use of such programs makes it possible to select the hashing algorithm that will produce the smallest number of collisions.

Direct files tend to be convenient to process. For example, many access functions involving direct files are for user inquiries. The user simply enters a record key. The system finds the desired information and prints or displays the record. There is no impact upon the content of the file.

Record updating in a direct file occurs *in place*. That is, after one or more fields are modified by the update program, the record is written back to its address in the file, replacing the original record. Thus, file updates are destructive to the original information contained in the records. For this reason, updating nearly always involves creation of transaction logs. For example, when a given record is updated, the original record and the transaction that was applied to it may be written to the log file. Thus, if the direct file is destroyed inadvertently, it becomes possible to reconstruct it using a backup copy of the direct file and the log file. Either magnetic tape or disk can be used to maintain log files.

When a record is to be deleted from a direct file, it is not removed physically but is flagged for deletion. In this case, a *flag* is a group of special characters written to a field appended to the record to indicate a specific condition. When these special characters are encountered during processing, they indicate that the record is not to be considered as part of the file. During random access, this flagged record is ignored by the systems software. During sequential processing of the file, however, all records are made available to the program. Thus, an application program that processes the file sequentially—say, a program to sort the direct file in sequence by key—must contain routines to check specifically for flagged records and skip such records during processing.

Over time, direct files require reorganization and rebuilding. The reorganization requirement may occur because a new application requires additional fields be appended to the records. More often, a direct file that has been in use for some time and has experienced many additions and deletions may develop greater incidences of collisions that impair efficiency in the use of the file.

One way to determine whether an existing direct file needs reorganization is to consider run times, or access times. Many operating systems include packages that keep track of access times and other operating statistics involving use of direct files. A sure way to tell that a file is losing efficiency is a steady, though usually gradual, increase in access time experienced by users. These increases in access times occur as the computer is forced to execute increasing numbers of SEEK functions to find a needed record. Statistical packages within operating systems, typically, are able to report on the average number of SEEK executions necessary for each access transaction.

If a new file is expected to require continual reorganization, a profitable practice may be to set up an *access count field* within each record. This count, then, is incremented by 1 for each access function. These access counts can be used to identify the high-activity records that will be loaded in the first pass when the new file is created. This practice can be particularly valuable if there is no way of knowing in advance, before a file is built initially, which records will experience the greatest activity.

Indexed-Sequential File Organization

Indexed-sequential organization can provide equally effective support for applications requiring both sequential and random access to disk files. Records are maintained in key sequences. At the same time, control software for the file system creates and maintains indexes that can be referenced to determine addresses of individual records for random access.

When an indexed-sequential file is created, records are loaded onto disk, in order, by record key. As is the case with sequential files, the records are written in a physical order that matches the logical order of the keys. The disk area that holds this file is called the *prime data area*. Initially, all records loaded into a file are recorded in this area, as shown in Figure 5-15. Separate from the prime data area is an overflow area that is used for the placement of records to accommodate file expansion. A common technique for designating an overflow area is to apportion a segment of the tracks available in each cylinder for this purpose. An alternative technique for handling file

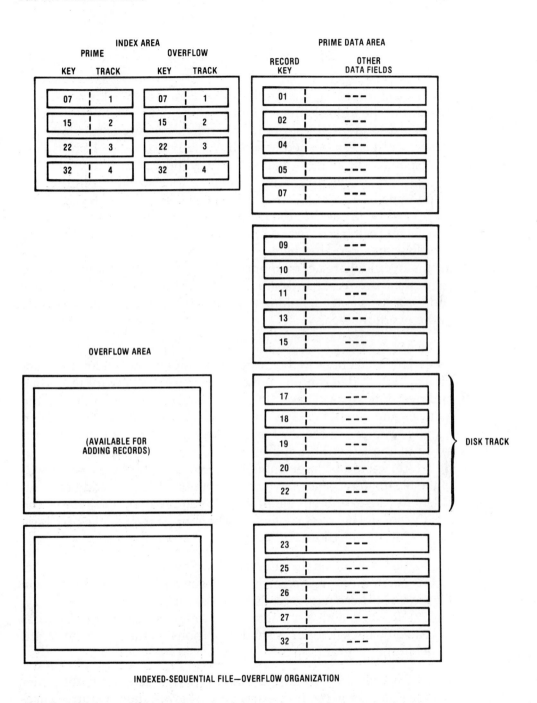

Figure 5-15. Organization pattern of records in an indexed-sequential file under the overflow approach.

expansion is to establish an independent overflow area on separate cylinders.

A third recording area indicated in Figure 5-15 is known as the *index area*. The index is the reference table used to identify, or point to, record locations in support of random access instructions. An index record is established for each block of records—which usually corresponds with a track—within every cylinder. This index is keyed, as indicated earlier, to the highest value for any record in the corresponding block. Note, in Figure 5-15, that separate index areas are set up for the prime and overflow data areas. When a file is loaded initially and can be contained entirely within the prime data area, the overflow area remains blank. Therefore, the initial index entries for the overflow area correspond with the track index values of the prime data area. As records are moved to the overflow area, the track addresses are changed, linking records in the prime data area to those in the overflow area.

One of the challenges in the design of an indexed-sequential file, obviously, lies in determining how much space to allow for the prime and overflow areas. These figures should be developed during the analysis and general design phase and refined as part of the design activity. The prime area will accommodate the record volume that exists at the time the application is initiated. The overflow area, therefore, should be large enough to provide for anticipated growth of the file up to the level at which the number of record additions and deletions causes enough inefficiency to require reorganization of the file. Obviously, if a file's content is to be static, it may not be necessary to provide for overflow.

Figure 5-16 shows the effects of overflow on the content of a file. In this illustration, new records have been added to those shown in Figure 5-15. These new records have key values of 03, 12, 14, 21, 24, and 30. Records with key values of 19 and 23 have been deleted, as indicated by flags (shown as shaded portions on the left of the record boxes in the diagram).

As records are added to the file, they are inserted within their normal sequence in the prime data area. This insertion, in turn, causes the transfer of displaced records into the overflow area. As records

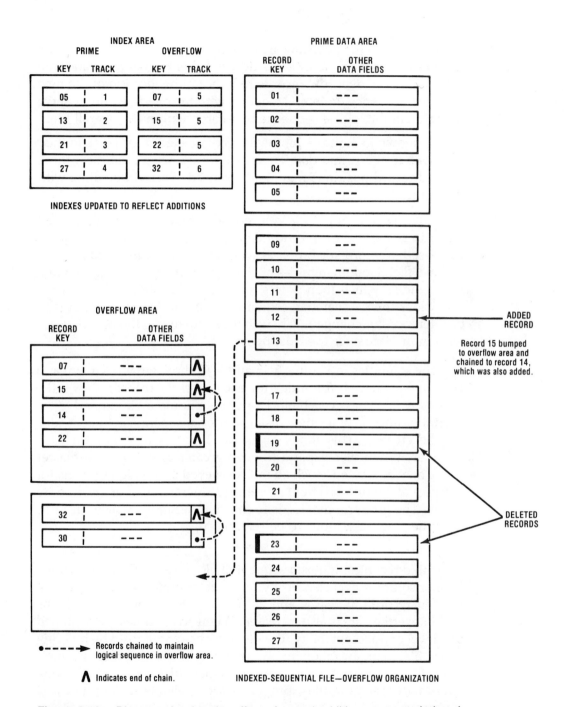

Figure 5-16. Diagram showing the effect of record additions upon an indexed-sequential file with an overflow organization approach.

are written into the overflow area, corresponding changes are made for these records in the index. Address fields known as *pointers* are recorded in the overflow records to indicate the location of the next record in sequence. Pointers, together with the overflow index, serve to maintain the logical sequence of records within the overall file.

As shown in Figure 5-16, the flagged, deleted records retain their positions in the prime data area. In indexed-sequential files, these deleted record positions may not be reused, as could be done with direct files.

Inquiries or changes for indexed-sequential files can be handled either in sequential or in random order. Sequential transactions are processed in record key order, as would be done with an ordinary sequential file. Random inquiries or change transactions are referred first to the index, which then points to the block location of the record. That block then is searched sequentially to find the record that is to be referenced or changed. When changes are made, the new data are written in place over the existing record.

The examples in Figures 5-15 and 5-16 show what is known as a *track index*. That is, the index for each block of records shows the highest-value key on each block or track. Should a file expand to exceed the capacity of a single disk cylinder, a separate cylinder index can be created. This index gives the highest-value key on each cylinder. When files expand to even greater dimensions, a master index may be created. Such an index is keyed to the highest value for a group of cylinders. Thus, reference is hierarchical, or ordered by levels. The master index leads to the cylinder index which, in turn, identifies the track index to be referenced.

As with direct files, indexed-sequential files lose efficiency over time. Response times experienced by users deteriorate as the number of deleted records increases and as the complexity of pointer chains within overflow areas also increases. When access times are degraded to a level of unacceptability, the file must be reorganized by building a new file. Reorganization involves copying all existing records, in sequence, into the prime data area of the new file, allowing for a new overflow capacity. Utility programs are available to handle these reorganizations.

An alternative method of building an indexed-sequential file is called the *free space* type of organization. The disk area that holds this file is called the data area. As shown in Figure 5-17, this area may be divided into blocks of records. A certain percentage of each block of records is left as a free space when the file is built. In addition, a certain number of blocks is left totally free. Initially, this free space contains no data records; it is used as space to add records to the file as the file undergoes change.

In addition to the data area, this indexed-sequential file is supported by an index area. Within the index area are records that associate keys with the block in which the corresponding records are stored. Initially, the index records contain the values for the highest keys in the blocks. As shown in Figure 5-17, the highest key in Block 1 is 04; the highest key in Block 2 is 09, and so on. Also, the index points out the blocks composed totally of free space.

As records are added to the file, they are inserted in sequence within their proper blocks. Records with higher keys are moved forward within the block, and free space is used to receive records "bumped" from their original locations. Figure 15-18 shows what happens when records are added to the file. Here, a record with the key 03 has been added to the first block. To make room for this addition, record 04 is moved forward into free space.

Over time, free space within a block becomes full. When the free space within a block is filled, half of the records in the block are moved into one of the free blocks. The result is the replacement of one full block with two blocks that are only half full. The insertion is made in the appropriate new block. Of course, creation of the new block requires the addition of a corresponding index entry. Figure 15-18 illustrates the organization of the indexed-sequential file after several records have been inserted.

When a record is deleted from the file, all of the following records are shifted upward, physically deleting the record and increasing the free space.

Updating of records occurs in place. The record is accessed, changes are made, and the record is rewritten to the file in the same location.

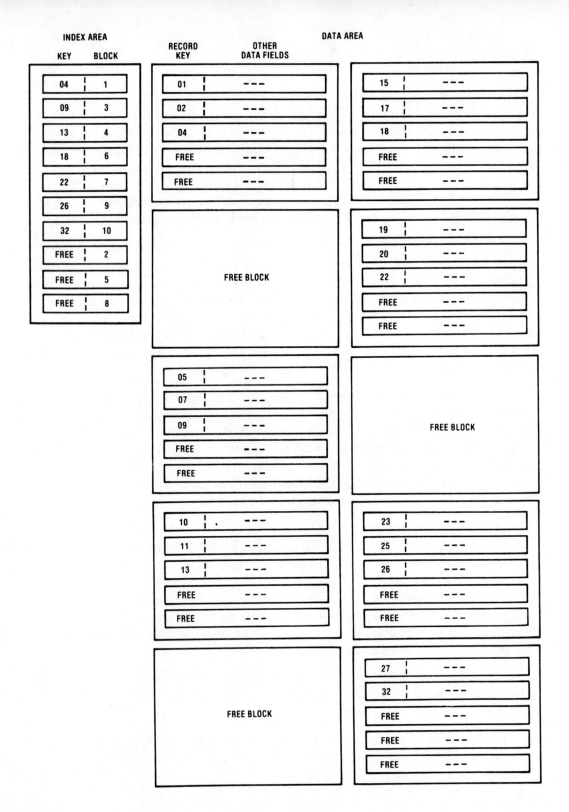

Figure 5-17. Organization pattern of records in an indexed-sequential file under the free-space approach.

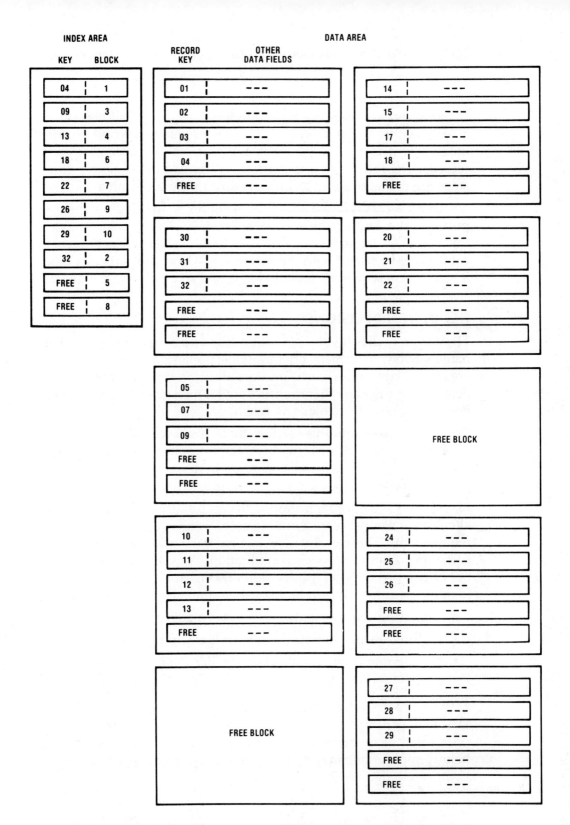

Figure 5-18. Diagram showing the effect of record additions upon an indexed-sequential file with a free-space organization approach.

Records in a free-space type of indexed-sequential file also can be accessed either sequentially or directly. For sequential access, the first index record is read to establish the address of the first block of records. The records in this block then are read in sequence. These steps are repeated for each index record in the index area.

Direct access is based on use of the indexes. In this case, the user provides the key of the record of interest. The system searches through the existing levels of indexes to locate the blocks of records containing the record with the search key. Then, the particular block is searched in sequential order to locate and access the desired record.

FILE DESIGN DECISIONS

The design of files involves identification and consideration of alternatives involving formats, organizational options, and controls. These areas of file design, in turn, produce decision criteria considered in determining the exact characteristics of files for a given application and deciding where those files will reside.

The discussion in this section, then, deals with the elements of final file design decisions. In effect, the list of topics below provides a checklist that can be used as a guide in decision making.

The decision-making process itself involves identifying alternatives for format, organization, and control—then evaluating these alternatives against corresponding costs, values, or compliance with requirements. These evaluations involve trade-offs that are unique in each organization and for each application. Thus, this discussion does not offer a prescription for decision making. Rather, the intent is to identify areas for concern and consideration. These areas include:

- Processing requirements
- Required response time
- Activity rate of the application
- Volatility
- Backup requirements
- File device capacity.

Processing Requirements

Data files, generally, support applications that will be processed either on a batch or on an interactive basis. In some cases, as discussed below, applications may involve combinations of batch and interactive processing.

Batch processing often involves the accumulation of an entire transaction file that is entered as a unit and processed against a master file. In most cases, the efficient organization method for such files will be sequential. For sequential processing, both the transaction and master files have to be in the same order. Matching of records then becomes rapid and efficient, as long as a high percentage of master file records is to be updated with each processing of the file.

A typical batch application is payroll as described earlier in this chapter. Other typical applications supported by sequential files include logs, backup or protection files, and archival files. A typical logging application, for example, would be electronic journaling. Electronic journaling is a chronological listing of all transactions processed within an installation. These files can be used as audit trails or for reconstruction of files under recovery procedures. A backup or archival file is a sequentially created copy of a master file. The original files themselves may or may not be sequential.

To illustrate use of interactive files, consider the airline reservation application. A reservation system requires access to information on individual flights on a direct basis. Therefore, there would be no way to anticipate the order in which transactions are presented to the system or to organize transactions so that they correspond with the sequence of a master file. Thus, with a lack of correspondence between the ordering of transactions and master file records, a direct file organization becomes the only way to support such an application.

Direct file organization also is appropriate for any type of file used primarily for reference. Typical applications can involve inventory files in wholesale or manufacturing organizations, securities quotation reporting, or even automated libraries.

In some instances, applications require a mixture of accessing and processing capabilities. For example, transactions might be processed sequentially against an indexed-sequential file. However, it also may be necessary to support an on-line reference application that uses the same file. Such a mixture of processing methods is implemented by many banks. Processing of checks written by depositors typically is done sequentially. The checks are sorted in account-number order, then these transaction files are processed against customer master files in batches. The same file, however, also can be referenced directly, at random, by bank officers and tellers who need status information.

Trade-offs in processing, therefore, consider such factors as the type of access (serial, sequential, or random) that must be supported, whether batch or on-line processing is required for conformance with processing cycles, and the types of correlations that must be established between files. These considerations become evident during the analysis and general design phase of the systems study and are used as the primary criteria for choosing a file organization and formatting method. In some cases, processing requirements dictate clearly the type of organization scheme that must be employed. In other cases, two or more options will exist: For example, where serial access can be used, the designer can implement the file as a purely serial file, a sequential file, a direct file, or an indexed-sequential file. In these instances, and assuming that necessary hardware is available, additional criteria are applied in choosing the appropriate organization method.

Required Response Time

Consideration of *response time* applies to direct access applications only. The volume of inquiries and the response time required must be taken into account in determining what equipment and what organizational approach to use.

For example, an airline reservation system represents a high-volume application in which there is a real, monetary value associated with rapid response time. A direct file organization with direct access provides the fastest method of getting at needed records. By contrast, if an indexed-sequential file were used for airline reservations, access time could be at least twice that for a direct file. This extra time under the indexed-sequential approach is needed because the computer must perform an extensive search of an index before it can even initiate a record access operation.

Activity Rate

The *activity rate* of an application refers to the frequency with which records are accessed by the application. For example, a payroll application would have a high activity rate, or *hit rate*, because most records are processed on every processing run. By comparison, checking account updating applications would have a lower hit rate because only a relatively small percentage of account holders write checks on any given day.

Normally, sequential processing and access methods are appropriate when hit rates are high because sequential processing represents the fastest method for accessing each record in a file. By contrast, occasional reference to records within a file would make on-line, direct access more efficient.

Volatility

The *volatility*, or rate of change and expansion, of master files should be taken into account in determining the organization to be used. If additions to, and deletions from, files are to be relatively great, it is often best to use a sequential organization plan. An excellent example can be seen in billing programs for transient classified advertising in newspapers. These advertisements are called in by telephone. Usually a classified ad runs for a few days and then is dropped. Its customers are one-time or occasional users. When a bill is paid, the customer name is dropped from the accounts receivable file. Thus, there are large volumes of additions to, and deletions from, the file every day. If such a file were organized for direct access, the entire file would have to be restructured almost every day. Under sequential processing, structuring is routine. That is, a sequential file is rewritten every time it is processed. Thus, additions or deletions present no problem; sequential files are volatile by nature.

Special user requirements, of course, can have an overriding effect on the selection of file organization method. To illustrate, highly volatile files are used by law enforcement agencies to record stolen cars or wanted persons. Content of these files changes rapidly. If batch processing were possible, this application would lend itself well to sequential files. However, on-line reference is a must. Therefore, the application itself dictates a direct access capability. Thus, even though greater processing efficiency might be attainable with sequential files, it is virtually necessary to use an indexed-sequential or direct file organization.

Backup Requirements

Every system needs backup and recovery procedures. If master files are destroyed or if an error occurs during update processing, there must be some method for restoring the files to their proper state. Backup files provide a starting point. Recovery procedures specify a plan for restoring the files.

Sequential files have an advantage in that such files automatically create backup files, because a new file is written during each update processing run. Thus, the input file and the transaction file become backup files for each newly created file.

If direct access files are used, special backup procedures must be developed. The protection is needed because master records are updated in place. Transactions are usually logged as they occur, but transaction records are only valuable as related to a current version of the master file. Master records may be written to an update log file before and after updating, providing a basis for file restoration if something goes wrong during the update processing. It is also good practice to make a backup copy of the master file before a direct update is performed.

File Device Capacity

While not a characteristic of the application itself, the proposed file organization should be reviewed to be sure that the file devices available can handle the files to be created. For example, there might be enough room on a disk file to handle a sequential file application. However, it might turn out that there isn't enough room to accommodate the space overhead of a direct or indexed-sequential file. Additional storage capacity—and corresponding additional cost—may be necessary. In general, one of the checkpoints that should be covered is to make sure that the storage devices to be used can accommodate the files to be created.

DATABASE MANAGEMENT SYSTEMS

The term *data management* refers to the hardware and software technology necessary for data organization, storage, retrieval, and presentation for processing. Data management facilities in most computer installations are provided by traditional *file management systems.* That is, data are organized within logical groupings called records; these records are grouped within sequential, direct, or indexed-sequential files; and files of similar record types are placed on secondary storage devices such as magnetic tape and disk for convenience of retrieval.

For the most part, file management systems are designed to allow access to data through applications programs. In fact, files frequently are organized to support specific applications written in a specific programming language. Data records and their relationships, therefore, often become application-dependent; that is, the file organization supports the viewpoint of a limited number of users who are serviced by the application. Nevertheless, the process of normalization is an attempt to generalize the file structure to make it more flexible and amenable to differing viewpoints and applications.

A *data management system,* on the other hand, is a collection of generalized file management software with the purpose of taking over many of the common processing functions that traditionally have been implemented through applications programs. These systems include report writers and report generators, sort packages, file creation and updating facilities, and basic inquiry processing and data manipulation functions. The purpose of data management systems is to allow users to perform many recurring data processing functions without the time and effort required to develop specialized programs. Data management software provides a convenient interface between the user and data organized under traditional file organization methods.

A *database management system (DBMS)* is a collection of data organization and access techniques that extend the capabilities of data management systems. In general, a DBMS is capable of integrating and managing data to serve the needs of a variety of users with a variety of viewpoints. A DBMS overcomes the constraints imposed in traditional file-oriented environments by organizing data independently of applications. A database management system supports three viewpoints on how data are stored in a database:

- The physical view of data refers to the way in which data actually are organized and stored on secondary storage devices. For the most part, direct and indexed file organization techniques are used along with methods for relating data maintained in physically separate files.

- The logical view of an entire database is called the *schema.* A schema is composed of the set of entities, attributes, and relationships that describe the logical organization of data necessary to support all users of the database. The schema is derived through normalization techniques and documented with data access diagrams.

- The logical view of a single user of a subset of the database is called the *subschema*. The subschema comprises the collection of attributes drawn from the schema to support the processing needs of a user or application program. Whereas a database has a single schema, it is capable of supporting any number of subschemas.

The data access diagram in Figure 5-19 repeats the schema from Figure 5-1 for one of the databases that could be defined to support a student records system. The entities within this database are STUDENTS, CLASSES, and GRADES, each with their own attributes, or data elements. Through various keys, relationships have been established between the entities. The schema has been designed to support user needs for accessing different combinations of attributes for various applications.

Figure 5-19 shows two potential subschemas that can be derived from the schema. Subschema A represents the collection of attributes required to produce class rosters. This subschema may represent the viewpoint of an application program that will generate the rosters at the beginning of a term. Subschema B, on the other hand, represents a different viewpoint. In this case, the combination of attributes can be used to produce grade reports that will be made available at the end of the term. Yet, in each case, a single schema is the basis for the application. It is not necessary that separate files be established to support the different applications.

The schema is a global view of the entities important to the organization, the attributes that characterize these entities and that will be made available to different users, and the relationships that must exist for maneuvering through the database and accessing the required attribute values. Definition of this schema may be one of the responsibilities of the systems analyst, working in conjunction with users and a database administrator. Once the schema and subschemas are defined and documented, a database administrator or other technical database specialist will format the schema for maintenance by the database management system software. It should be noted that a DBMS, in and of itself, does not provide assurance of quality data management. In a sense, a database management system is simply another file organization and access method. However, it does provide extended potential for managing data resources if the logical

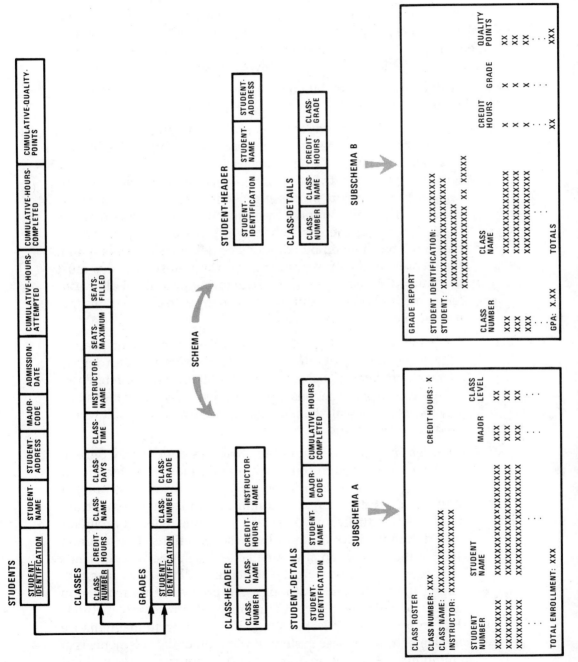

Figure 5-19. Schemas and subschemas to support processing within a student record system.

viewpoints represented in schemas and subschemas are of high quality. Thus, the technical aspects of database management systems are of secondary importance. What is important in a database environment, as well as in a file environment, is the appropriateness and accuracy with which these logical views are modeled.

With an appropriate data model and the availability of database software, several benefits can accrue. These include:

- *Data independence* means that the physical organization of data is not tied directly to specific applications programs or user needs. The data are insulated from changes in processing requirements. For example, in a traditional file environment, files may be organized to support processing needs that are in effect at the moment. Data are bound physically to the programs that use them. Whenever there is a change in processing requirements, files must be changed by additions or deletions of fields within the records, and all programs that process those files also must be changed. In addition, the restructuring of a file to allow different methods of access propagates changes to all programs that use the file.

  Database technology, however, allows convenient adaptability to change. Programs define data needs through logical references to database subschemas rather than through references to physical file organizations. Thus, processing changes mean that only the program with the particular subschema viewpoint would need to be changed.

- *Nonredundancy* means that entity attributes appear only once in the database. Although some redundancy may be preferable for enhancing database performance, the goal is to keep the number of identical attributes at a minimum. In a file environment, redundancy is virtually unavoidable in providing different programs with different views of data. In these situations, if a program changes the value of an attribute in one file, special arrangements must be made for updating the attribute in all the other files containing the same field. Although the process of normalization helps in avoiding excessive redundancy, file processing methods still can produce duplications of fields that lead to maintenance problems. In a database system, replication of data elements is at a minimum. Applications can share common attributes without being bound together by common physical views of data. Thus, programs can operate independently.

- *Flexibility of access* means that extraction and processing of data contained within a database need not be limited to specialized applications programs. In addition, the database can be referenced through higher-level, nonprocedural languages that give the nontechnical user access to data. Normally, database management systems are supported by query languages, report writers, and other generalized software that permit natural-language-type inquiries of the database. In effect, database technology allows as many different applications as can be modeled through subschemas.

- *Integrity* refers to the preservation of a high degree of consistency and quality of data. By its very nature, a database system maintains data integrity. The characteristics of data independence and nonredundancy help assure control over data resources. The DBMS helps reduce the need for users to coordinate their use of data that will be changed, replaced, or linked with other data through new relationships.

- A database system helps provide *security* of data resources. Through elaborate software mechanisms, the system can establish accessibility limits on data and permit only authorized users to gain access to the data defined by the users' personal subschemas. In effect, a DBMS places data resource control into the hands of data administrators and out of the hands of technical persons with the know-how to access the data.

- *Performance* of the database system represents a trade-off in three areas: application processing efficiency, processing overhead, and disk storage space.

A database presents multiple logical views, or subschemas, of the data to multiple users. Yet, these views are all based on a single set of physical files that implement the overall schema. It is not surprising, then, that some application programs might run very efficiently, while others do not. Normally, the underlying physical files are "tuned" so that they provide the most rapid and efficient support to the most critical programs. In this way, the applications that are run most efficiently or have the most stringent response time requirements receive the better service.

The very ability of a database to provide different views of the data to different users, as well as different access paths for different applications, creates a great deal of "behind-the-scenes" complexity. Data maintained in database systems typically require significantly more storage space than would be used in traditional file processing environments. There is also a heavy processing overhead required to maintain the various access paths as the database is updated. These disadvantages can be more than offset, however, by the long-term flexibility provided by a database environment. As business needs change and new application requirements arise, such requirements often can be accommodated by simply developing a new logical schema, rather than by restructuring the entire database and the programs that use it.

Within a database environment, determination of data organization and access is primarily a logical, rather than a physical, consideration. During systems analysis activities, the database schema is derived and modeled as a collection of entities, attributes, and relationships. This schema may have to be integrated with other schemas representing other systems in the organization. During design, specialists such as database administrators assist in creating and formatting the physical database through *data description* or *data definition languages*. These special languages, peculiar to each individual DBMS, communicate the logical data structures to the DBMS so that the data can be organized on devices to allow access according to various subschemas, or views.

DATA MODELING FOR FILES

Even with traditional file environments, database-style techniques for organizing data are applicable and produce similar benefits. The process of normalizing files to reduce redundancy, to eliminate the need for variable-length records, and to permit access through primary keys can produce systems that emulate database environments. Although processing inefficiencies may result for any particular application, system efficiency will be enhanced. The integrity of the data will improve and adaptability of the system to future needs will increase.

Generally, the collection of data managed by an organization remains fairly stable. Applications that process the data, however, change and expand. If a system design—especially its file structure—is modeled after those applications, a change in processing requirements will dictate a change in the file structure, with corresponding modifications required in all software that accesses those structures. If, on the other hand, a system design is anchored to the data structure rather than to the processing structure, the impact of changes is reduced. Files organized according to business needs rather than to computer processing needs serve long-term organizational goals as well as immediate data processing objectives.

Thus, determination of file organization methods should be preceded by careful modeling of the data. This modeling, in turn, should borrow from schema development techniques that are applied in database environments. The result will be a file structure that is generally applicable to both the current processing environment and to planned changes and contingencies.

Summary

In building and using computer information systems, files are the major focal point. It has been said that an organization's collection of files—its data base—models the organization itself.

Within the systems development life cycle, the inputs for file design come from the outputs of the analysis and general design phase.

File design involves the process of normalization, which produces data structures that can be packaged physically as attribute files or as correlative files.

By analyzing data stores and their input and output data flows as depicted in data flow diagrams, identification of entities and attributes can take place. The process of normalization distills this information into sets of attributes and relationships that are characterized by nonredundancy and simplicity of access.

A second method for deriving an organizational data structure begins by soliciting the information from managers and other persons in the organization. Normalization is applied to the resulting structure in deriving the final design that will be implemented as files or databases.

File access methods include physical sequential (or serial) access, logical sequential access, and direct (or random) access.

File organization methods include sequential, direct, and indexed-sequential.

Storage media and devices may be categorized as sequential access devices or as direct access devices. In general, sequential access usually involves magnetic tape and direct access typically involves magnetic disk.

A distinction is made between logical and physical records. File design involves determination of the blocking factor to be used in creation of files. Files may be blocked or unblocked. Also, design decisions may depend upon whether records are to be of fixed or of variable lengths. Blocking decisions relate to the size of blocks and the gaps that occur between blocks. Blocking factors differ for magnetic tape and for magnetic disk implementations.

Direct and indexed-sequential files tend to become inefficient over time and must be reorganized.

Areas of concern in the process of file design include processing requirements, response time, activity rate, volatility, backup requirements, and file device capacity.

Beyond data management systems, database management software allows for multiple views, based on subschemas, of a single database, based on a schema. With an appropriate data model and the availability of database software, the benefits that result include data independence, nonredundancy, flexibility of access, integrity, security, and performance (system throughput). Similar results can accrue within file processing environments if the techniques of schema development are applied.

Within a database environment, determination of data organization and access is primarily a logical consideration.

Key Terms

1. normalization
2. primary key
3. attribute file
4. correlative file
5. secondary key
6. data model
7. file access
8. physical sequential access
9. serial access
10. logical sequential access
11. direct access
12. random access
13. file organization
14. sequential file organization
15. direct file organization
16. address
17. composite key
18. indexed-sequential file organization
19. index
20. sequential access device
21. direct access device
22. read/write head
23. transport speed
24. inches per second (ips)
25. parity bit
26. check bit
27. track
28. recording density
29. bytes per inch (bpi)
30. disk cartridge
31. disk pack
32. diskette
33. cylinder
34. sector
35. hard disk
36. megabyte (MB)
37. access time
38. head positioning
39. transfer rate
40. kilobyte (KB)
41. rotational delay
42. millisecond (ms)
43. logical record
44. physical record
45. block
46. unblocked
47. interblock gap (IBG)
48. interrecord gap (IRG)
49. count-data format
50. data area
51. count area
52. count-key-data format
53. key area

54. buffer
55. fixed-length record
56. variable-length record
57. fixed-block architecture (FBA)
58. blocking factor
59. relative file organization
60. hashing algorithm
61. relative position
62. randomizing function
63. collision
64. synonym
65. consecutive spill
66. progressive overflow
67. 80-20 rule
68. home record
69. two-pass loading
70. update in place
71. flag
72. access count field
73. prime data area
74. index area
75. pointer
76. track index
77. free space
78. response time
79. activity rate
80. hit rate
81. volatility
82. data management
83. file management system
84. data management system
85. database management system (DBMS)
86. schema
87. subschema
88. data independence
89. nonredundancy
90. flexibility of access
91. integrity
92. security
93. performance
94. data description
95. data definition language

Review/Discussion Questions

1. What is the basic relationship between an organization and its data files?

2. What is the process of normalization?

3. What are the three principal methods of file access?

4. What are the three principal methods of file organization?

5. Into what two categories may storage devices be grouped and what specific type of hardware typically is associated with each?

6. What is meant by the blocking factor with reference to files?

7. Which types of files tend to become inefficient over time and how must this inefficiency be corrected?

8. How are hashed files built?

9. What are six key concerns in the process of file design?

10. What is the basic function of database management software?

Project Assignments

The Appendix presents three case studies that can be used in support of project assignments. All three scenarios provide functional system specifications that would result from systems analysis. Included are data flow diagrams, data dictionaries, and process narratives that would result from systems analysis. Continuing the work begun in Chapters 3 and 4, assignments that relate to the content of this chapter will be found in *PART 2: Transition into Systems Design.*

II DETAILED DESIGN AND IMPLEMENTATION PHASE

The Detailed Design and Implementation Phase of the systems development life cycle represents the primary focus of this book. The work of this phase—and the content of this part—begin with a key transition. The project moves from analysis to design—from an emphasis on communication with and responsiveness to users into a technical concentration on design and implementation of a workable solution to an identified problem.

The activities of the Detailed Design and Implementation Phase are reviewed in brief chapters. Over and above these important procedural milestones, there are also vital chapters on the skills associated with systems design and implementation. These chapters provide a basis you can use to build an important foundation of knowledge and capabilities that relate to application software. These software chapters stress the basic principles of application design, design alternatives and their evaluation, adoption of and adherence to application software strategies within a CIS organization, and strategies that can be followed in designing and performing tests of application software.

At the completion of the Detailed Design and Implementation Phase, a project team has a system that is fully developed and tested. Users have been trained and have "bought" the new system. The work of the project team is drawing to a close and users are prepared to assume ownership responsibilities associated with the system.

TECHNICAL DESIGN 6

LEARNING OBJECTIVES

On completing reading and other learning assignments for this chapter, you should be able to:

- ☐ Tell why technical design is a pivotal activity within the systems development life cycle.
- ☐ Describe what system design specifications must be arrived at before technical design can begin.
- ☐ Give the two principal objectives of the technical design activity.
- ☐ Describe the scope of the technical design activity.
- ☐ Name the end products of technical design and describe what is contained in each.
- ☐ Describe the steps in the process of technical design.
- ☐ Tell what personnel are involved in this activity.

ACTIVITY DESCRIPTION

Technical design is a pivotal activity: The framework for the remainder of the project—and for the structure of the system itself—is formed at this point. As described earlier, a transition from analysis to design takes place at the beginning of the detailed design and implementation phase. This transition is implemented during Activity 7. Members of the project team receive copies of the new system design specification (prepared in Activity 5) as their input. The major output of technical design, a detailed design specification, serves as a blueprint for the system to be implemented and installed as a production system.

This activity, following principles presented in Chapters 1 and 2, emphasizes a separation between overall systems design and detailed design for the technical portions of the system. This relationship between systems design and technical design serves to structure the activity in that a system design must be in place before technical designs can be developed and implemented. Consider:

- Processing volumes must be defined and quantified before hardware specifications can be set.

- Decisions must be made about the mix of on-line and batch processing within a system before application software can be designed and programs written.

- The processing cycles for portions of the system—daily, weekly, monthly, and possibly quarterly—also are necessary inputs for application software design.

- Policy decisions about centralization or distribution of both processing and data will shape all aspects of the system—particularly to establish whether and what data communication capabilities must be implemented.

All of these issues are addressed during systems design, which is an essentially transitional activity. That is, systems design considers information gathered and initial design decisions made during the analysis and general design phase as a basis for establishing the framework for detailed design.

There are sound reasons for reevaluating the general system design specifications at this point. For one thing, the emphasis has shifted. During the analysis and general design phase, the objective was to carry the feasibility analysis far enough for it to be updated with reasonable accuracy. The revised feasibility analysis, then, permits the management steering committee to make a major ''go/no-go'' decision on continuing the project. However, in this activity, at the start of the detailed design and implementation phase, the need to prepare detailed design specifications puts a different emphasis on the study of system specifications. In addition, as pointed out in Chapter 2, the CIS professionals and user managers all have gained sophistication—both technically and organizationally—in the process of systems analysis.

Thus, for a variety of reasons, the general level of systems design that took place in the analysis and general design phase may be modified at the start of this activity. This modification can result from identifying new opportunities, from recognizing cost-saving methods, or from a variety of technical reasons. The point is that the technical design activity becomes the arena in which desire meets reality. The product of systems analysis, in effect, documents user desires. In systems design, these desires must be modified in keeping with such realities as requirements of other systems used within the same organization, physical constraints of hardware, budgets, changes in the economy, and many other factors.

Technical design, as described in the Process section later in this chapter, is a two-level activity. Initially, an overall systems design and a general data base design are devised as part of a series of events involving the review and challenging of the new system design specification. This effort produces a set of technical specifications and guidelines that, in turn, drive a series of tasks dedicated to detailed design of specific portions of the new system. Specific areas and products of detailed design are covered further in the Process section of this chapter.

OBJECTIVES

Two major objectives should be met during this activity:

- The transition from analysis to design—as a preparation for implementation of the new system—should be bridged successfully.
- A detailed technical design for the new system should be produced and carried forward as a basis for design and implementation activities that follow.

These objectives are critical because the technical design activity is the point within the life cycle at which quality is built into the new system. There is an old adage that you can't add quality after the fact, nor can you inspect quality into a product. Rather, quality must be an integral, organic part of the process from which a new system is derived.

One measure of quality lies in the extent of useful life that the new system will enjoy. This determination of useful life, in turn, depends upon the maintainability of the system.

All computer information systems, no matter what application they handle, exist in a dynamic environment. It is axiomatic that any new system, no matter how sound its initial design, will be modified throughout its life. The degrees of urgency and necessity for these modifications may vary, but the basic need is universal. Therefore, a major measure of quality—and a major objective of the technical design activity—lies in the extent to which future systems maintenance requirements can be anticipated and long-term flexibility can be assured by the overall design.

Another measure of quality for the work done in this activity is that the documentation generated—particularly the detailed design specification—must be precise and complete. In any sizeable CIS project, a number of people who have not been associated with the project previously will be brought in at this point. This is particularly true for programmers. A number of programmers may be assigned to prepare a variety of subsystems or program modules. Precise and complete definitions are essential: The work of many contributors must fit together so that the individual work products can function eventually as an integrated system. Put another way, without module-to-module compatibility, there can be no effective computer information system. Specific requirements include:

- A precisely defined data base (down to the level of storage formats, field sizes, access methods, and other specifications) must be specified.

- Inputs and outputs must be defined down to levels that include exact spacing, the content of standard labels and messages, and the precise terminal displays that occur for on-line portions of the system.

- Program definitions must encompass both the processing within individual modules and the passing of control and data among modules.

SCOPE

The extent and scope of Activity 7 are illustrated in the Gantt chart in Figure 6-1. Clearly, this activity leads the detailed design and implementation phase; none of the other activities is launched until technical design is well along. All of the other activities either continue or do not start until after Activity 7 is concluded.

The working scope of this activity involves a reconfirmation and possible modification of the new system design specification produced during analysis and general design and a transition from the new system design specification to a detailed design specification.

Overlapping and interdependency are particularly strong between the technical design activity and Activities 8 and 9, involving the preparation of test specifications and the writing of programs. The first parts of the system to be designed can move forward to the latter two activities even while design continues on other portions of the system.

Activity 10, relating to user training, is largely independent of the others in this phase in that requirements and response patterns in user training are different from those associated with systems and technical design. User training will, of course, depend in part on the results produced during the programming activity and on the terminal dialogs established during the technical design activity.

Activity 11, involving the testing of the complete system by users and operations personnel, cannot begin until all of the other activities have been completed.

END PRODUCTS

Three end products result from the technical design activity:

- The detailed design specification
- Computer operations documentation
- Specifications for special programs to be used for file conversion.

Detailed Design Specification

The *detailed design specification* includes and also extends the new system design specification developed at the close of the analysis and general design phase. As described in Chapter 1, both an overlapping and a transition exist between the end of the analysis and general

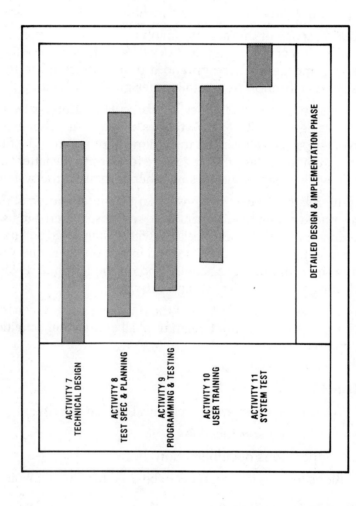

Figure 6-1. Gantt chart illustrating the parallel nature of the activities of the Detailed Design and Implementation phase.

design phase and the beginning of the detailed design and implementation phase.

When the technical design activity is completed, all of the documentation required for technical implementation of the new system should be completed and in place. All of this documentation is incorporated in a detailed design specification that can become quite extensive. Content of this detailed design specification can include:

- An overview narrative describes the system's purpose, goals, and objectives—as well as the basic logical functions that must be performed.

- Processing descriptions include a context diagram and a hierarchical set of data flow diagrams. Diagram 0 should identify the major subsystems within the system being developed. Lower-level diagrams should indicate the physical packaging decisions that have been made for the new system.

- Annotated systems flowcharts should be provided for each job stream. In the process, program identifications are established and all necessary procedure language or job control language (JCL) statements are written.

- Individual programs are designed and specified through use of structure charts. Structure charts are created for all major job steps in the computerized portion of the system.

- A program inventory, or listing, is prepared to cover all programs in the system. This inventory is organized according to program identification and name, the job stream or streams in which each program occurs, and the external programs (if any) that call each particular program.

- Program specifications are prepared. These specifications include detailed descriptions of modules, of interfaces among modules as specified on the structure charts, and of inputs and outputs. In addition, processing descriptions for these modules and program components are prepared. These descriptions are executed at the level of structure charts and pseudocode.

- Specifications are included for backup requirements and recovery procedures for all files in the system.

- A description is prepared of the audit trail and logging requirements to be incorporated in the new system.
- Data dictionary definitions are updated as necessary to support the processing specifications.
- Precise specifications for data base or file definitions are prepared. These specifications include access methods or paths, record layouts, storage formats, edit criteria, and other details.
- A catalog is developed to show all outputs to be delivered to the user—either printed or displayed. A description sheet is prepared for each output. Precise layout charts are prepared for all outputs. If preprinted forms are to be used, final designs are prepared.
- A catalog is prepared that lists all inputs, including a descriptive document sheet for each. Layout charts for input records and file designs for any preprinted forms are prepared.
- User interfaces with the system are reviewed and validated or modified as necessary. Precise terminal dialogs for all on-line functions are created.
- Performance criteria that are critical to either computer or manual processing are documented. These performance criteria can include response times, volumes, or other quantifiable specifications.
- Security and control measures aimed at limiting access to either equipment or files are documented. (Security and control measures that deal with processing are included within processing specifications.)
- Policy considerations associated with the new system are reviewed. Any unresolved decisions are carried forward for resolution.

Computer Operations Documentation

Computer operations documentation comprises instructions on the setting up and execution of programs for each application. These instructions can include brief narratives describing the processing of systems and subsystems for the benefit of computer operators.

Operations manuals should include estimates of processing volumes and run times. For each job stream, a descriptive sheet should be prepared that covers the name of the job stream, input file requirements, set-up instructions, outputs, data control provisions, backup procedures, recovery and restart instructions, and any special requirements.

Conversion Programs

Conversion programs may be necessary in situations in which files are to be converted from an existing computer system to a new one. Very often, it is necessary to prepare special programs, used one time only, to accept existing files as inputs, restructure those files, and write the files onto the media that will support the new system. One of the end products of this activity is a set of technical specifications for any such conversion programs.

THE PROCESS

Chapter 1 introduces the process-oriented relationship between the major tasks of this activity. Initial efforts are directed toward overall system design. Data base modeling and design tasks overlap these efforts. The results of these system and general data base design efforts then lead to a series of more detailed technical design steps.

Work occurs at two levels of detail. Distinguishing between these two levels makes it possible to structure the transition from the end products of analysis to the inputs required for detailed technical design.

The systems design process itself is described in Chapter 2. As described there, the process actually was begun during the analysis and general design phase, with results presented in the culminating document of that phase—the new system design specification. At the start of this technical design activity, those system design considerations are reviewed and modified following the process described in Chapter 2. These revised specifications form the basis for the detailed technical design steps that follow.

It is worth stressing that the process for general systems design is not as well structured nor as widely accepted as are the processes for either systems analysis or for software or program design. The problem is that presently there is no formal, generally accepted methodology for the transition from systems analysis to the specification of systems design criteria.

Detailed design issues are covered in greater detail in other chapters. However, to make clear the scope and focus of the technical design activities, the following sections review some of the key concepts associated with each of the areas that comprise detailed technical design.

Human-Machine Interface Design

There are fairly straightforward procedures that can be followed in analyzing and meeting needs for interfaces between people and computing equipment. It is important to understand what is done during this activity. Rough sketches of all input documents, output reports, and on-line screens were produced during analysis and are input to this activity. The job here is to turn the rough sketches into precise formats. Design includes specific content of standard messages and labels as well as detailed specifications for format and spacing.

On-line interfaces present a special concern. In this activity, the designer takes the basic input and output screens that were identified earlier and builds on-line conversations based on those displays.

Although the emphasis usually is placed on these on-line conversations, it is important to remember that the human-machine interface also may involve submission of batch inputs and production of printed reports. It is also important to specify all of the detailed procedures that the user must follow within the scope of the system. Specifications for on-line conversations feed the detailed software design and programming tasks. Specifications for manual user procedures feed the user training activity that normally occurs in parallel with the programming and testing activity. Among the important human-machine design considerations are:

- Readability of reports or screen formats should be evaluated for each delivered item. Spacing should enhance user recognition and understanding. Data elements should be presented in logical groups corresponding with the way the receiver actually will use the information. As a general rule, the normal reading sequence—top-to-bottom and left-to-right—should be followed.

To the extent possible, the overall design of reports and displays should be kept uniform throughout a system. This uniformity can enhance their readability and understandability, especially for the occasional user.

- The amount of data presented to a user at any given time should be considered. On displays, for example, data content should not be overloaded to a level that inhibits recognition or reaction time. In both reports and displays, emphasis should be on delivering what is needed, not all that is known. The key to determining content is the purpose or intended use of the output.

- Terminology should match the normal words, phrases, or terms that are natural to a user and should be standardized throughout the system. Avoid use of codes or abbreviations when designing outputs for occasional users.

- Messages displayed on terminals should have common or similar formats for easy recognition. Diagnostic messages should be stated clearly and be meaningful to the user.

- Don't assume that a user will be familiar with rules or procedures after he or she has heard them once. As user procedures are defined, outlines are prepared for user manuals and training materials that will be developed later during the user training activity. For on-line portions of systems, much of this material can be provided in the form of ''help'' screens that can be called up at the user's request. These help screens should be self-contained and easily accessible to the inexperienced user and should be capable of being bypassed by the knowledgeable user.

- The level of sophistication of users should drive the content and design of interactive conversations and guide the selection of instructional aids or other options that are made available.

Detailed File Design

A logical data model to support the system was created during the analysis and general design phase. This model serves as input to the general data base design tasks. If the system is to be supplied by traditional computer files, the result is a specification of the data content, file organization, and access keys for each of the files specified. If database management software is to be used, the database administration group will specify the logical view of the data—the view as seen by the programmer and as supported by the particular database software package to be used. This group also will specify the necessary changes to the physical database in order to support the estimation of implementation costs. Those results now are reviewed and refined in light of the system design decisions reached at the start of this activity. In general, major changes should not be required.

After completion of the general data base design tasks, the detailed file design is fairly straightforward. Precise record definitions are specified for each file. These definitions include exact record layouts, storage formats and lengths, primary and secondary keys, and so on.

Both general data base and detailed file design considerations are covered in greater depth in Chapter 5.

Network Design

Basic decisions in the area of network design, increasingly, concern the degree of centralization or distribution of processing functions and data resources. Flexibility and creativity can pay big dividends both in terms of the cost of equipment and communications facilities and in terms of system responsiveness and business services provided.

Most of the detailed network design tasks will be performed by communications software and hardware specialists working in parallel with the project team during this activity. It is the responsibility of the project team to provide the necessary input to these support people concerning business requirements as they affect network design. This input includes:

- Outline the geographic distribution of sites and the system functions to be supported at each site. Many of these sites probably will be local. Some may be remote.

- Specify the type of processing by function and site. Which function will be batch and which will be on-line?

- For each function and site, specify the response time required. The time will depend on the nature of the business. Pricing routines used with a bar code reader in a grocery checkout application require responses in less than a second. An airline reservation system can tolerate responses within several seconds, while response to queries in a nationwide medical diagnosis system could be several minutes or longer. Response time measured in days may be adequate for other applications. For example, maintenance schedules for vehicles in a nationwide distribution system could be distributed weekly.

- When remote sites are involved, it is particularly important to analyze where the different data elements are used in addition to the required response times. This analysis will affect decisions about which data elements to keep at a central location and which to distribute to remote locations. Clearly, these decisions have an impact on the network design.

- Normally, some communication network capability will be in place already. Communication needs for the system under development should be tied into the larger picture of existing communications among local and remote sites. Comparing the communication needs of this system with those of other existing systems will help communications specialists determine the adequacy of current equipment and network support.

Systems designers on the project team must work closely with communications support personnel to develop an understanding of communications requirements and impacts. Network design decisions will impact directly both data base and software design decisions—and, hence, the ultimate cost to implement and run the system. Trade-off decisions involving cost and level of service certainly will be necessary.

Application Software Design

It is highly significant that the design of application programs and the writing and testing of those programs comprise separate activities within the systems development life cycle. This positioning stresses the importance of assuring thoroughness and quality in executing and evaluating designs before program coding even begins. The design tasks done during this activity concentrate on creating overall designs for each of the application programs at the structure chart level. The emphasis is on creating tight, functional modules and clearly defining the data relationships among them. Algorithm design to implement individual modules normally is done as part of the later programming and testing activity.

In the design of application software, extensive sets of tools are available for identifying modules that are related closely to business functions, for determining the data relationships that should exist among those modules, and for evaluating the correctness and maintainability of the program designs as documented in the resulting structure charts. These program development principles and techniques are covered in later chapters.

At this point, attention should be paid to the relationship between the design of application programs and the preparation of test specifications and test plans for those programs. Ultimately, programs are evaluated by the results produced when test data are processed. Thus, the activities for deriving test data and the design of programs can and should overlap, as shown in Figure 6-1. However, as is pointed out later, although the tasks of software design, test specification, programming, and testing may all overlap in time, these tasks should not overlap in personnel. That is, test specification and testing should be done by an independent set of people—not by those responsible for creating or implementing the design.

A number of concerns should be addressed throughout the process of application software design. These concerns include:

- Always remember that a program is an implementation of a plan to solve a business problem. It can be tempting, as the work becomes detailed and technical, to concentrate on program efficiencies and refinements that may be implemented only at the expense of the business objectives to be met. Truly elegant program designs are useless if those designs are not ready in time or, worse, if they do not solve the specified business problem. Software designers should restrict all impulses to deal with technical challenges for their own sake, even though the technical aspects of a system may seem more intriguing. In short, never lose sight of the business objectives for any given application.

- Almost invariably during application software design, some opportunities for system enhancement will be uncovered. It can be tempting to modify existing designs in the belief that these changes will make the system significantly better. However, such decisions should not be made at this stage. The function of an application software designer is to meet existing specifications. If additional opportunities are uncovered, those opportunities should be carried forward into the list of enhancements to be considered in the review phase.

Each application software program should be designed to accomplish as much as possible within defined specifications. Many options are available in the structuring and execution of program designs. That is, there always will be two or more ways to implement any given set of specifications. It is the designer's responsibility to select the option that presents the best long-range value for the system and its users. Thus, it would be best to choose an option that is most compatible with the perceived future needs and directions of the system. The designer should avoid options that would work equally well at present but might present greater difficulties when the system is modified or expanded in the future.

- Any temptation to rush or short-change the program design step should be put down. Many programmers are far more comfortable writing code than they are designing programs. As a result, there can be a temptation to rush to coding. Almost invariably, a design derived hurriedly will be low in quality and will produce more problems and higher costs. The design step is the point at which efficiency and maintainability are built into and assured for programs. The higher the quality of design, the simpler program coding will become. Thus, challenges and satisfactions should be sought in the design process. In thinking of the process of programming as an overall entity, the actual writing of code, if designs are done effectively, becomes just a little higher than a routine clerical function.

PERSONNEL INVOLVED

The makeup of the project team shifts dramatically during the technical design activity. Up to this point in the project, there has been a mixture of users and CIS professionals. At this point, the active membership of the team is largely technical. Depending upon the size of the project, a number of designers and programmers should be added to the project team. Some systems analysts may be reassigned to other projects.

Even though users may not be involved actively in day-to-day design tasks, it is important to keep lines of communication and consultation open. One method for doing this is to hold regular, formal status meetings conducted by the project leader and attended by key user managers and supervisors of technical activities that are currently in process.

The purpose of such meetings is not to solve problems. Rather, the purpose is to keep the various user areas informed of activity and progress. Typically, topics would include:

- Review the major tasks completed since the last meeting.
- Review the major tasks currently in process.
- Outline the work scheduled to be completed before the next meeting.
- Discuss any scheduling problems.

A 30- to 60-minute meeting held weekly or bi-weekly usually is sufficient to meet the objective.

Bear in mind that users have been involved actively in project development for perhaps six or nine months. Bear in mind also that this is their system and they are beginning to anticipate its benefits, perhaps with some impatience as they become more aware of the shortcomings of methods they are using presently. It would be counterproductive to cut them off entirely during the technical phases and activities of a project. Remember that, although the project team is mainly technical, users are still on call and should be consulted during the technical design activity.

Technical Support Considerations

Certain projects may require technical support from outside the project team. Two common situations are noted below. These situations are significant because of the coordination that must occur between the project team and technical support capabilities that these specialists can provide.

Database considerations.   If the computer system on which the new application will run utilizes database management software, there will be special concerns and considerations connected with integrating the data requirements and outputs of the new system with existing databases. In such a situation, a database analyst typically will begin to work with the project team during the last portions of the analysis and general design phase.

The database administration group may be involved and would be responsible for the physical database design. Later, this group will handle the creation of the physical database during the conversion and installation activities. The database group also will be involved during testing to oversee program efficiency in terms of database accesses and access paths.

Hardware and system software concerns. Special technical considerations will arise if the system under development requires new items of computer equipment or new systems software capabilities. To put this requirement in perspective, there probably will be no special concern unless significant new hardware or systems software purchases are required. However, if the computer installation is being altered to accommodate the new application, a technical specialist would begin work with the team late in the analysis and general design phase. This person would become a key member of the team during technical design. The technical services group would be responsible for the acquisition, testing, and acceptance of new hardware and systems software.

Summary

The major output of technical design, a detailed design specification, serves as a framework for the system to be implemented and made operational.

Technical design is a two-stage activity. Initially, an overall systems design is devised as part of a series of events involving the review and challenging of the new system design specification. This effort produces a series of technical specifications and guidelines that, in turn, drive a series of tasks dedicated to detailed design of specific portions of the new system.

Two major objectives of technical design include the transition from analysis to design and the production of a detailed technical design.

Specific requirements of technical design include a precisely defined data base, precise physical design of inputs and outputs, and program definitions that encompass both the processing within individual modules and the passing of control and processing among modules.

The working scope of this activity involves transition from the new system design specification to a detailed design specification, with possible modification as part of the transition process.

The three end products that result from the technical design activity are the detailed design specification, computer operations documentation, and specifications for special programs to be used for file conversion.

The process of technical design includes human-machine interface design, data base design, and design of application software.

Although the makeup of the project team during this activity is mainly technical, users are still on call and should be consulted.

Technical support considerations include database considerations and hardware and systems software concerns.

Key Terms

1. detailed design specification
2. computer operations documentation
3. conversion program

Review/Discussion Questions

1. Why may technical design be considered a pivotal activity within the SDLC?

2. What system design specifications are necessary inputs to system design?

3. What are the principal objectives of the technical design activity?

4. What specific requirements exist for the output of the technical design activity?

5. What is the scope of the technical design activity?

6. What are the end products of technical design?

7. What are the main elements of the detailed design specification?

8. What is contained in computer operations documentation?

9. What is the purpose of conversion programs?

10. What are the main steps in the process of technical design?

11. What pitfalls may be encountered in writing program code?

12. What personnel are involved in technical design?

Project Assignments

The Appendix presents three case studies that can be used in support of project assignments. All three scenarios provide functional system specifications that would result from systems analysis. Included are data flow diagrams, data dictionaries, and process narratives that would result from systems analysis. Assignments that relate to the content of this chapter will be found in *PART3: Detailed Systems Design*.

7 FOUNDATIONS OF SOFTWARE DESIGN

LEARNING OBJECTIVES

On completing reading and other learning assignments for this chapter, you should be able to:

- ☐ Explain the purpose and scope of software design.
- ☐ Tell how software may be structured by statements, modules, and program control structures.
- ☐ Describe the attributes of modules in terms of function, logic, and interfaces.
- ☐ Explain the use of program control structures for sequence, repetition, and selection.
- ☐ Describe graphic representation tools used in software design.
- ☐ Explain the methods of structuring software, including the algorithmic approach and the hierarchical approach.
- ☐ Describe the use of software design strategies, including functional decomposition, data flow, and data structure.
- ☐ Describe software design evaluation criteria, including coupling, cohesion, and heuristics.

SOFTWARE DESIGN DEFINITION

A *design*, any design, is a representation of an object to be constructed. A *software design* is a representation of programs to be written.

In effect, a software design is a paper model representing the structural and functional characteristics of some unit of software, such as a complete program or a module that will be part of a program, or a set of integrated programs. The design serves to answer a series of questions about the unit of software to be constructed:

- What functional components (modules) are required?

- What structural relationships exist among these components?

- What detailed processing components comprise the modules?

- What control interfaces are required among modules?

- What data interfaces must exist among modules?

Software design is one step in the preparation of a total system design. Systems designs deal with overall requirements and results in specification of the input, processing, output, and data resources that will interact to provide the needs stated as a result of analysis. In a sense, a systems design can be likened to an architect's rendering of a building to be constructed. The viewer can tell what the building will look like. But, before construction can begin, sets of extremely detailed plans and specifications must be developed. These plans serve as the basis for actual construction, or implementation, of the design.

A user specification for a new system plays the same role and serves the same basic purpose as an architect's rendering for a new building. That is, a specification document provides an overall picture of what an information system will look like and what it will do. Beyond that, a series of detailed technical designs must be produced for a system under development. One of the types of detailed designs that must be produced is a file design, as described in Chapter 5. Another requirement is a design for the application software needed to implement the system. The foundations of software design are covered in this chapter. The chapters that follow deal with criteria for evaluating designs and strategies for producing designs.

THE PURPOSE OF SOFTWARE DESIGN

Systems analysis takes a complex business problem and breaks it down into a system of data and processing components. Systems design takes this resulting logical system and organizes it into jobs, job steps, and individual interactive procedures. Software design, then, takes one of these system components—the processing function—and bridges the gap between its broad, general specifications and the myriad lines of program code that fulfill those specifications.

Thus, software design is the process of taking the specifications for a job, job step, or on-line procedure and translating those specifications into a hierarchical organization of modules. The basis for the specifications is documentation in the form of data flow diagrams, data dictionary entries, and other documentation produced during analysis.

As in any phase of technically oriented design, software design takes an overall assignment that is too complex to deal with in its entirety and breaks that job down into a series of smaller parts that can be understood and handled individually. The skills and methods for identifying what those smaller parts are, how extensive each of the parts is to be, and how all of the parts relate to one another, is the crux of software design.

As computers have become more generally available and have grown in capacity and capability, the systems that can be planned and implemented have become more extensive. Thus, complexity of applications is, in a sense, an outgrowth of the success achieved by the builders of computers and by developers of system software. The extended capabilities of hardware and software have, in turn, been passed along into the application area.

Unfortunately, trends in applications software development have not kept pace with trends in the hardware and systems software areas. Design methodologies that were sufficient when hardware and software constraints limited the application of the computer to rather simple data processing systems are no longer adequate for building the systems of today. There is an order of magnitude difference between designing and developing a basic accounting system of 5,000 to 10,000 lines of code and a management information system of, say, 300,000 lines of code. The degree of complexity inherent in the latter type of system places analytical demands on the designer that were not even considered in earlier systems.

Another difference between the systems being developed today and those of a decade ago relates to the types of users who operate and are serviced by the systems. Data processing systems traditionally have been batch-oriented systems run by computer professionals. All data entry, processing, and output took place within the confines of the computer room, which was considered off-limits to the eventual user of the information produced. Within such a highly controlled environment, data processing was the responsibility of the computer specialist who designed systems more in accordance with machine constraints than with user needs. Nowadays, however, the clients of information systems have become more knowledgeable about computer processing. They request—often demand—increasingly complex processing services and expect those services to be delivered. Long past are the days when computer professionals could hide behind the excuse, "It can't be done!"

Computer power also has moved out of the computer room and onto the shop and office floors. Users are performing more of the data entry and data processing operations that once were closeted within the computer department. Therefore, systems today must be designed for use by nontechnical people. Thus, input and output facilities become increasingly important, and human-machine interfaces may become the focal points for design. Extensive editing and other preprocessing functions must be built onto the front ends of applications and more flexible, "user friendly" formats must be designed into the outputs.

Business itself has become more competitive and dependent upon computer power for survival. Although the computer still performs valuable bookkeeping and record-keeping functions, these services are overshadowed in importance by the need to meet competition. The use of management information systems and growth in the use of decision support systems signals the integration of the computer into the management structure of many organizations. These types of systems perform valuable management control services, business simulations, and performance projections that require access to comprehensive databases of information and elaborate software to search, extract, and manipulate the data. The effectiveness of software designs in this environment is judged, in the final analysis, on the ability of the organization to reach its goals, not just on the accuracy and timeliness with which business records are maintained.

These and other developments have made software design an exacting and demanding skill. Software design now must proceed with an eye on flexibility, generalizability, and maintainability. In this environment, software can become complex, and change is inevitable. Today's systems increasingly are susceptible to changes in user needs and to changes dictated by a volatile business and technical environment. Thus, software design has become an extremely dynamic area of endeavor. There are no firm, positive prescriptions for how best to design applications software. The field has, as yet, written no "cookbooks" about how to develop these products. However, certain trends are clear.

One of these trends is marked by a term that has been applied with some acceptance to the entire area of application software development—*software engineering*. In effect, this acceptance means that software developers are attempting to apply the disciplined, relatively standardized procedures found in the engineering of structural, mechanical, electrical, and electronic systems to the construction of software products.

In principle, the approach is based on the use of functionally independent, integrated components. That is, software design attempts to subdivide problem solutions into software modules, or pieces of code, that implement specific, identifiable processing components organized to produce minimal connections among those components. In effect, software design principles recognize that most systems are much too complicated to be considered in detail in their entirety. Thus, a large program is broken down into smaller and smaller modules. At succeeding levels within this hierarchy of modules, functions and interconnections are defined. Further, design proceeds downward from a relatively broad definition of processing requirements to exacting operational details at the lower levels. This process brings order and simplicity to design. Each component of the software system becomes a functional identity. It stands apart from other components in terms of detailed design considerations; yet, it interfaces easily with other modular components. Such characteristics contribute to flexibility in design, ease of implementation, and reduced problems in long-term maintenance.

Another dimension to software design lies in communicating solutions to the people involved in implementing those solutions. In effect, the paperwork products of application software designers are the blueprints for software systems in the same way that blueprints developed by construction specialists are the guides followed by plumbers, electricians, and other craft specialists. Graphical techniques have been developed both to guide the design process and to communicate design solutions clearly to the implementers of software systems.

THE SCOPE OF SOFTWARE DESIGN

Application software design encompasses all of the activities needed to produce structural designs for the collections of programs that will implement a system. The starting point for software design is the set of systems flowcharts that defines the programs and data files to be used. The systems flowcharts also specify how the programs and data files will be integrated within a job or processing procedure. Before application software design begins, data files have been designed and major input and output requirements have been determined. In essence, the objective at this point is to figure out how to turn the defined inputs into desired outputs.

The scope and extent of activities involved in going from systems flowcharts to a set of finished programs will vary widely with the nature and types of programs involved. In some instances, a systems flowchart delineates a program to an extent that application software design becomes relatively routine. This occurs in situations, for example, in which there are many precedents—and much experience within a programming group—for designing and developing certain types of programs. For example, existing modules may be reworked and hung together to form the needed program. Again, an analogy with construction is appropriate. Some houses are mass-produced and, in effect, wheeled in from a factory to place on a prepared foundation. In other cases, every aspect of a home is custom tailored. For programs that are opening new areas of information processing applications or using new technologies, custom tailoring—specific, innovative design—becomes necessary.

In either case, whether design is routine or innovative, the primary motivation of application software design remains the same.

That is, application software must be cost-effective. In this context, cost-effectiveness means that programs do the required job accurately and reliably, that results meet specifications, that computer and human resources are used efficiently (within reasonable cost parameters), and that results can be enhanced and maintained easily throughout the operational life of the system.

An application software package is going to be around for a long time, possibly longer than the people who develop it. Therefore, quality and cost-effectiveness standards must be met. Specifically, cost-effective application software must:

- *Work correctly.* It goes without saying that a software product must work. Obviously, a program that produces incorrect output has no value to the organization. Thus, the first criterion is correctness.

- *Meet specifications.* Besides working correctly, the software must conform with expectations established by and for users and documented in the user-approved systems specification. The designer and programmer have little, if any, latitude in modifying those specifications. Any proposed changes should be approved by users and reflected as new specifications that cannot be changed unilaterally.

- *Be reliable.* The software product must work correctly and meet specifications over time and under changing circumstances. It must be dependable to produce results consistently for as long as it remains in production. Reliability also implies *robustness.* That is, the system must be able to deal with expected and unexpected situations inevitably arising from use by nontechnical users.

- *Be maintainable.* Software products have relatively long useful lives over which changes in requirements are inevitable. Across the operational life of a system, more effort and time likely will go into modifying the system to meet these changing requirements than went into its original development. Thus, the software design must contribute to convenience and ease in performing maintenance work.

- *Be easy to use.* Programs developed for use by managers, workers, and other nontechnical personnel must be easy to use and not require technical expertise. Systems must conform as nearly as possible to the normal work patterns of people rather than require people to adapt totally to the technology. Helpfulness, consideration, and tolerance for human error are watchwords for the designer.

- *Be easy to implement and test.* Without compromising the above criteria, the software must be designed to make efficient use of development time and resources. It must be structured for ease of programming and testing. The designer must recognize that development teams, rather than an individual, typically will be involved in implementation. Further, the designer must consider that multiple programs and data files will require interfacing and coordination.

- *Use computer resources efficiently.* Although not as important as it was when computer hardware was relatively expensive and the need to optimize use of those resources was critical, the need to use the technology efficiently is still an important consideration. This is especially so with small computer systems such as minicomputers and microcomputers. Specifically, efficiency relates to CPU and memory usage, as well as to auxiliary storage space utilization and access.

It should be evident from the above list that cost-effectiveness is related primarily to human factors rather than to technological ones. The expenses involved in developing and using an information system result mainly from the cost of personnel required to analyze, design, implement, use, and maintain the system. Similarly, the effectiveness of a system is primarily contingent upon the time, effort, and success people have in using it. Therefore, quality and cost-effectiveness standards for software encompass hardware, software, and human resource contributions.

THE STRUCTURE OF SOFTWARE

It is probably more accurate to say that software is constructed, or built, than to say that it is written. For example, the systems analyst prepares a set of detailed specifications documenting the data and the major processing functions of a system; the systems designer defines the overall physical parameters for the system, identifying the programs and data files necessary to implement it as a computer-based system; next, the software designer prepares the blueprints to be followed in implementing the programs, specifying the software components and their relationships; and finally, the programmer constructs those components and interfaces utilizing programming languages and other technical tools of the trade. Thus, software undergoes an extensive development process from conceptualization through specification, design, and implementation. Whether software is developed by a team of specialists or by one person working in isolation, the steps in the software development process remain virtually unchanged.

The activities performed by the software designer have as their main purpose the definition and structuring of the components that will comprise the set of programs required to make a system design operational. Thus, the software designer considers the question of which program functions should be interconnected and in what ways, to meet processing specifications. The resulting design, then, becomes the blueprint for the programmer, who elaborates the design into sets of computer operations that perform requisite input, processing, output, and storage functions.

At the most fundamental level, software is constructed from two basic components. These are:

- Statements
- Modules.

These two elements, or *primitives*, are the basic building blocks from which applications programs are built. Program *statements* are the elementary computer operations that are selected and organized within *modules* that, in turn, perform problem-oriented processing functions. At the software design level, the main design component is the module. The task of the designer includes determining software functions and the manner in which those functions are related logically. Taken together, these software functions—packaged as modules—comprise the program.

At the programming level, the main design components are statements. The programmer selects and organizes statements within modules to carry out the processing functions identified during software design. To restate the analogy—the software designer is the architect of the system, and the programmer is the carpenter.

Statements

To restate the definition, a statement is a primitive element in a programming language. Statements form the lowest level of operation that can be carried out under that language. Languages, in turn, differ in the sets of primitives they provide and also in the extent to which the actual machine functions performed are made apparent to the user or programmer. The higher the level of the programming language, the less apparent the details of statement execution at the machine level will be to those who write or deal with programs. That is, the more problem-oriented a language, the further removed will be the person who writes the program from the actual physical functions of the computer.

To illustrate, the COBOL language uses relatively high-level primitives. When an ADD instruction is used in a COBOL statement in conjunction with identified fields of data, machine-language code is generated. This code causes the computer to find the data fields, add them, and deposit the sum in a location described by the statement. Several machine-level instructions are produced for each COBOL instruction. In applying COBOL or any other higher-level language, the user is removed from concern with machine details. Thus, software construction primitives are more problem-oriented than machine-oriented.

In building program modules, groups of statements are arranged in a structure that causes the computer to complete a processing sequence, accepting inputs and producing prescribed outputs. There is nothing inherently "good" or "bad" about statements. Rather, it is their structure—the way they are arranged and the relationships among those statements—that introduces a quality dimension to design.

Modules

As defined above, a module is a collection, or organized grouping, of source code statements. Statements are organized within modules so that the modules can be linked to form programs. Characteristics of a program module include:

- A module consists of a group of statements that are physically contiguous and that are executed as a unit.

- The group of statements that form a module have identifiable beginnings and endings.

- In most cases, the group of statements has a single entry point and a single exit.

- The group of statements that form a module can be referenced collectively within programs by a specific *mnemonic,* or assigned name.

Modules are identified by standard programming terms, such as a COBOL section, paragraph, or subprogram; a BASIC subroutine; a PL/1 procedure or task; or a FORTRAN function or subroutine. Thus, a module can be either a separately compiled program or subprogram or an identifiable internal procedure within them.

Attributes of modules.   All modules have three basic attributes:

- Function
- Logic
- Interfaces.

A module's *function* is the data transformation that takes place when the module is executed. A function can be described as the *black-box* behavior of a module. In other words, the function of a module consists of the results that can be discerned externally without looking inside the module. The nature of this examination is implied by the term black box. Functions of a black box are deduced by an understanding of its inputs and outputs.

In developing software designs, a module is represented by a labeled box describing its function. The name assigned to each module both describes its function and encompasses the names of subordinate functions that contribute to execution of that module.

The collection of modules that represents a design usually is documented within a *structure chart.* Functional labeling and connections for a group of modules within a segment of a program structure chart are shown in Figure 7-1. This chart shows a module and a group of subordinate modules that implement an integrated function, GET VALID TRANSACTION RECORD. To implement this high-level module, there are a series of submodules. One of these submodules performs the function of getting the record; the other four edit the record, field by field.

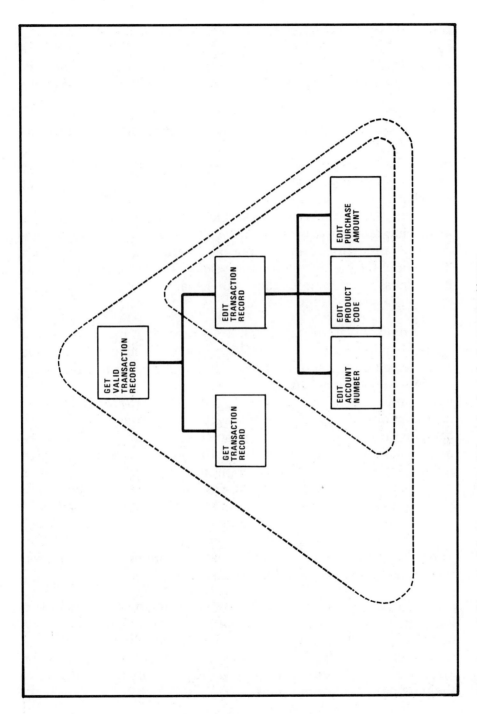

Figure 7-1. Software modules are represented on structure charts as named boxes, labeled to indicate the functions performed. As indicated by the dotted lines, the module name refers to the function of the module itself *and* to the composite functions of any subordinate modules.

A module's *logic* is a description of, or specification for, the actual processing that takes place within the module. The logical description of a module has been likened to a *white box,* or clear box. With the logic exposed, it becomes possible to understand the inner workings of a module.

The distinction is made between the function of a module and its logic because the two attributes are developed at different times, usually by different people. At the systems design level, the term module is used in reference to a processing box on a systems flowchart. This symbol indicates an entire program with broad functional purposes. At this level of design, there is little concern with the processing details necessary to implement the module. Of primary importance are its input and output file interfaces and the black-box behavior of the module within the system. The module is known by its business processing function rather than by the processing logic that comprises it.

At the software design level, a reference to a module is also a reference to its function. The problem-oriented function of the module as described on the systems flowchart is expanded into a hierarchy of processing-oriented functions that accomplish its purpose. Normally, concern with the module's logic is limited to consideration of its activation sequence. Little or no interest is taken in the detailed computer operations required to implement its function. For example, the activation sequence would be the primary concern if the program were a common, routine application with no complicated or novel logic. On the other hand, the designer may expand the design to the processing logic level if warranted by the complexity of the application. Pseudocode notation, decision tables, or decision trees would be used in describing and documenting the module's logic. Thus, the software designer works primarily with black-box modules, incorporating logic descriptions as necessary.

At the programming level, concern is mainly with the module's logic. Processing algorithms are designed to transform the inputs into outputs, and these algorithms are translated into statements. Design, therefore, centers on the statement as the basic building block of algorithms. The module's function is the given. By the same token, the module's logic is the result.

Interfaces are the connections between modules. Module interfaces serve to establish paths across which data are transferred from one module to another. For example, in a COBOL program, the transfer of control among modules takes place on execution of a PERFORM statement, a subroutine CALL, or a GOTO instruction. Transfer of data, then, occurs through identification of the data elements to be passed from one processing module to another, or through reference to a data structure common to both modules.

Connections between modules, therefore, serve two main purposes:

- Transferring control from one module to another
- Passing data from one module to another.

The effectiveness of a software design depends largely on how effectively control is transferred between modules and the method used to pass data from one module to the next.

Program Control Structures

Software design takes place at two levels: At the logical level, the processing modules required for an application are defined and are organized into a hierarchy. This hierarchy represents the problem-related connections among the software components. It documents the processing activities that will be implemented and shows how these activities are related to one another in terms of their data interfaces. A structure chart prepared to describe this problem solution provides documentation of the overall logical structure of the software.

Once a logical solution is devised, it must be adapted for computer processing. Thus, at the physical, or implementation, level, consideration is given to the execution structure of the software. This consideration involves determining the sequencing of computer operations within modules.

Any given module, and any set of statements within it, will be governed by one of three basic types of control structures:

- Sequence
- Repetition (iteration)
- Selection.

These three methods for relating modules and statements within modules represent the minimum set of logical control structures that are necessary and sufficient for controlling the activation sequence of processing components within a program.

Sequence. The activation *sequence* of program modules is controlled through use of PERFORM or CALL statements within the source program or, alternately, by the physical placement of modules to correspond with execution sequence. In the structure chart drawn to indicate the execution structure, modules usually are organized to reflect this sequence. That is, the top-down, left-to-right ordering of the modules parallels the sequence in which the modules are executed. Likewise, processing operations within a module normally are carried out in the physical sequence in which they appear. Unless it is instructed otherwise, the computer will execute statements in this top-down order. Sequences, therefore, represent the natural order of processing as viewed from the standpoint of the computer.

Repetition. Another valid type of control structure for modules and statements is a *repetition*. A repetition is the continued activation of a module or a set of statements for as long as a stipulated condition exists. With each execution of a repetition, a condition test is first applied. If the condition for ending the loop has not occurred, the processing is repeated. When eventually the condition is found to exist, repetitive processing terminates and control passes to the next statement in sequence.

Selection. A *selection* control structure implements processing decisions. That is, based on comparative tests applied to data being processed, control is passed to any one of two or more modules or sets of statements. Typically, a selection is invoked with an IF command. Selection control structures often are referred to as *case constructs* whenever three or more alternative processing activities are included within the structure. Some programming languages have command structures specifically utilizing CASE designations. In other languages, case constructs are contrived by using multiple IF statements or nested IF-THEN-ELSE statements.

The importance of recognizing this difference between the logical, problem-related structure of a piece of software and its physical, execution control structure cannot be overemphasized. Initial design activity is at the logical level. The designer attempts to model a software structure that mirrors the structure of the problem to be solved. At this point, little if any consideration is given to the possible execution structure. Modules are designed to correspond with processing functions implied in the problem structure; connections between modules are modeled after the relationships existing in the problem structure.

Once the logical structure of the solution is determined, it then is packaged for implementation. The physical constraints of the computer system are taken into account as the design is modified to include execution control structures.

To manage the complexity of the development process, the designer must proceed systematically from logical to physical design: By working initially at the logical level, the designer is able to abstract only the essential features of the system as building blocks. A logical solution structure is one that parallels the problem structure without introducing complications based on physical execution. As the development process moves into the programming stage, this logical model becomes the guideline for implementation. A physical model of the software is realized by packaging statements within program modules and by adding necessary control logic to the structure. If this logical-to-physical process is followed, the final software product will have a structure that closely resembles the structure of the original problem. This structure facilitates implementation, testing, and long-term maintenance of the product.

Graphic Representation of Software

Typically, modules or groups of modules are represented within structure charts. Structure charts, in turn, are used as a means of communication about software designs. At this time, there are no generally accepted standards for the notations or conventions used to represent modules and the interconnections among modules within structure charts. However, the differences are such that anyone who can read one style of notation should have no trouble deciphering and dealing with others. To illustrate, four popular notation methods are presented in Figures 7-2 through 7-5.

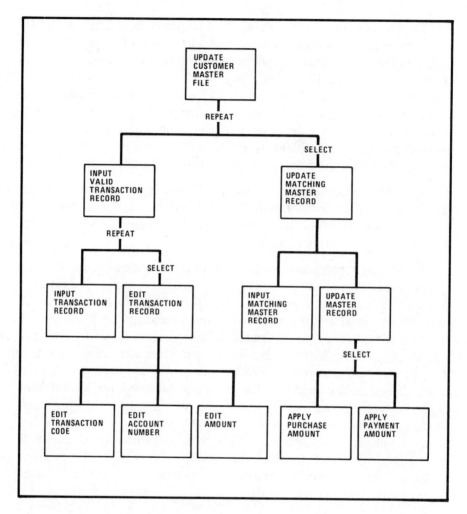

Figure 7-2. A traditional structure chart indicates the main functional modules in a program and their hierarchical relationships with one another.

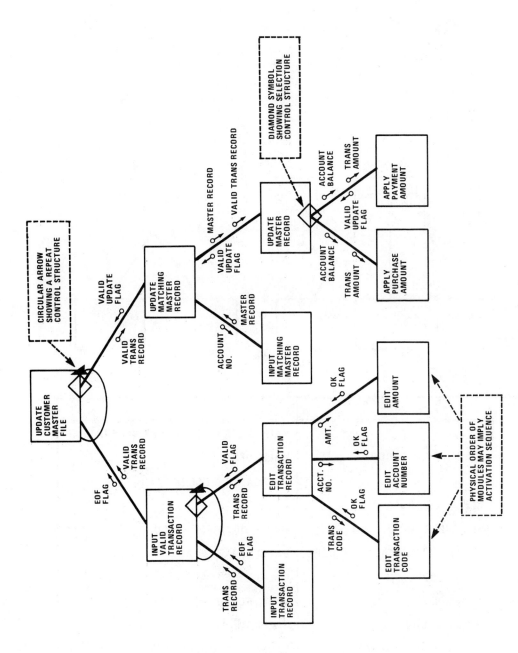

Figure 7-3. One system of notation for structure charts illustrates the modules in a program and their hierarchical relationships, and optionally includes identification of data passed between them and the types of control structures used to relate them.

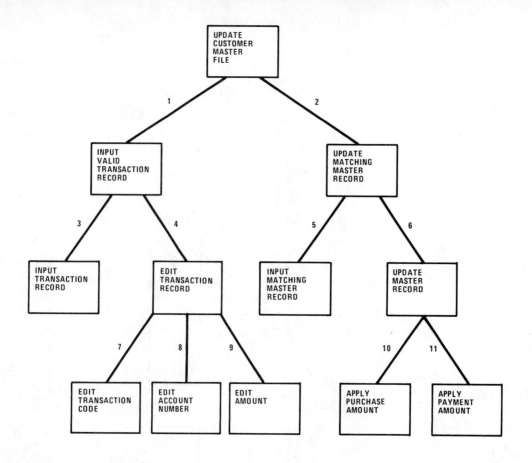

IN		OUT
1	–	VALID TRANSACTION RECORD / EOF FLAG
2	VALID TRANSACTION RECORD	VALID-UPDATE FLAG
3	–	TRANSACTION RECORD / EOF FLAG
4	TRANSACTION RECORD	VALID-RECORD FLAG
5	ACCOUNT NUMBER	MATCHING MASTER RECORD
6	VALID TRANS RECORD, MASTER RECORD	VALID-UPDATE FLAG
7	TRANSACTION CODE	VALID-CODE FLAG
8	ACCOUNT NUMBER	VALID-ACCOUNT-NUMBER FLAG
9	TRANSACTION AMOUNT	VALID-AMOUNT FLAG
10	TRANSACTION AMOUNT, BALANCE	VALID-UPDATE FLAG
11	TRANSACTION AMOUNT, BALANCE	VALID-UPDATE FLAG

Figure 7-4. An alternate style of structure chart optionally is supported by a table of input and output interfaces between modules.

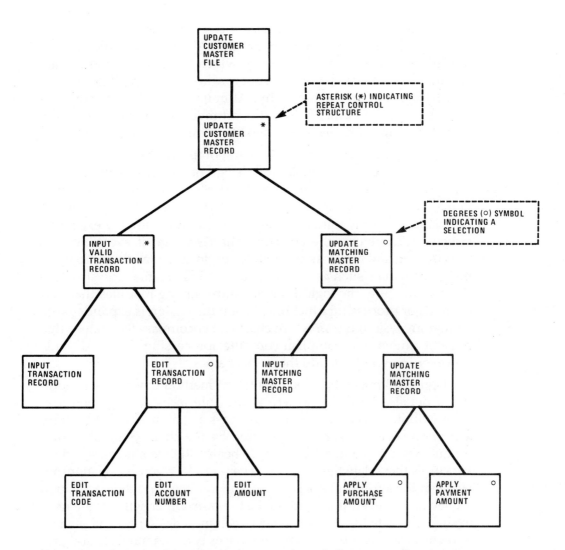

Figure 7-5. Structure chart using alternate notation to indicate connections between modules and repeat and select control structures.

Figure 7-2 presents a traditional structure chart, or hierarchy chart, that often is used to document program designs. Modules are arranged in an activation hierarchy in which lower-level modules are executed from their immediate superordinates, the modules on the next higher level of the hierarchy. Although the order of activation cannot be inferred directly from the arrangement of modules, there is often an implied sequence from top to bottom and from left to right.

The illustration in Figure 7-2 again points out the distinction that must be made between the logical design of a piece of software and the final physical design that can be implemented on a computer. Logical design considers the hierarchical structure of processing functions with only minor concern given to modeling the execution structure. Therefore, the structure chart seldom, if ever, includes initialization and termination modules or documentation of programming techniques such as ''look-ahead'' reads to a file. The structure chart shows only the logical relationships among modules that perform major processing functions. When the design is expanded into a program design, the structure chart will encompass techniques that must be added to accomplish computer processing. Initial designs, however, may not include these mechanics.

Figure 7-3 presents a documentation method that provides additional details about the structure and interfaces among modules. Within limitations that do not encompass actual machine execution, activation sequences are documented on the chart. Symbols representing repetition and selection have been added to show the control structures that relate the modules. There can be a general relationship between the left-to-right ordering of the modules and their activation sequence; however, this need not be the case. Other symbols are used to show the data that are passed between modules. The small, named arrowheads document the major interfaces between modules and provide confirmation that proper inputs are available to the module and that expected outputs can be produced. If a particular module will be implemented as a subprogram, the notations refer to explicit data arguments that are passed. If the module will be implemented as an internal subroutine, the named data interfaces refer to data elements within the global data structure that must be available for use.

The structure chart in Figure 7-4 contains the same basic information as the one in Figure 7-3. In this case, data interfaces between modules are documented in a separate table of input/output parameters. The table then is cross-referenced to the structure chart. Again, the chart documents the logical organization of processing modules, not necessarily the physical organization.

Figure 7-5 is another variation. Here, the logical activation sequence of modules is documented with other symbols. The asterisk (*) within the upper-level module indicates that the function will be repeated; the degree symbols (°) in the lower-level boxes imply a selection control structure in which one of the two modules will be activated depending on a condition test in their superordinate module.

As design moves from a consideration of the logical structure of the software to integration of computer processing techniques, either the same or other notational systems may be used. For example, structure charts can be expanded easily to include initialization and termination routines. In fact, structure charts can be taken to a level of detail from which code can be produced directly. Usually, however, there is a change to a more convenient and explicit notation, such as pseudocode.

Pseudocoding provides a method of expressing module logic in a form relatively independent from any particular programming language. The pseudocode formally documents the computer operations and control structures necessary to implement the modules. Processing is presented in sufficient detail for virtually automatic conversion into a programming language on a line-by-line basis. Figure 7-6 shows this type of pseudocode documentation. Some designers even prefer to use pseudocode during all phases of design. That is, the notation is used to document the overall logical design of the software system, as well as the transitional steps through to coding. Figure 7-7 shows this form of pseudocode notation used to document a software structure.

No matter what kind of graphic notation scheme is used, all have a common purpose and feature. That is, all of these methods present programs and program modules in terms of hierarchical structures. Further, structured presentations are modeled on a top-down basis, from a logical view of the software components and their relationship to a physical view that can be translated directly into code. Any method that presents these program design concepts clearly is acceptable. The method is less important than the thought that goes into it and the ideas that are communicated.

UPDATE CUSTOMER MASTER FILE
Set Valid-Trans-Flag = " "
Perform INPUT VALID TRANSACTION RECORD
    until Valid-Trans-Flag = "Y"
    or EOF-Trans-File
If not EOF-Trans-File
    Perform UPDATE MATCHING MASTER RECORD

INPUT VALID TRANSACTION RECORD
Perform INPUT VALID TRANSACTION RECORD
If not EOF-Trans-File
    Perform EDIT TRANSACTION RECORD

UPDATE MATCHING MASTER RECORD
Perform INPUT MATCHING MASTER RECORD
Perform UPDATE MASTER RECORD

INPUT TRANSACTION RECORD
Read Transaction-File
    at end
        Set EOF-Trans-File

EDIT TRANSACTION RECORD
Perform EDIT TRANSACTION CODE
Perform EDIT ACCOUNT NUMBER
Perform EDIT AMOUNT

INPUT MATCHING MASTER RECORD
Move Account-Number to Master-Key
Read Customer-Master-File

UPDATE MASTER RECORD
If Purchase
    Perform APPLY PURCHASE AMOUNT
Else
    Perform APPLY PAYMENT AMOUNT

EDIT TRANSACTION CODE
If Trans-Code not equal "PR" and not equal "PY"
    Set Valid-Trans-Flag = "N"
Else
    Set Valid-Trans-Flag = "Y"

EDIT ACCOUNT NUMBER
Move Account-Number to Master-Key
Read Customer-Master-File
    Invalid Key
        Set Valid-Trans-Flag = "N"
    If Valid-Trans-Flag = " "
        Set Valid-Trans-Flag = "Y"

EDIT AMOUNT
If Amount < = $\varnothing$
    Set Valid-Trans-Flag = "N"
Else
    Set Valid-Trans-Flag = "Y"

APPLY PURCHASE AMOUNT
Add Amount to Customer-Balance

APPLY PAYMENT AMOUNT
Subtract Amount from Customer-Balance

Figure 7-6. Pseudocode notation for describing module processing.

UPDATE CUSTOMER MASTER FILE
REPEAT: Until EOF-Trans-File
  Set Valid-Trans-Flag = " "
  REPEAT: until Valid-Trans-Flag = "Y"
    or EOF-Trans File
    INPUT VALID TRANSACTION RECORD
      INPUT TRANSACTION RECORD
        Read Transaction-File
        at end
          Set EOF-Trans-File
      SELECT: on not EOF-Trans-File
        EDIT TRANSACTION RECORD
          EDIT TRANSACTION CODE
            If Trans-Code not = "PR" and not = "PY"
              Set Valid-Trans-Flag = "N"
            Else
              Set Valid-Trans-Flag = "Y"
          EDIT ACCOUNT NUMBER
            Move Account-Number to Master-Key
            Read Customer-Master-File
            Invalid-Key
              Set Valid-Trans-Flag = "N"
            Else
              Set Valid-Trans-Flag = "Y"
          EDIT AMOUNT
            If Amount < = Ø
              Set Valid-Trans-Flag = "N"
            Else
              Set Valid-Trans-Flag = "Y"
      END SELECT
  END REPEAT
  SELECT: on not EOF-Trans-File
    UPDATE MATCHING MASTER RECORD
      INPUT MATCHING MASTER RECORD
        Move Account-Number to Master-Key
        Read Customer-Master-File
      UPDATE MASTER RECORD
        SELECT: on Trans-Code
          APPLY PURCHASE AMOUNT
            Add Amount to Customer-Balance
          APPLY PAYMENT AMOUNT
            Subtract Amount from Customer-Balance
        END SELECT
  END REPEAT

Figure 7-7. Alternate method for specifying pseudocode in describing a software design.

METHODS OF STRUCTURING SOFTWARE

Software components—modules and statements—are organized within structures to implement systems designs. Statements are packaged together within modules that are interconnected to perform consecutive processing on data as the data flow throughout the system. The primary challenge of software design is in deriving the structure, or relationships, among these components. The effectiveness with which this can be done contributes directly to the quality and cost effectiveness of the system from an implementation and maintenance standpoint.

There are two basic approaches used in determining the structure of software:

- The algorithmic approach
- The hierarchical approach.

Software designers and programmers should understand and be able to form needed relationships between these two approaches. That is, each approach has its place within the creation of software structures. Typically, the hierarchical approach is used to establish the overall *architectural design,* or logical structure, for the software. Program functions are *decomposed,* or broken down, into major processing functions that, in turn, are decomposed, or *partitioned* into detailed processing functions. Modules are defined in a top-down, hierarchical fashion through increasing levels of detail. The end result of this partitioning process is a hierarchy of modules that are interconnected to provide paths across which control and data are transferred from one module to another.

Once the logical, hierarchical structure of processing modules is defined, the *algorithmic approach* is used to derive an *execution structure,* or specific processing statements and sequences, for implementing the design on a computer. Detailed processing algorithms are designed to structure the statements within modules. Thus, it is important to understand the separate approaches and to be aware of the situations in which each is the proper method of design. It should be recognized that these are complementary approaches, each with its own benefits and appropriate uses.

Algorithmic approach. The algorithmic approach is the classical technique for program development. It can be viewed as basically a bottom-up approach. That is, the focus is on the program statement as the basic building block of design, and the object is to arrange statements in a proper execution sequence. The approach is fundamentally linear. Program design becomes a process of defining which statements should be ordered in what sequence so that required processing is carried out at the right times. The program, in effect, is regarded as a chronology of events describing what happens first, second, third, and so on. Often, program flowcharts are used to document the resulting design.

In terms of overall structural design, the algorithmic approach requires a high degree of precision. It is critically important that each statement appear in the precise order required by its time sequencing. Thus, software design based on the algorithmic approach can be extremely difficult when applied to complex or especially large processing requirements.

Certain consequences result from this sequential, linear approach. First, the procedural, rather than structural, aspects of the program are emphasized. It causes the designer or programmer to be machine-oriented rather than problem-oriented—to think in terms of computer operations and their effects rather than in terms of problem structures and solutions. The approach emphasizes how the computer will operate as opposed to what the overall structure is of the problem to be solved. Consequently, the algorithmic approach seldom will lead to an understanding of the problem and its solution.

Second, the algorithmic approach tends to lead to the creation of *monolithic* programs, or programs that behave as single, tightly-cemented blocks of code. Each statement is related to all other statements by the mere fact that its positioning within the program is important to the functioning of the other statements. Such programs are difficult to implement, modify, and maintain since sections of code are not functionally independent but are related physically and chronologically. The effects of a change in a single function may have to be traced throughout the entire program, since statements are packaged according to time sequence rather than as isolated functions.

Third, a section of code of an algorithmic-structured program is an implementation of the program's logic as well as its processing. The physical order in which the statements appear in the program is as important as their functions. In fact, the ordering often is dependent on the function. For example, input statements must precede processing statements physically. Processing statements, in turn, must appear before output statements. Accordingly, statements are related in time as well as in function. Though this approach is the most straightforward, it also would be possible to order the statements in a different sequence and cause processing control to jump from one part of the program to the other. However, the transfers of control among the different functions would be difficult to track and would obscure program logic.

The fourth consequence of using an algorithmic approach to overall software design is that a program is not really designed *a priori*, or before the fact. The final structural design is not realized until after the program is written. Furthermore, the resulting structure is related more to the machine steps necessary to accomplish the processing than it is to the problem structure. Thus, maintenance of the software becomes more difficult because changes in the problem structure cannot be matched directly with corresponding changes required in the program structure.

Finally, since the focus is on the program statement, the problem solution is linked closely to the programming language used. With this approach, it is difficult to design a language-independent solution. The methodology forces the programmer to think about which statements appear in what sequence, and the statements required will depend upon which language is chosen.

In summary, algorithmic approaches tend to be better suited to describing the internal logic of modules rather than to deriving total software designs. The scope of a module should be limited to a single, well-defined task; as a result, many of the consequences described above are acceptable. The sequencing of statements should be left as a follow-up activity to the task of designing the overall architecture of the software. Once this structure of modules is determined, algorithms can be created for implementing the structure. At this point, attention must be focused on designing the chronology of events that will implement the solution on a computer.

Hierarchical approach. Rather than focusing on the statement as the basic building block for programs, the hierarchical approach concentrates on the module. The structuring of a program is accomplished by identifying a hierarchical relationship of subproblems within the overall problem. In effect, this approach makes it possible to frame solutions that parallel the basic structures of the problems themselves. Putting it another way, the hierarchical approach is problem oriented rather than processing, or machine, oriented. Emphasis is on problem solving functions rather than on the step-by-step logic of computer operations. Eventually, algorithms will have to be developed for program statements. But, in a hierarchical approach to design, it is possible to avoid dealing with this level of detail at the design stage.

There are several consequences of using the hierarchical approach to software design, most of which are corollaries of the shortcomings of the algorithmic approach. First, the hierarchical approach leads to an understanding of the problem, its structure, and its logical solution. The focus is on the module as the basic building block of software. Consequently, attention is concentrated upon the overall organization of major processing functions. The complications of integrating machine details within the problem solution are effectively removed. The designer can concentrate on modeling a solution structure based on the problem structure, and, in the process, equate the two.

Second, the hierarchical approach leads to programs that do not behave as single, monolithic entities. Programs contain functionally independent modules that are not strongly related to one another. Thus, individual processing functions are relatively independent so that a change in any one module has minimum impact on other modules. This characteristic makes implementation, maintenance, and modification relatively easy tasks. Also, since the program structure is tied to the problem structure, changes in the nature of the problem that will require program changes can be traced easily to the corresponding modules.

Third, there is a separation of control logic from processing logic within hierarchically designed systems. Except within the individual modules themselves, there is no significance attached to the physical ordering of components. The activation sequence within a program is determined primarily by its overall logical control structure rather than by the physical ordering of statements. Therefore, implementation and testing is simplified. It is possible to design, code, and test major interfaces between modules (normally, a major source of program bugs) prior to testing actual processing results.

Finally, the structure of the software is evident before the program is written. The solution to the problem is known before consideration is given to its implementation details, thus making a clear distinction between architectural design, processing design, and coding. A hierarchical approach enforces consideration of only the essential elements of a problem solution at appropriate levels of design. The designer, therefore, can put off consideration of implementation details until the structure of the software is firm.

As indicated, each of the design approaches is appropriate at different phases of the design task. Hierarchical structuring is the preferred method for developing an overall logical design for a program. This approach leads to an understanding of the problem structure and to the development of a solution structure that parallels it. The modular components of the solutions are relatively independent, allowing design to continue separately within each module. There is reduced concern about the effects one module has upon other modules in the system. After this overall structure has been designed, algorithmic approaches are applied in devising the internal processing logic of each module. At this point, the precise ordering of statements within modules is important. However, by this time the mass of details that would have had to be considered for the program as a whole have been localized to independent modules. The reduced complexity allows use of algorithmic approaches for sequencing the statements within modules.

In terms of documentation technique, therefore, structure charts are appropriate tools for analyzing and describing the architecture of the software system. Program flowcharts and pseudocode become useful in designing and documenting the processing logic within the modules of the system. Each of these tools enforces a somewhat different approach. By their nature, structure charts require that a hierarchical method of thinking be applied. Flowcharts, on the other hand, are compatible with the linear thought process applied during detailed module design. Pseudocode is appropriate for identifying the sequence of detailed processing operations that comprise a module, once its position within the hierarchy has been established. However, effective designs can be produced with pseudocode only, since this technique allows expression of both hierarchical designs and linear sequences.

SOFTWARE DESIGN STRATEGIES

To this point, the discussions in this chapter have covered some general approaches to program design and implementation, as well as the elements and tools used in building programs. The application of these approaches and tools, in turn, must be directed by an overall strategy. Broadly, three design strategies can be identified for the design and development of programs:

- Functional decomposition
- Data flow
- Data structure.

Both parallels and differences exist among these three strategies. The strategies are described briefly below and are dealt with in greater depth in Chapter 9.

Functional Decomposition

Functional decomposition approaches software design by partitioning, or decomposing, a complex problem, breaking it down into a series of simpler, individually solvable subproblems.

Functional decomposition follows a top-down approach through what are sometimes called *levels of abstraction.* That is, efforts begin by isolating a top-level, or global, module that represents an entire program. This module represents the highest level of abstraction within the program. Succeeding levels of modules to implement subordinate functions represent lower levels of abstraction. At each level of abstraction, the level of detail increases. Ultimately, identification of levels brings modules down to some predefined amount of coding that can be coped with readily. This approach to the elaboration of design often is called *stepwise refinement.* The modules identified through stepwise refinement are organized and related so that they can be implemented through structured programming constructs.

Data Flow

The *data flow* strategy employs a technique for modeling a problem on the basis of the data flowing through the system. This data flow model, in turn, becomes the basis for both analysis and design of a solution. Analysis is performed through data flow diagramming. Single processes and groups of processes within a data flow diagram are regarded as black boxes that can be composed through preparation of structure charts. The data flow strategy employs the concept of levels of abstraction, much like functional decomposition. However, in this case, there are explicit criteria for partitioning a system into modules. The result is a design having a structure that parallels the problem structure, as defined by the data transformations that take place.

Data Structure

Data structure approaches also are based upon the need for partitioning of complex problems into a series of simpler subproblems. In these approaches, the real key to program design and development rests with the structures of the data elements to be processed. The collections of fields within records within files provide a model of the organization they represent. Thus, modeling based on data structures attempts to develop a software structure that parallels the structure of data processed.

A common denominator of these three strategies is to approach the design of programs by partitioning them into a set of modules. This subdivision is carried down to a level at which program modules may be managed and implemented through simple collections of statements. The statements, in turn, are related through the minimum set of program control structures. There is no best, single strategy to follow in designing and implementing application software. Thus, any of these strategies is workable, with each especially effective with different types of problems. Success as a software designer means being familiar with the basic tenets of all of these strategies.

SOFTWARE DESIGN EVALUATION CRITERIA

Although there is no best design strategy, there is a set of criteria that can be applied to evaluate application software designs, regardless of whether functional decomposition, data flow, or data structure approaches are used. These criteria include:

- Coupling
- Cohesion
- Design heuristics.

Coupling

Coupling is a measure of the strength of the connections among the modules of a system. Coupling, in effect, describes the degree of interdependence between modules, indicating the degree to which it will be necessary to study and deal with one module to understand the content and processing of another. Good design seeks to minimize coupling between modules in order to improve the maintainability of a system.

Cohesion

Cohesion is a measure of the interrelatedness of the statements that describe the processing logic of a module. The standards of cohesion measure the effectiveness of partitioning a system into modules that are functionally tight and contain only those statements necessary for implementing a single, independent processing task. A high degree of cohesion implies a low degree of coupling and, consequently, improves the maintainability of the system.

Design Heuristics

As compared with the rather formal evaluations applied under coupling and cohesion criteria, it is also possible to evaluate software designs by applying some broad rules of thumb, or *heuristics*. These heuristics are applied in evaluating such characteristics as the distribution of decision making within the control structure of a system, the size of modules, and the number of submodules controlled by each superordinate module.

These criteria for evaluating software designs should be understood before techniques for selection of design strategies are reviewed. Therefore, these evaluative criteria, which are applied during the design process, are covered in the following chapter. Then, the design strategies are presented.

Summary

A software design is a representation of programs to be written and specifies program modules, structure of modules, detailed processing within modules, control interfaces among modules, and data interfaces among modules.

Software design is the process of taking the specifications for a job, job step, or on-line procedure and translating those specifications into a hierarchical organization of modules. The basis for the specifications is documentation in the form of data flow diagrams and supporting data dictionary entries.

In applying software design principles, a large program is broken down into smaller and smaller modules. At succeeding levels within this hierarchy of modules, functions and interconnections are defined. Further, design proceeds downward from a relatively broad definition of processing requirements to exacting operational details at the lower levels.

The starting point for software design is the set of systems flowcharts that defines the programs and data files to be used. The systems flowcharts also specify how the programs and data files will be integrated within a job or processing procedure.

Application software must work correctly, meet specifications, be reliable, be maintainable, be easy to use, be easy to implement and test, and use computer resources efficiently.

Statements and modules are the two primitives from which applications programs are built.

Statements are programming commands that are contained within program modules. Attributes of modules include function, logic, and interfaces.

A module's function is the data transformation it performs. A module's logic is a description of the actual processing that takes place within the module. Interfaces are connections between modules. These connections cause transfer of both control and data from one module to another.

Program control structures that determine the order of execution include sequence, repetition, and selection.

The hierarchical relationship among modules in a top-down design may be documented in a structure chart, for which there are several graphical conventions. Pseudocode may be derived from structure charts to document module logic or to develop complete designs.

Software may be structured through algorithmic or through hierarchical approaches. An algorithmic approach focuses on the sequence of statements and generally results in a monolithic program. A hierarchical approach focuses on the structure of modules, which, in turn, tends to parallel the structure of the overall problem being solved. Typically, a hierarchical approach is used to design the structure and an algorithmic approach is used to design the internal workings of modules.

Design strategies may include functional decomposition, data flow, and data structure.

Software design evaluation criteria include coupling, cohesion, and design heuristics. A high degree of cohesion, or interrelatedness between modules, implies a low degree of coupling and improves the maintainability of a system.

Heuristics are applied in evaluating such characteristics as the distribution of decision making within the control structure of a system, the size of modules, and the number of submodules controlled by each superordinate module.

Key Terms

1. design
2. software design
3. software engineering
4. robustness
5. primitive
6. statement
7. module
8. mnemonic
9. function
10. black box
11. structure chart
12. logic
13. white box
14. interface
15. sequence
16. repetition
17. selection
18. case construct
19. architectural design
20. decompose
21. partition
22. algorithmic approach
23. execution structure
24. monolithic
25. functional decomposition
26. level of abstraction
27. stepwise refinement
28. data flow
29. data structure
30. coupling
31. cohesion
32. heuristic

Review/Discussion Questions

1. What is the purpose of software design, and what does the scope of software design include?

2. How are statements and modules related to software structure?

3. What are three attributes of modules, and how is each defined?

4. What is the minimum set of program control structures, and how are these structures related to statements and modules?

5. What graphic representation tools may be employed in software design?

6. Despite differing conventions for drawing structure charts, what do these conventions have in common?

7. What is the relationship of pseudocode to structure charts and to program statements?

8. What is the algorithmic approach to software design, and in what circumstances might it be applied?

9. What is the hierarchical approach to software design, and in what circumstances might it be applied?

10. What are three principal software design strategies, and how may each be used?

11. What is the relationship between cohesion and coupling when speaking of program modules?

12. How may heuristics be applied in evaluating a software design?

Project Assignments

The Appendix presents three case studies that can be used in support of project assignments. All three scenarios provide functional system specifications that would result from systems analysis. Included are data flow diagrams, data dictionaries, and process narratives that would result from systems analysis. Continuing the work begun in Chapter 6, assignments that relate to the content of this chapter will be found in *PART 3: Detailed Systems Design*.

SYSTEMS DESIGN SKILLS:

EVALUATING SOFTWARE DESIGN 8

LEARNING OBJECTIVES

On completing reading and other learning assignments for this chapter, you should be able to:

- [ ] Describe the need for and the characteristics of quality software.
- [ ] Explain the criteria of coupling and cohesion in the evaluation of software design decisions relating to the partitioning of modules within a program.
- [ ] Describe how transfers of data and transfers of control between modules may be implemented.
- [ ] Tell how cohesion and coupling are interrelated and what impact this relationship has on design of software modules.
- [ ] Give the levels of coupling and the levels of cohesion and tell how these criteria may be applied.
- [ ] Tell what heuristics may be applied in making software design decisions.

THE NEED FOR EVALUATION CRITERIA

In the previous chapter, it was pointed out that applications software, in general, is becoming more complex. Complications arise because of several factors, including higher expectations of users, the increased capabilities of hardware and systems software, and the integration of multiple processing routines within on-line systems. For these and other reasons, today's systems of software demand greater skill in their construction and increased attention to their quality.

Possibly the most important of the quality dimensions of software is maintainability. Applications systems will be around for a long time. It is not unusual to find, even today, production systems that were written fifteen years ago. Thus, systems must be developed with an eye on their future impacts. These systems must be designed so that changing business requirements can be integrated easily within the ongoing software products, and the systems must be adaptable to changing hardware/software environments.

Quality software, therefore, exhibits two main characteristics: It is easy to implement and test, and it is easy to maintain and modify. These are qualities that must be built into the software from its inception. Thus, criteria for assuring quality in software must be applied early in the development process, during initial logical design of the system. At this point, and regardless of the design strategy used, the criteria are applied to the various iterations on the design to improve its implementation and maintenance impacts. As a result, considerable time and effort must be applied at the front-end of the development process. It is especially important that the designer resist pressures to produce a design quickly and proceed with implementation. Granted, an abbreviated design process may lead to earlier products; however, such products seldom will be the easiest products to implement, test, and maintain. Over the life of the system, the time saved on design will be absorbed quickly in the time it will take to revise the system to meet changing requirements.

As surveyed in the previous chapter, there are two main criteria used in evaluating software design:

- *Coupling* is a measure of the degree to which the component modules of a software system are interrelated. For ease of implementation and maintenance, the strength of these interconnections should be at a minimum.

- *Cohesion* is a measure of the degree of interrelatedness of the processing components, or statements, within a module. Implementation and maintenance are facilitated to the extent that the statements all relate to a single, well-defined function.

The criteria of coupling and cohesion are applied to each design decision as the hierarchy of modules is expanded from top to bottom. As each module is factored into its subordinates, consideration is given to the components of the modules and to the interconnections among them. The motivation is to develop a software product that has a high degree of cohesion among the elements within modules and a low degree of coupling among modules. Software with these two qualities generally will be easier to implement and test, and will be easier to modify and maintain over the operational life of the system. The reason is that modules are highly independent; a change in one module will affect a minimum number of other modules.

In addition to coupling and cohesion, there are other factors that can be applied in evaluating a design. These include the distribution of decision making within modules of the system, the span of control exhibited by superordinate modules, the general usefulness of detailed processing modules within different portions of the system, the location of decision making effects within the system, and the physical sizes of modules. All of these criteria are discussed in this chapter.

It will become evident that these criteria do not represent formal measurements, or *metrics*, that can be applied to a design. Rather, they can be best classified as *heuristics*, or rules of thumb derived from experience. There is still a need for considerable judgment to be applied in evaluating a design on the basis of these criteria. Yet, in the absence of quantitative measures of software quality, the guidelines presented here are appropriate and valuable.

COUPLING

Again, coupling is a measure of the strength of connections between modules in a system. The strength of coupling, high or low, is viewed as the probability that, in designing, implementing, or modifying one module, the characteristics or contents of another module will have to be taken into account. Thus, the more free-standing a module, the lower its coupling will be with other modules; the greater the interdependence between modules, the higher the level of coupling.

The existence and extent of coupling within a software system is influenced by four factors:

- Type of connection between modules
- Complexity of the interface between modules
- Type of information flow across the interface
- Binding time of connections.

Each of these factors is discussed below, and examples are given to illustrate their effects on coupling.

Type of Connection between Modules

A *connection* within a program is a reference by an element within one module to the name, or *identifier,* of another module. Connections occur, for example, with execution of program instructions that CALL a subprogram or subroutine, or with PERFORM or GOTO instructions. In these instances, control is transferred from the module containing the *branching* statement, or selection, to the module named or identified in the instruction. In general, connections between modules serve two purposes:

- Transferring of control
- Passing of data.

That is, upon execution of the branching instruction, program control is sent to the module named in the instruction, and any data required by the receiving module is made available for processing. When a subroutine CALL is issued, the data are passed explicitly as *arguments* within the CALL command. They are received as *parameters* by the called module and become available for processing. In the case of PERFORM or GOTO instructions, the data are available within a global data structure accessible by all of the program's internal modules.

Transfers of control.   Transfers of control between modules may be either *conditional* or *unconditional.* In a conditional transfer, control is transferred to and returns routinely from the subordinate module. An example would be a PERFORMed paragraph in COBOL. The statement, PERFORM MOD-A, transfers program control to the module named MOD-A. Statements within this module are executed, and when the next paragraph is encountered within the execution sequence, control returns automatically to the statement following the PERFORM. There is no requirement that the activated module contain branching instructions to send control back explicitly to the originating module.

The same situation exists with called subroutines. When the CALL instruction is executed, control branches immediately to the named module. Upon encountering a RETURN instruction within the subroutine, the system returns control directly to the calling module, continuing on with the statement following the CALL.

A characteristic of a conditional transfer, then, is that a single connection between modules serves a dual purpose: It serves as the path across which control is sent from one module to another, and it becomes the path through which control is returned. Figure 8-1A shows graphically these control paths for modules involved in a subprogram CALL; Figure 8-1B illustrates the control characteristics for a PERFORMed paragraph in COBOL.

An unconditional transfer of control is one in which the receiving module, in turn, determines the next assignment of control. Control may or may not return to the originating module. At minimum, two or more connections result from an unconditional transfer. A path must be defined for sending control to the subordinate module, and a second path must be defined for returning control.

Unconditional transfers are implemented with GOTO statements. A module issues a GOTO command, in which the receiving module is named or otherwise referenced. In COBOL, for example, the GOTO statement contains the name of the paragraph to which control is to be transferred. In BASIC or FORTRAN, however, the reference is to the line number of the initial statement in the receiving module.

Once control has been passed to the receiving module, execution continues in sequence until module processing is completed. However, in this case, there is no automatic return to the originating module. A second GOTO command must be included within the receiving module to accomplish this return. Furthermore, a second identifier, or *label*, must be added within the originating module to serve as a referent for the GOTO statement within the receiving module. Figure 8-2 shows graphically the paths that are defined with

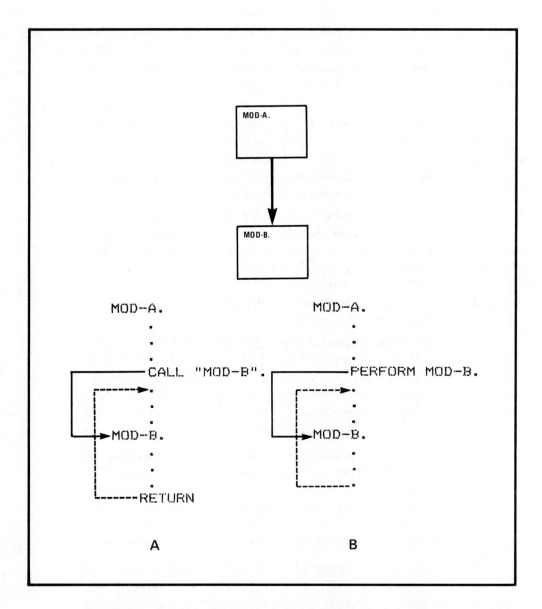

Figure 8-1. Conditional transfers of control define a single interface across which control is transferred to and returned automatically from a module.

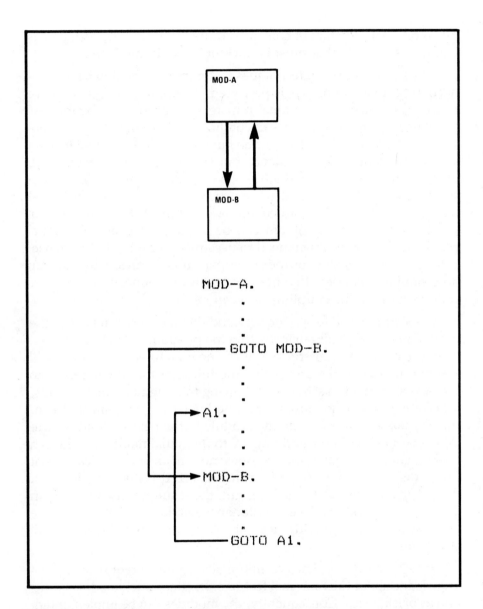

Figure 8-2. Unconditional transfers of control require two or more separate connections to establish paths to and from modules.

unconditional transfers of control and the kinds of *referents,* such as identifiers or calls, that must be included within modules.

In terms of coupling, techniques that minimize the number of connections between modules is preferred. Therefore, designs that are based on the use of conditional transfers of control will exhibit less coupling than designs that use unconditional transfers. One of the obvious reasons this is so relates to the number of paths required to send and receive control. In conditional transfers, implemented with CALL or PERFORM commands that invoke modules, a single connection serves both to transfer and to return control. With unconditional transfers, at least two connections are required. Since all control procedures within a program can be implemented as conditional transfers, there is really no need to use any other type. In this sense, unconditional transfers introduce superfluous connections that can and should be avoided. In other words, unconditional transfers serve only to increase the coupling between modules.

A second reason for preferring conditional transfers relates to the need to maintain predictability within the program's control structure. With use of GOTO constructs, there is no guarantee that control will return routinely to the originating module, nor that it will return to the next statement in sequence following the original transfer. Thus, it becomes easy to get lost within the logic of a program that continually passes control from one module to the next without returning control immediately and directly to the initial module. Although use of unconditional transfers does not mean automatically that programs will be developed in a disorganized fashion, it does not discourage this result. In the long run, the resulting discontinuity in program logic will produce maintenance headaches that could have been avoided easily with designs based on conditional control structures.

Finally, modules that are minimally coupled require the least knowledge about their internal features. In effect, such modules function as black boxes. Consequently, the modules can be implemented, tested, and maintained more easily and can be integrated within new applications with the minimum of trouble.

As an example, consider the transfers of control shown in Figure 8-2. In this case, the receiving module, MOD-B, must know something about the internal structure of the originating module, MOD-A. In particular, MOD-B must be aware of the existence of label A1 so that it can return control to this location. Thus, any changes made in MOD-A could impact directly the design of MOD-B. If, for some reason, MOD-A required restructuring such that control should not be returned to A1, then MOD-B itself would require rewriting. The two modules are coupled tightly; there is a high probability that changes in one module would require corresponding changes in the other. With conditional transfers, on the other hand, this type of coupling is nonexistent. If processing changes are made in the originating module, there is no effect in the receiving module. Changes are isolated in the calling module, which handles any and all processing changes subsequent to the execution of its subordinates.

Transfers of data.   In addition to providing paths for the transfer of control between modules, connections serve as data paths, or interfaces. Data can be provided to a receiving module in one of two ways: In one case, data are passed explicitly through argument lists, as is common with subprogram calls. Alternately, data are made available implicitly, without specific reference, by being a part of a global data structure. The DATA DIVISION in a COBOL program is an example of global data.

Within subprogram calls, as illustrated in Figure 8-3, the receiving module has access only to the data elements that are passed to it. For example, in the following COBOL statement, a subprogram named SUB is called and is provided access to three data fields named DATA-1, DATA-2, and DATA-3 within the calling program:

CALL "SUB" USING DATA-1, DATA-2, DATA-3

Control is transferred to the named subprogram, which receives the data through parameters listed in the statement:

PROCEDURE DIVISION USING DATA-A, DATA-B, DATA-C

Although the names in the argument list do not have to match those in the parameter list, there still must be correspondence in the number of data items and their formats. The important point is that the subprogram has access only to those data items specifically named in the argument list and passed to it.

Such conditional transfers of control are said to be fully *parameterized*. A connection causing minimal coupling is made between

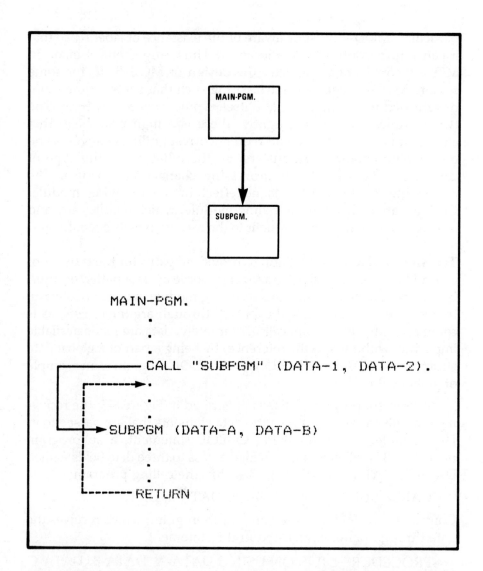

Figure 8-3. A subprogram CALL serves as a conditional transfer of control and defines a path between modules across which data are passed and received.

modules. This connection serves four distinct functions with relation to the receiving module: the transfer of control, the return of control, the passing of data, and the return of processed data. Thus, this type of connection is sufficient to realize all functions necessary for passing control and data within a modular system. In addition, it represents the minimal connection required, and, therefore, offers the least amount of coupling possible.

Nonparameterized conditional transfers of control, such as those provided with PERFORM statements in COBOL, offer benefits that are similar. In these cases, data are not passed specifically through argument lists but are extracted by the receiving module from a global data pool.

The use of global data by program modules raises some complications, however. Although minimal connection between modules may exist for transfers of control, coupling problems can arise because global data are common to all modules. Any module in the program may change the value of any data element in the global data pool. Further, such changes may occur regardless of whether the module, logically, should have access to the data item. As a result, there is not sufficient protection given to DATA DIVISION elements to ensure that they will not be changed inadvertently. Thus, strictly speaking, all modules that have access to the global data structure are coupled, since a change produced by any one module has potential impact on all other modules that share the data item. The same consequences can arise with transfers of control through GOTO statements, or unconditional transfers.

Some design guidelines can be stated that contribute to reduced coupling. In general, the designer should elect fully parameterized, conditional transfers of control and data. Of course, this rule may not be practical for all possible cases of coupling. Within a COBOL program, for instance, the design and coding overhead involved in implementing subprograms for every paragraph would be particularly bothersome. In such cases, careful definition and use of the global data structure should keep coupling problems at a minimum. However, in large systems in which several major processing functions are required, use of subprograms should be the rule. Use of subprograms is particularly helpful if two or more development teams are involved in the project and will work independently on software for the separate functions. In addition, these minimal control structures often will be used for developing general-purpose modules that will appear in several different systems.

Complexity of the Interface

The second dimension of coupling is the *complexity* of the interface. For this purpose, complexity refers to the characteristics of the data being passed from one module to another. From an operational standpoint, complexity relates to the amount of data being passed from one module to another and to the structure of those data.

In general, the number of data elements contained within an argument list in a subroutine call provides evidence of complexity. The larger the number of data items, the more complex the interface. Thus, a subroutine call that contains 25 arguments is, generally, more complex than one with 10 arguments. Further, such increased complexity contributes an increased degree of coupling between modules. Where data are passed implicitly from one module to another, the degree of complexity and coupling can be inferred by the number of data elements referenced within the receiving module.

This type of coupling cannot be evaluated only on the count of the number of data items passed. The evaluation of complexity should consider the structure of the data as well. For example, an argument list that contains 25 table elements that are being passed as a group to a subprogram would not be a particularly complex interface. Similarly, passing a single record that contains 25 data fields would not be considered complex as long as the argument list made reference only to the entire record and not to its fields individually.

In summary, the designer must be aware of the potential for increased coupling when modules are designed. It is likely that excessive numbers of independent data elements being passed between modules are evidence of improper partitioning. Recall that a goal of design is to produce functionally independent modules. Complex interfaces can point out situations in which modules are performing more than one function or in which partitioning has not been carried out fully.

Information Flow Across the Interface

There are two types of control that relate to the connections between modules. *Active control* refers to those situations in which modules are connected through conditional or unconditional transfers and only the data involved in actual processing are transferred across the interface. A second type of control is referred to as *coordination*. Coordination occurs when one module also passes information to direct the processing that takes place in another module.

To illustrate, consider a payroll program in which a single module performs all payroll calculations. Different sets of calculations are required for salaried and for hourly workers. When this module is called, it must be informed of the type of calculation to perform, as well as receive the record that will be involved in the processing. Thus, the module requires both data and a *flag*, or software switch, to indicate whether the record is for a salaried or for an hourly employee. This flag represents control information, or coordination. The calling module is involving itself in the internal processing details of the called module by directing the type of calculation to be performed. Thus, coordination couples the modules strongly because neither is functionally independent of the other. Figure 8-4A shows this form of coordination between the two modules in the payroll example.

Also, coordination is superfluous control. For example, in the payroll illustration, the calling module evaluates the record type and sets the flag to indicate whether a salaried or an hourly record is being passed. Then, in the called module, this flag is tested again, and one of the two processing routines is executed. Thus, the same test is applied in both modules. If, on the other hand, a separate module were defined for each of the salary and hourly calculations, there would be no need for a flag. The calling module would make the test and send the record to the appropriate subordinate module. Figure 8-4B illustrates the design of the modules to eliminate the use of the flag.

The determination of whether a program flag represents either data or control information depends upon the intent of the module passing the flag. Usually, if a flag is passed from a superordinate to a subordinate module, it is a form of coordination. Information is being passed to tell the receiving module what kind of processing to perform. A return flag from a subordinate module, however, is usually not coordination but is simply an indication of the results of processing. For instance, an end-of-file flag passed from an input module to its superordinate module represents data, not control. The superordinate is being told what took place and can perform whatever subsequent processing is necessary.

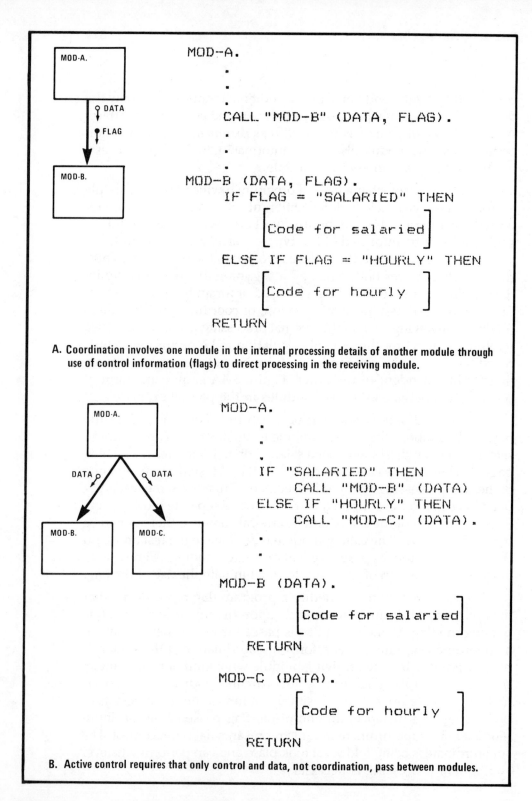

```
MOD-A.
    .
    .
    .
    CALL "MOD-B" (DATA, FLAG).
    .
    .
    .
MOD-B (DATA, FLAG).
    IF FLAG = "SALARIED" THEN
        ⎡Code for salaried⎤
    ELSE IF FLAG = "HOURLY" THEN
        ⎡Code for hourly  ⎤
    RETURN
```

A. Coordination involves one module in the internal processing details of another module through use of control information (flags) to direct processing in the receiving module.

```
MOD-A.
    .
    .
    .
    IF "SALARIED" THEN
        CALL "MOD-B" (DATA)
    ELSE IF "HOURLY" THEN
        CALL "MOD-C" (DATA).
    .
    .
    .
MOD-B (DATA).
        ⎡Code for salaried⎤
    RETURN

MOD-C (DATA).
        ⎡Code for hourly  ⎤
    RETURN
```

B. Active control requires that only control and data, not coordination, pass between modules.

Figure 8-4. Coordination versus active control.

In most cases, the need for coordination arises because the designer has not done an adequate job of partitioning. Lower-level modules have not been broken down into functionally independent subordinate modules. Thus, modules with more than one function must be passed information on which of the processing activities to perform. Usually, coordination can be avoided by making sure that lower-level modules implement a single function.

Binding Time of Connections

Binding is the process for resolving, or fixing, values of data items in a program. Binding takes place, in a COBOL program for example, when the VALUE clause is used to establish an initial value for a field described in a PICTURE clause. Also, when a program table is dimensioned to a specified number of elements, this value is bound to the program and becomes an integral part of it. Another instance of binding occurs when a program loop is defined and the upper limit of the controlling subscript is specified as a constant. In all of these cases, a constant value is embedded within the code. Thus, if changes in processing requirements dictate changes in these constants, the program must be rewritten and recompiled. In general, the appearance of constants within the source code impairs the flexibility of a program and causes maintenance problems.

In addition, when binding takes place early in the implementation phase (such as when the program is coded), there is an increase in coupling. All modules that make reference to the constant value cannot function independently. All modules that reference the constant are related, or coupled, through its value. For example, if one module defines a table length of a specified value, all modules that process the table must be aware of and adapt processing to that value. If, for some reason, the table length changes, all modules that process the table likewise must be changed.

Binding can take place at any of four points in the development of a program. Values can be fixed during program coding, during compilation, during linkage editing, or at execution time. For example, consider the need to establish a program table of federal withholding amounts that will be used in a payroll system. The table will be built within the program that calculates weekly payroll. The programming requirement, then, is to establish the table length and to assign withholding values to the table elements. If the program is to be coded in COBOL, the table size and values can be bound to the program at any one of four different times:

- *At coding time.* The table-length value can be coded as a constant in the OCCURS clause that defines the table. Table values are coded through VALUE clauses. In this case, the table values become part of the program. Thus, when the next tax year rolls around, the program will have to be recoded with the new withholding values. Also, if different withholding categories are established, all modules that make reference to the table will require recoding, since processing is tied to the specified table length.

- *At compilation time.* The source code containing table specifications can be placed in a source statement library and added to the program through use of COPY facilities during compilation. In this case, changes can be made to the library code and inserted into the program when the library is compiled. Possibly, all modules that perform table processing can be grouped within this library so that they are readily available when changes are required.

- *At linkage editing time.* Table definitions and processing instructions can be isolated also within separately compiled subprograms that are retained within object program, or *image*, libraries. The binding of values then would take place when the subprogram is linked with the main program. Changes to the program would require only that the subprogram be changed. Other modules could remain in object code versions without the need for recoding or recompilation.

- *At execution time.* The previous three options require that all or part of the program be rewritten at the source code level when changes occur. This last option requires no recoding, recompilation, or relinkage. In this case, table dimensions are defined as variables within the program. The OCCURS clause establishes a range of values for the table size, the upper limit being a value that is not likely to require changing. Table values would be maintained in a separate data file. When the program is executed, the table would be loaded from the file, with a record counter serving to tally the number of records and to establish the dimensions of the table. From a maintenance standpoint, the program would not require modification to adapt to a changing number of records in the file. The dimensions of the table and the upper limits of all loop subscripts used to reference it would vary automatically according to the number of records in the file. Of course, the data file itself would require maintenance. It is not unusual for a reference file to be used in several different programs. Thus, if all the programs are bound to current file dimensions, a change in the file size would require changes to all the programs. If given the choice, it is nearly always better to opt for file maintenance over program maintenance.

In general, binding should take place as late as possible. The earlier that values are bound into a program, the less flexible the program will be and the greater the likelihood that revision and recompilation will be needed. Of course, consideration must be given to the likelihood that constant values must be changed. In some cases, a constant may never be expected to change. For example, a formula that calculates a monthly sales average by dividing yearly sales by the number of months could be coded safely with a divisor of 12. Unless someone changed the number of months in a year, the program containing this formula never would need changing. Therefore, judgments must be made in considering the trade-offs between program maintenance requirements and the consequences of maintaining constant values external to the program.

Levels of Coupling

The various forms of coupling described above can be summarized within four distinct categories. This classification scheme provides a mechanism for evaluating design techniques on the basis of their effects on coupling. The four levels of coupling, ranked from worst to best, are:

- Common coupling
- Control coupling
- Stamp coupling
- Data coupling.

Common coupling occurs when two or more modules share access to the same data items from a common pool. Modules lack independence because they are, in effect, linked together by the data structure they all share. In COBOL, exclusive use of the PERFORM statement rather than the subprogram CALL statement to transfer control implies common coupling. All modules share the same data pool as described in the DATA DIVISION. Common coupling also results from early binding of data values. This makes for ineffective design because, as explained, any change in a constant value requires revision and recompilation of all modules that reference the value. Thus, all such modules cannot function independently. Common coupling is rated as the most severe form of coupling because even minor changes in data values can necessitate extensive changes in the software.

Control coupling occurs when processing coordination—for example, the passing of a control flag—appears between two modules. This type of control requires that one module be aware of the internal processing details of the other module. Thus, a change made in one of the modules propagates corresponding changes in the other module. Again, the modules cannot function independently.

Stamp coupling relates to the complexity of the interfaces between modules. It occurs, for example, when a calling module passes an entire data record to a subordinate module but only selected fields are required by the subordinate module. Passing unneeded data elements increases the complexity of the interface. For example, if the record size in the calling module changes, the subordinate module also will need to be changed, even though none of the data elements used by the called module are affected. Increased complexity, in turn, decreases the independence of the modules, since they are tied to the same data structure, which may not be related strongly to the problem structure. The subordinate module cannot be reused in other programs because of its dependence upon the specific structure. Stamp coupling is similar to common coupling, except that a global data structure is not involved. Nonetheless, a module is given access to data elements for which it has no logical need.

Data coupling should be the goal of design. Data coupling exists in modules connected through active control and which transfer only needed data items. Ideally, fully parameterized, conditional transfers of control and data should be used. These types of connections offer the least amount of coupling between modules.

These criteria are not offered as exacting standards. Rather, this discussion surveys coupling consequences of design alternatives. It is not assumed that data coupling will be the best alternative in all cases, nor that techniques that lead to common, control, or stamp coupling are to be avoided totally. It is possible that, in some circumstances, design techniques that produce higher degrees of coupling will be the best alternative, given the practical constraints within which the designer works. As designs are being developed, trade-offs among the coupling consequences of various techniques should be explored. At least, the designer should be able to justify the alternative selected in light of other viable solutions.

COHESION

Cohesion is a measure of the degree to which processing elements, or statements, within a module are interrelated, referring to the degree to which statements contribute toward carrying out a single problem-oriented function. Whereas coupling focuses on the connections between modules, cohesion centers on the connections between elements within modules. Ideally, a module should have maximum cohesion, or strength of internal connection, among its statements.

Seven levels of cohesion can be identified. In order from lowest to highest, these levels include:

- Coincidental
- Logical
- Temporal (classical)
- Procedural
- Communicational
- Sequential
- Functional.

At the low end of this scale would be a module in which none of the statements pertain to a single, clearly identifiable function; at the high end would be a module in which all of the statements pertain only to a single problem-oriented function. In practice, it is unlikely that the degree of cohesion will be this clear-cut. Instead, modules will exhibit varying degrees of cohesion, and the designer should look for opportunities to increase or enhance the relationships among module elements.

Coincidental Cohesion

Coincidental cohesion implies that there is little or no relationship among the elements of a module. In effect, the module consists of a random set of statements.

It is not likely that a purely coincidental level of cohesion actually would exist. To produce such a module, a designer would have to designate module functions totally by accident. Yet, it is possible to design modules that exhibit levels of cohesion that approach this low end of the scale. An example would be a situation in which the designer was motivated by concern for memory efficiency. After the program is coded, the designer might recognize recurring patterns of statements. So, to avoid this coding redundancy, these groups of statements are packaged together within a single module. The elements of the module are not related to any particular problem function but simply through the fact that they appear together in different portions of the code.

Another instance of coincidental cohesion might occur if an attempt is made to modularize a program that was designed originally as a nonstructured program. In an effort to devise a modular structure, groups of statements are lifted from the program and packaged together.

The problem with modules that have coincidental cohesion, of course, is that they have no relationship to the structure of the problem to which they apply. Therefore, if there are changes in processing requirements, it becomes difficult to trace the corresponding code. Also, if the code within a module is shared by two or more processing routines, a change that applies to one of the routines may cause erroneous processing for the other routine, thus requiring *patches*, or isolated software fixes, within the code. In general, a module with coincidental cohesion does not function as a black box; the designer must have detailed knowledge of its internal workings.

Logical Cohesion

Logical cohesion is said to exist when the elements of a module can be identified as falling within the same class of similar functions. For example, a module that performs two or more editing functions has logical cohesion. Similarly, a module that outputs several different types of error messages in response to a control flag passed to it is logically cohesive. In these cases, the module performs two or more processing functions that are similar in nature.

Logical cohesion can result from an incomplete partitioning of a program. For example, the system is decomposed down to the level at which an editing function is identified. Since, on the surface, record editing appears to be a single, cohesive function, no further analysis of requirements takes place to identify possible subfunctions. Also, logical cohesion may result from attempts to avoid coding redundancy. Using the editing situation as an example, the designer may find that there are several instances in which code can be shared among modules. For two or more fields to be validated, the identical numeric test may be applied. Therefore, the designer might decide to combine all of the edits that happen to involve a numeric check within the same module. The numeric check is coded only once; however, as a result, otherwise unrelated edits are packaged together. A preferable approach would be to create a single numeric-check edit module that could be called by the individual edit modules as necessary.

Several consequences of logical cohesion are evident. First, the modules cannot function as pure black boxes. Because more than one function is represented, the calling module must be aware of the internal logic of the called module. Often, control flags have to be passed to the module to select which of the multiple functions apply at a particular time within the execution sequence. Therefore, modules with logical cohesion require coordination control that, in turn, increases their coupling. Another problem with logically cohesive modules relates to maintenance requirements. If code is shared by the functions within the module, changes in the code for one function may impact another function inadvertently. Even if this possibility is considered, patches would be required that are unrelated to the problem structure.

Logical cohesion is avoided easily through complete partitioning of the program. At each level of design, the designer looks for identifiable subfunctions in which a single, specific processing activity is applied to a single, specific element of data. Even so, there still may be occasions in which the designer will opt for logical cohesion in a module to avoid other design problems. For example, it may be advantageous to package all error message output routines within a single module. This packaging would allow all editing routines to share the same output module to reduce coding redundancy; it also would provide flexibility in generating error messages, which could be maintained in a file external to the program. Changes in message content would require updating of the file rather than making changes to the program.

In other cases, it might be beneficial to include all input or output functions that pertain to a single file within a single module—to isolate program functions that interface directly with the external hardware/software environment, for example. If different file organization techniques are employed or different hardware devices are used, the greatest impact usually will be upon input/output functions. In isolating these functions within logically cohesive modules, locating and changing the affected statements becomes a relatively easy task.

In summary, there are trade-offs involved in evaluating the benefits and disadvantages of design decisions. Logical cohesion, for one, should not necessarily be avoided. However, the designer must be aware of the impact that such a decision will have on ease of implementation and especially on ease of maintenance of the modules involved.

Temporal (Classical) Cohesion

A module with *temporal cohesion* is one in which the elements are related in time. Statements appear within the module because they all represent functions that must take place within the same time frame during program execution. Examples occur in initialization and termination modules. At the beginning of the program, files are opened, counters are initialized, flags are set, and the first input record is read. Since all of these functions take place chronologically prior to the main processing function, these temporally related functions may be included within a single module that is called to perform the housekeeping tasks at the beginning of execution.

Of course, an ordered, time-related presentation of program functions is necessary for computer processing. A computer is a sequential machine that executes instructions in a time sequence. Therefore, functions, such as initializing constant values and opening files, must occur prior to any processing that involves those functions. However, there is no requirement that the functions be physically packaged together. For example, functions related to processing a given file can be grouped so that a read module can handle its own initialization and termination functions. Upon first entry into this module, a test is made to determine if the file is open. If not, the open routine is called, and a flag is set to indicate file status. Subsequently, on an end-of-file condition, the read module calls a subordinate module that closes the file. In this situation, functions related to a given file are localized within a separate set of modules rather than being disbursed throughout the program.

Time-related modules have low cohesion by the fact that more than one function is included. The degree of cohesion is rated higher than coincidental and logical cohesion because implementation and maintenance problems may not be as severe. A case in point would be a need to modify a program to either include or delete accumulators. Accumulators or counters by nature require initialization and are generally used in printing totals. This fact would cue the maintenance programmer to look for such occurrences in an initialization module and in a termination module. These occurrences would not be problematic; however, it still might be more convenient if all references were localized.

Procedural Cohesion

Procedural cohesion occurs when the elements within a module are grouped on the basis of the flow of control within the program. That is, possibly unrelated functions are packaged together in the same module simply because control is transferred from one module to the other. Usually, this type of cohesion results from packaging functions together that appear within the same procedural unit of a program. A procedural unit refers to the body of code that comprises a sequence, repetition, or selection control structure.

For example, consider an update program in which addition, change, and deletion transactions are processed. Depending on the transaction code, the program will dispatch control to one of three processing modules. If all processing relating to each of the transactions is packaged within a single module, that module has procedural cohesion; its elements are related because they occur within the same decision structure of the program. Another example would be the consolidation of all processing functions that take place within a program loop. Here, input, processing, and output statements would be grouped under the rationale that they all are performed within the repeat control structure.

Typically, modules with procedural cohesion result from a direct translation of a program flowchart into a modular structure. Because a flowchart documents the procedural aspects of a program, processing functions are tied directly to the control structures within which they appear. An almost unavoidable consequence of modularizing a flowchart, therefore, is a structure of activities related by this flow of control. As with other methods that result in low module cohesion, partitioning on the basis of control procedures can lead to implementation and maintenance difficulties: Modules contain more than one function, the specific order of execution of statements within modules become critically important, and data passed from one function to another are related more to the execution sequence of the program than to problem-related transformations.

Communicational Cohesion

Communicational cohesion is found in a module that performs multiple processing functions involving the same set of input or output data. This type of cohesion results from considering all of the things that can be done with a given item of data and assigning these activities to the same module. For instance, communicational cohesion would occur in a module that assembles a transaction record from several different sources, displays the transaction on the terminal, and writes a copy to an audit trail. Another example might be a module that calculates a student's grade point average, formats and prints the figure on a report, and updates the student record.

In each of these cases, the module performs multiple functions; yet, the functions are related to data transformations that are, in turn, problem related. All of the elements within the module pertain to the processing of a single data structure. Thus, communicational cohesion represents the lowest level at which relationships among processing elements are problem related rather than associated by machine processing characteristics.

Communicational cohesion is problematic to the extent that multiple functions are included within the module. Normally, however, all such functions are performed and are related closely to a data stream. If that data stream changes, the chances are good that all functions will need to be changed.

Sequential Cohesion

When a module performs multiple processing functions on the same data item *and* those functions represent successive transformations on the data, *sequential cohesion* results. In effect, the output from one function becomes the input for the next function. An example of a module with sequential strength would be one in which a table is searched for a particular value, the value is extracted from the table, and the value is used in a series of computations. Sequential cohesion can result from thinking procedurally about a problem; however, it differs from procedural cohesion in that the processing steps that are packaged together follow the flow of data through the system rather than the flow of control.

The only real problem with modules having sequential strength is that more than one function appears within them. However, unlike other such modules, control flags that increase module coupling usually are not needed. The functions are not selected individually by the calling module but are executed in the normal processing sequence. Attendant problems can be resolved easily by further partitioning of the module into two or more subordinate modules.

Functional Cohesion

A module with *functional cohesion* applies a single transformation to a single item of data. Functional cohesion is rated as the highest level of cohesion because each element within the module contributes only to a specific transformation. Modules with functional strength are preferred because they behave as classical black boxes. That is, for each item of input to a module, there is a known and predictable output. Further, the module can be used without knowledge of its internal workings. None of the maintenance problems associated with other levels of cohesion are evident in functionally cohesive modules.

A straightforward test of whether a module has functional cohesion often can be made by assigning a name to the module. If the module can be described as performing a single transformation on a single data structure, it probably has functional strength. For example, the following module names suggest functional cohesion:

COMPUTE GRADE POINT AVERAGE

INPUT TRANSACTION RECORD

SEARCH TAX TABLE

PRODUCE PAYROLL REPORT

At first glance, the last module name in the above list—PRODUCE PAYROLL REPORT—might not appear to suggest functional cohesion. To produce such a report, a complete program would be required. Thus, it might be implied that the module must contain all of the input, processing, and output instructions required to carry out the function. However, if proper design techniques were applied, this module would be a top-level control module that contains only control statements and calls to lower-level modules that perform the processing. Recall that the name of a module implies the functions of all subordinate modules. Therefore, as a control module, PRODUCE PAYROLL REPORT has functional strength.

Information Hiding

Software design normally will be motivated by the objective of defining hierarchical structures of functionally cohesive modules. In most cases, such a design will contribute to ease of program implementation and maintenance.

In other cases, however, this type of design will not be the best in terms of module independence. Consider a program in which the following three functional-strength modules appear:

BUILD TAX TABLE

SEARCH TAX TABLE ON GROSS AMOUNT

EXTRACT WITHOLDING AMOUNT FROM TABLE

These three functions may be scattered throughout the program, depending on the sequence in which they are activated. Although the modules are functionally independent from the standpoint of the problem structure, they are all dependent on the structure of the table. If the format of the table changes, all three modules probably would have to be changed.

A solution to the potential maintenance problems associated with modules that are functionally cohesive but interrelated through some information structure lies in the concept of *information hiding*. That is, functions that depend on an information structure or other system resource that is not problem related are isolated, or hidden, within a single module. The structure or resource is made transparent to the designer or user of the software. Changes in the processing environment, therefore, produce software changes only in the informational-strength module and do not require changes in the structure or processing logic of the program.

Figure 8-5 illustrates the use of an informational-strength module for the table-processing situation described above. Here, a single

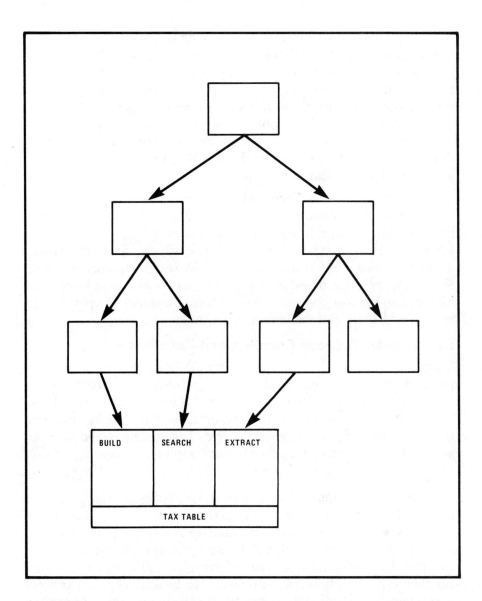

Figure 8-5. An informational-strength module has multiple entry points and no overlapping code for the multiple functions performed.

module, coded as a subprogram, contains all processing related to building, searching, and extracting information from a tax table. The table itself is defined within the subprogram. The subprogram contains a separate entry point for each of the processing functions. Superordinate modules call the subprogram and specify the entry points for the desired functions. Within the subprogram, there is no overlapping code, and each function includes a distinct exit point. If changes occur in the structure of the tax table, the processing changes are isolated within this single module.

A module with informational cohesion results from the packaging together of two or more functional-strength modules. Of course, this solution lessens functional cohesion and increases coupling. However, the system is maintained more easily to cope with changes in the processing environment. A general design strategy might be to develop a system based on the use of functional-strength modules and then to look for occasions to reduce environmental dependencies by defining informational-strength modules.

Relationship Between Coupling and Cohesion

Coupling and cohesion are closely interrelated. On one extreme, a program might be composed of one module. There would be no coupling within the system; however, the module also would rate very low in cohesion. At the other extreme, each module in a program might be composed of a single statement. This system would rank not only high in cohesion but also high in coupling. Of course, neither of these designs is a good solution.

The designer should be guided by the goal of establishing a proper balance between coupling and cohesion. Just what is the proper balance and how to achieve it are questions for which no definitive answer can be given. Usually, if the designer aims for modules with functional cohesion, in which functions relate to the structure of the problem under study, a proper balance can be achieved. The process is still largely judgmental; however, the criteria presented here can guide the designer's evaluation.

OTHER EVALUATIVE CRITERIA

In addition to the main evaluation criteria of coupling and cohesion, other design guidelines can be applied. These include:

- Distribution of decision making
- Span of control (fan-out)
- Fan-in
- Scope of effect and scope of control
- Module size.

Distribution of Decision Making

The modules that comprise the structure chart of any program will fall into two broad categories: control or detail processing. If the program is well-designed, decision making and control processing will be applied by modules at the top levels of the hierarchy. Lower-level modules then will carry out detailed processing. This distribution of decision making is shown in Figure 8-6. In this example, the shading in the modules represents the proportion of elements that provide program control functions. As indicated, the incidences of control decline at lower levels of the structure.

Span of Control (Fan-Out)

Span of control, also referred to as *fan-out*, refers to the number of modules that are immediately subordinate to a module. A high span of control can indicate improper partitioning of a system into upper-, intermediate-, and lower-level modules. Figure 8-7A shows a high span of control with wide fan-out, indicating that probably too many modules are subordinate to and under direct control of the higher-level module. Such a configuration indicates that partitioning of a function has not proceeded hierarchically through intermediate-level abstractions. A possible problem with this design is that the higher-level module has too many functions to oversee, making the logic of that module highly complex. High fan-out is not always an indication of poor design; however, it does present evidence of a possible source of problems that should be examined.

By comparison, Figure 8-7B illustrates the opposite extreme. The span of control is possibly too low, indicating that partitioning has gone forward with too much vigor. The excessive control built into this structure means that intermediate-level modules are likely to contain trivial processing functions that can be absorbed into higher-level modules, leaving a configuration in which two detail modules are under the top-level module.

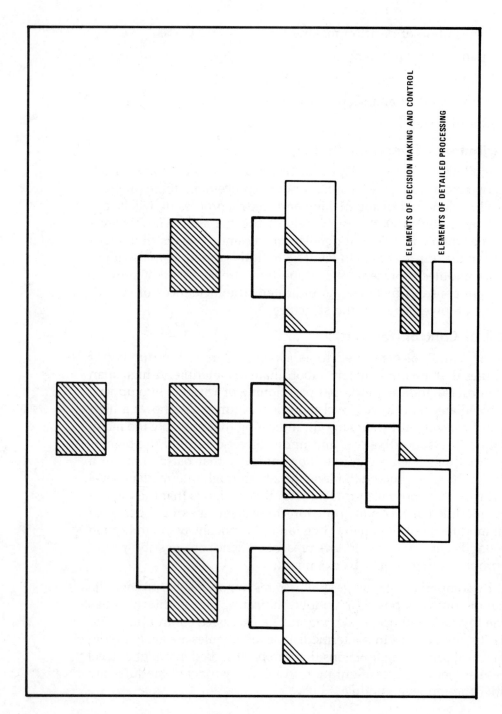

ELEMENTS OF DECISION MAKING AND CONTROL

ELEMENTS OF DETAILED PROCESSING

Figure 8-6.   Distribution of decision-making within modular systems.

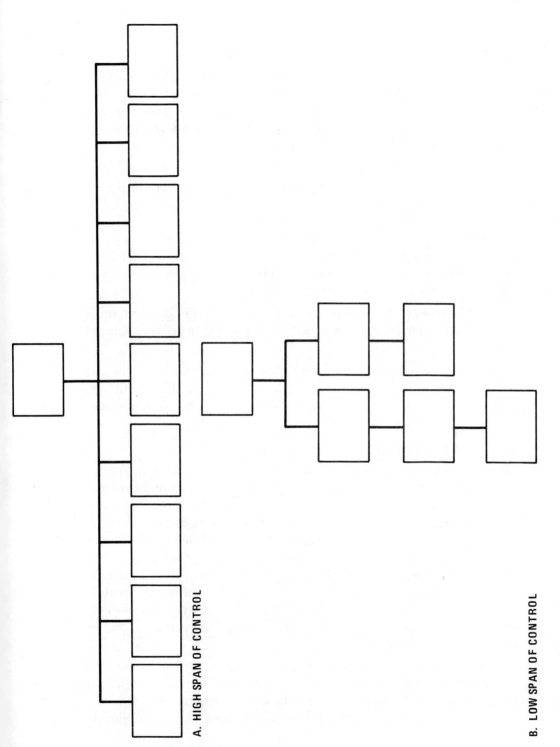

A. HIGH SPAN OF CONTROL

B. LOW SPAN OF CONTROL

Figure 8-7. Span of control in a system indicates the effectiveness of top-down partitioning across multiple levels of abstraction.

Fan-In

Fan-in is a measure of the number of higher-level modules that call upon a lower-level module. High fan-in is desirable; it indicates that the called module is highly functional and can be used without modification in several parts of the system. Figure 8-8 shows a segment of a structure chart in which the shaded module has high fan-in.

As a general rule, it is desirable to have a lower-level module called by two or more higher-level modules. However, fan-in should not be achieved at any cost. For example, fan-in can be achieved by packaging several related functions within the same module. Various higher-level modules, then, would call this routine and select a particular processing function through use of a control flag. Thus, a module of low cohesion (perhaps logical cohesion) is created and coupled tightly to the modules that call it.

The increased coupling resulting from this design decision must be balanced against the convenience of having such a general-purpose module.

It is impractical to list those situations in which fan-in is the best solution and those circumstances in which it probably should be avoided. Designer judgments should be made on the basis of trade-offs of coupling and cohesion.

Scope of Effect and Scope of Control

The *scope of effect* of a decision is the total group of modules containing conditional processing based upon that decision. In other words, the scope of effect includes all modules affected by a selection control structure. By comparison, the *scope of control* of a decision extends to all subordinate modules, as well as to the module containing the decision. Figure 8-9 shows a structure chart in which the scope of effect can cut across the branches of a structure chart. This is, basically, an undesirable result, which requires setting a flag within one module that must be passed to, retested by, and acted upon by another module. The flag introduces excessive coupling between the modules affected by the decision.

A much sounder approach, shown in Figure 8-10, is to include the scope of effect within the scope of control. That is, modules affected by a decision are subordinate to the module in which the decision is made. In this case, no control flags are required, and coupling is

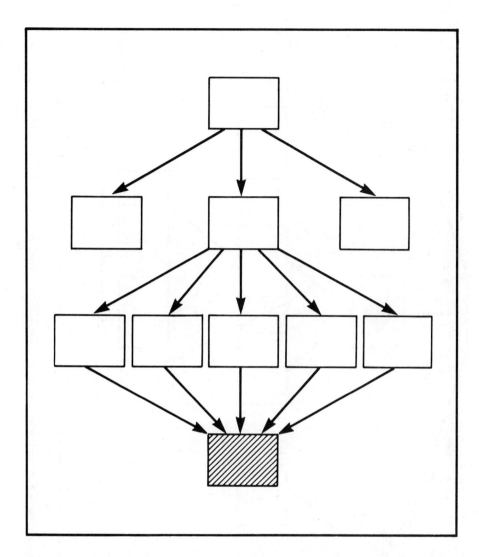

Figure 8-8. High fan-in refers to the use of a single, lower-level module by two or more higher-level modules.

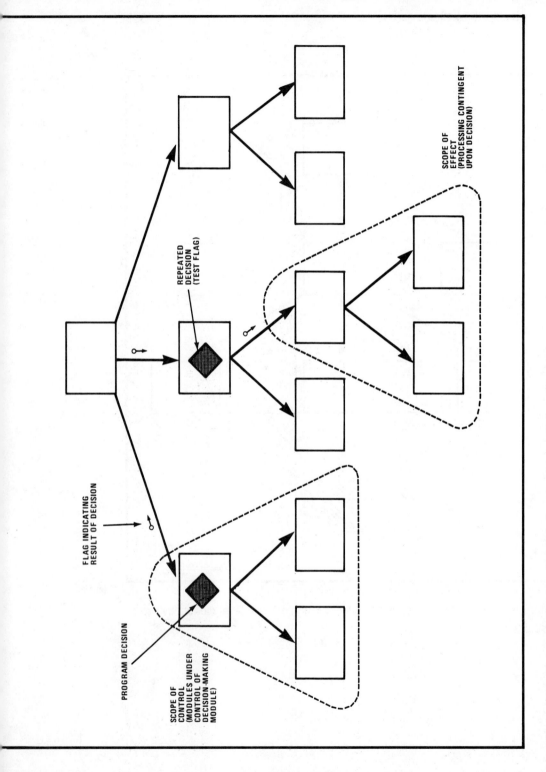

REPEATED
DECISION
(TEST FLAG)

FLAG INDICATING
RESULT OF DECISION

PROGRAM DECISION

SCOPE OF
CONTROL
(MODULES UNDER
CONTROL OF
DECISION-MAKING
MODULE)

SCOPE OF
EFFECT
(PROCESSING CONTINGENT
UPON DECISION)

Figure 8-9.   Scope of effect and scope of control in a modular system.

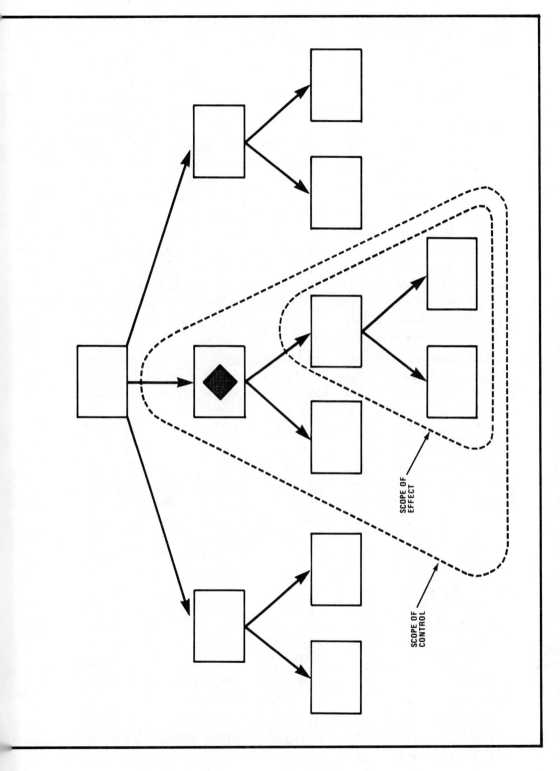

Figure 8-10. The scope of effect should be a subset of the scope of control to avoid use of unnecessary flags and their retesting within parts of the program.

reduced. The decision-making module calls upon the appropriate processing module directly.

Module Size

As a general rule, large modules often result from an incomplete breakdown of a problem into appropriate subproblems. Thus, one of the measures of effectiveness of design lies in the size of individual modules. Several rules of thumb have been suggested for evaluating module size. One of these holds that no module should be longer than 50 lines of source code. This length corresponds with the number of lines that can be printed on a single page of computer printout. Another suggested size limitation is that no module should contain more than 30 statements. This limit is suggested by studies indicating that a programmer's comprehension of a module drops sharply if it contains more than 30 lines. Individual CIS organizations are known to enforce module sizes ranging up to 500 lines.

The number of lines appropriate within any module, for any programmer, is highly judgmental. However, the principle is that as the number of statements in a module grows, it pays to inspect it to determine whether it contains more than one function. Excessively large (or small) modules may point out problems affecting coupling and cohesion.

The evaluation criteria reviewed above may be applied in judging the quality of software designs. Such criteria can be applied after a design has been completed; they also can serve as guidelines during the development process. As yet, these criteria do not have the formality of quantitative approaches to quality assessment. It is not possible to apply them as software metrics to rate coupling and cohesion with precision. However, the criteria can promote an awareness of design trade-offs and the consequences of design decisions. Also, this approach can result in software designs that are highly cohesive and that exhibit low coupling.

Summary

Quality software exhibits two main characteristics: It is easy to implement and test, and it is easy to maintain and modify.

To implement quality, the criteria of coupling and cohesion are applied to each design decision as the hierarchy of modules is expanded from top to bottom.

Coupling, or the connection between modules, is influenced by the type of connection between modules, the complexity of the interface between modules, the type of information flow across the interface, and the binding time of connections.

Connections between modules serve to pass control or to transfer data. Data are passed as arguments from one module and are used as parameters within the receiving module. Transfers of control may be conditional or unconditional. Transfers of data may be fully parameterized (passed explicitly as arguments) or may be non-parameterized (shared within a global data pool).

Complexity of data transferred may have an effect on coupling. Coordination between modules results when both data and processing instructions or flags must be passed.

Binding is the process of fixing the values of data items within a program. In general, binding should occur as late as possible, preferably at execution time.

Levels of coupling, ranked from worst to best, include: common coupling, control coupling, stamp coupling, and data coupling.

Levels of cohesion, ranked from lowest to highest, include: coincidental, logical, temporal (classical), procedural, communicational, sequential, and functional.

In applying the principle of information hiding, functions that depend on an information structure or other system resource that is not problem related are hidden within a single module. Changes in the processing environment produce software changes only in the informational-strength module.

Other evaluation criteria include the distribution of decision making within modules of the system, the span of control exhibited by superordinate modules, the general usefulness of detailed processing modules within different portions of the system, the location of decision making effects within the system, and the physical sizes of modules.

Key Terms

1. coupling
2. cohesion
3. metric
4. heuristic
5. connection
6. identifier
7. branching
8. argument
9. parameter
10. conditional
11. unconditional
12. label
13. referent
14. parameterized
15. nonparameterized
16. complexity
17. active control
18. coordination
19. flag
20. binding
21. image

22. common coupling
23. control coupling
24. stamp coupling
25. data coupling
26. coincidental cohesion
27. patch
28. logical cohesion
29. temporal cohesion
30. procedural cohesion
31. communicational cohesion
32. sequential cohesion
33. functional cohesion
34. information hiding
35. span of control
36. fan-out
37. fan-in
38. scope of effect
39. scope of control

Review/Discussion Questions

1. What are the two principal characteristics of quality software?

2. What two related criteria may be used in making design decisions related to the partitioning of modules, and what are the definitions of each?

3. How are the criteria given in answer to the preceding question interrelated?

4. What types of connections may exist between modules?

5. What is the difference between parameterized and non-parameterized transfer of data between modules?

6. What is the difference between conditional and unconditional transfer of control between modules?

7. What is binding, and when should it occur?

8. What is complexity, and how may it affect coupling?

9. What are the levels of coupling?

10. What are the levels of cohesion?

11. What is meant by information hiding?

12. What heuristics may be applied in evaluating software design decisions?

Project Assignments

The Appendix presents three case studies that can be used in support of project assignments. All three scenarios provide functional system specifications that would result from systems analysis. Included are data flow diagrams, data dictionaries, and process narratives that would result from systems analysis. Continuing the work begun in Chapters 6 and 7, assignments that relate to the content of this chapter will be found in *PART 3: Detailed Systems Design*.

9 SOFTWARE DESIGN STRATEGIES

LEARNING OBJECTIVES

On completing reading and other learning assignments for this chapter, you should be able to:

☐ State the objective of software design.

☐ Tell how the criteria of coupling and cohesion may be applied in guiding software design decisions.

☐ Describe design approaches that involve functional decomposition, data flow, and data structure.

☐ Describe the relative strengths and weaknesses of the three software design strategies presented in this chapter.

☐ Tell how design approaches may be applied, resulting in a single, coordinated software design strategy.

SOFTWARE DESIGN APPROACHES

Chapter 7 deals with foundational knowledge related to the building of software products. The discussion points out that software is composed of statements organized into modules that, in turn, are structured hierarchically into complete, operational programs. In the following chapter, criteria are presented for evaluating the resulting designs, focusing primarily on the coupling and cohesion effects of different design decisions. In this chapter, these topics come together as strategies are considered for developing applications software.

The questions to be answered are: "How are designs produced that effectively package statements into modules and into programs? At the same time, what design decisions are necessary to minimize coupling and maximize cohesion?"

It is stressed in earlier discussions that, as yet, there is no single set of guidelines—no cookbook—covering a uniform, step-by-step method for developing quality software. Rather, a series of different approaches is evolving. Experience shows that each of these approaches has different advantages and disadvantages and that some of these characteristics actually make comparisons difficult. Thus, the presentations here are offered as guidelines for thinking about problems that may be encountered and solutions that will unfold. No prescriptions for automatically developing high-quality software are offered.

Each of the approaches for developing software designs can be classified readily into one of three general categories:

- Functional decomposition
- Data flow approaches
- Data structure approaches.

These general strategies approach software design from slightly different perspectives. *Functional decomposition*, for example, focuses on the identification of processing functions through a process of top-down partitioning. Abstract functions are decomposed into more detailed functions, proceeding successively down the program hierarchy. *Data flow approaches*, although they promote a similar top-down strategy, look to identify functional components by focusing on the transformations applied to data as they flow through the system. A *data structure approach*, on the other hand, considers the logical structure of the data to be processed by the program as the framework for deriving a parallel program structure. In all cases, and despite these differences, the design motivation is the same—to produce quality software that is easy to implement and test, easy to modify, and easy to maintain.

For these design approaches, a uniform set of assumptions is applied. It is assumed that, before design begins, a complete set of system specifications has been prepared. These specifications include statements of user requirements, including definitions of output reports, input documents, transaction documents, files, and databases, and any other data sources and destinations. A further assumption is that a systems flowchart has been prepared. This flowchart describes one or more job steps in a batch job stream or in a single on-line program that may include several on-line procedures selected through menu choices. The systems flowchart indicates the broad relationships that exist among input, processing, and output functions. Also included would be detailed specifications of the processing to be applied. These processing specifications would encompass a multi-level subset of data flow diagrams with accompanying process specifications for the lowest-level bubbles, or data transformations.

The three approaches described below are representative of three general strategies. Several variations on these approaches have been popularized and marketed as complete design methodologies. Yet, nearly all of these specific techniques have their roots in the approaches surveyed here.

FUNCTIONAL DECOMPOSITION STRATEGIES

Functional decomposition represented the first attempt, during the 1970s, to produce a set of procedures that could be followed in developing quality software products. At that time, there was an awareness of the benefits that could accrue from using structured programming techniques. By applying these standards, the task of programming was made more manageable at the coding level. So, functional decomposition applied the same rationale at the design level. That is, not only could the concept of a minimum set of control structures guide the structuring of code, but also it had application to the design and structuring of modules. By breaking a program down into modular components and relating those modules with sequence, repetition, and selection control structures, order could be brought to the design task.

Prior to functional decomposition techniques, attempts at establishing modularity in programs were outgrowths of flowcharting, or algorithmic, approaches to design. That is, programs were designed in a linear fashion, beginning with input and following chronologically through the processing steps and transfers of control up to output.

In breaking up flowchart organizations of programs into modules, subdivisions were rather arbitrary. Modules were defined as often by size criteria related to hardware constraints as by their individual processing functions. Consequently, no real progress was made in developing designs that were minimally sensitive to change. Maintenance was a continuing problem because the structure of the software bore little resemblance to the structures of the problems to which that software applied.

Functional decomposition was a first attempt to deal directly with this correlation at the design level. It presented a strategy for systematizing the development process, providing a problem-based rationale for decomposing programs into modules and into statements, and for relating modules with standard program control structures. The procedures of functional decomposition are also called *stepwise refinement* in recognition of the fact that the partitioning process carries the design through increasing levels of detail, beginning with an abstract statement of the program function and ending with a set of operational details that implement the function.

Functional Decomposition Techniques

Functional decomposition begins with identification of a top-level module representing the overall function of the program to be developed. From there, submodules are derived. At the second level, modules representing major subfunctions that comprise the overall program function are defined. These subfunctions are defined at an additional level of detail beyond the program function. As these and successive levels of modules are identified, additional levels of processing details required to implement the program on a computer are added. Finally, at the bottom level of the resulting hierarchy, modules representing computer processing functions are introduced.

Program partitions, or modules, then, represent processing tasks defined at several *levels of abstraction*. That is, at the top levels, processing details are ignored. Instead, concern is with identifying the business processing functions that must be accomplished and the relationship between those functions. The top-level module is primarily a control module, accomplishing the processing functions by calls to second-level modules. At subsequent levels, these business functions are decomposed into computer processing functions. Thus, functional decomposition has the effect of breaking down a large, complex job into a series of smaller, more manageable tasks. Tasks at successively lower levels contain more processing details than those tasks at higher levels.

Obviously, considerable judgment must be applied in the decomposition process. The designer is faced with the problem of determining which functions should be partitioned into what subfunctions. Also, in working downward through the levels of partitioning, the context of design changes. At the top levels, decomposition is carried out in light of the business functions that are to be performed. However, at lower levels, these business functions must be translated into computer processing techniques. Therefore, there is a transformation of the design from basically a problem-related structure to a machine-related structure.

Functional Decomposition Example

The process of functional decomposition can be illustrated by the following example. An on-line updating program is to be designed. This program will input transaction records interactively and post amounts contained within the records to corresponding master records. As in most on-line systems, a transaction log will be maintained. That is, a summary of the transaction will be written to a file that will serve as an audit trail of evidence that the transaction has been posted against the master. The flowchart in Figure 9-1 describes the procedure in terms of its major input and output files.

Functional decomposition begins with the definition of a top-level module that represents the entire updating procedure. This module would be a control module that calls upon subordinates to carry out detailed processing. First-level partitioning then would identify the major processing functions comprising this overall function. At this point, however, functional decomposition offers little guidance on exactly how to partition. The criterion is simply that subordinate functions should be identified and modules defined to carry out the processing.

The experienced designer will recognize that in the typical on-line update system, three common functions appear. First, the transaction record will have to be input and edited to ensure that only valid transactions are applied as updates to the master file. Second, the actual posting of amounts to the file will take place. Finally, a journal record

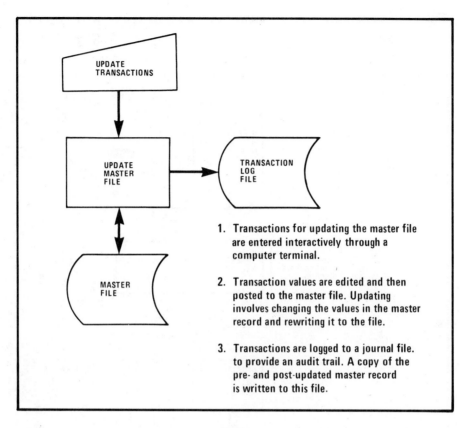

1. Transactions for updating the master file are entered interactively through a computer terminal.

2. Transaction values are edited and then posted to the master file. Updating involves changing the values in the master record and rewriting it to the file.

3. Transactions are logged to a journal file. to provide an audit trail. A copy of the pre- and post-updated master record is written to this file.

Figure 9-1. Annotated systems flowchart for master file updating.

will be written to the log file. Therefore, a first cut at a structure chart for this program might appear as shown in Figure 9-2A. The top-level module, UPDATE-MASTER-FILE, calls upon three subordinate modules—INPUT-VALID-TRANSACTION-RECORD, POST-TRANSACTION-AMOUNTS, and WRITE-TRANSACTION-LOG—to perform the major processing tasks in the system.

In thinking about the computer processing steps necessary to implement these functions, it becomes apparent that the structure shown in Figure 9-2A will have to be modified. The modules POST-TRANSACTION-RECORD and WRITE-TRANSACTION-LOG will not be activated for every input transaction; these modules will be executed for all transaction records except the last one. The last transaction, of course, will be a signal from the terminal operator indicating that no more transactions are to be input and that processing is complete. This suggests placing these two modules subordinate to a

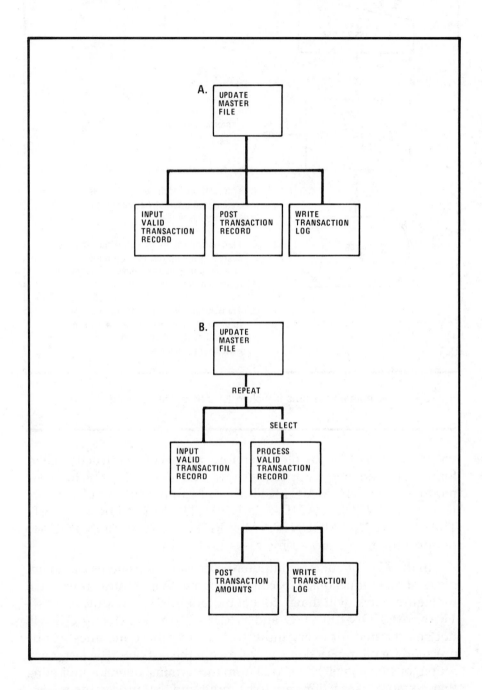

Figure 9-2. Structure charts for initial partitioning of master file update program.

control module, PROCESS-VALID-TRANSACTION-RECORD, which, in turn, will be governed by a selection control structure. That is, processing will take place only on input of a valid transaction record that is not an end-of-processing indicator.

These considerations are reflected in the structure chart shown in Figure 9-2B. Here, the modules have been reorganized to represent the structure under which processing will take place. The major program control structures have been superimposed over the problem structure to indicate how the functions must be presented to the computer for processing.

The next step in the design process is to decompose these initial modules into their subordinates. Design has not yet reached a level at which it can be translated conveniently into program code. Additional refinement is necessary as shown in Figure 9-3.

Consider the INPUT-VALID-TRANSACTION-RECORD module. This module will require partitioning into two subordinate modules. It is assumed that transaction input will take place on-line. An operator seated at a terminal will enter transaction values that will be edited and then used in updating a master record.

For each transaction, the program will display a data entry screen. Labels will be displayed to guide the operator through the data entry session. For example, prompts will be given for the master record key that identifies the record to be updated along with prompts for each of the data fields that represent update amounts. Therefore, routines will be needed to display the initial data entry screen and to input the transaction fields. These requirements are shown in Figure 9-3 as modules DISPLAY-INPUT-SCREEN and INPUT-VALID-TRANS-ACTION-FIELDS.

The INPUT-VALID-TRANSACTION-FIELDS module also has been partitioned into two submodules—ACCEPT-VALID-MASTER-KEY and ACCEPT-REMAINING-VALID-FIELDS. This partitioning is necessitated by the structure of data fields that comprise the transaction record. The transaction record must contain a key that corresponds with one of the records in the master file. Therefore, one of the requirements is to input and validate the master key. This operation involves moving the key that is input to a master file search-key data area and accessing the master file. If the corresponding master

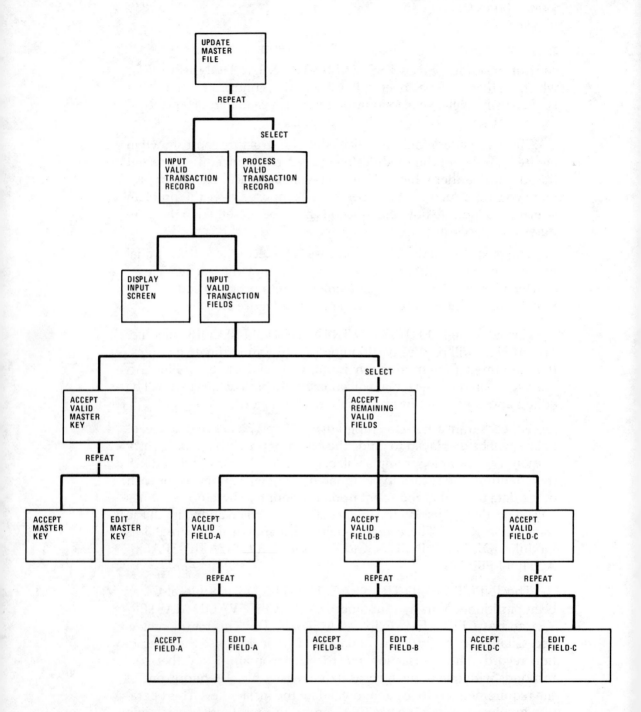

Figure 9-3. Structure chart for partitioning of INPUT-VALID-TRANSACTION-RECORD procedure of master file update program.

record is in the file, the remaining transaction fields are input. If there is no matching key, an "invalid key" condition exists, in which case the program prompts for another key value. Thus, the routines to input the key are repeated until either a valid key is input or the operator decides to terminate the updating session. It is assumed that the end of the session is indicated when the operator bypasses entry of a key, leaving the field with a null, or blank, value.

If a valid master key is input, the module named ACCEPT-REMAINING-VALID-FIELDS is called. This module directs the acceptance and editing of each of the data fields that contain updating values. In the current example, it is assumed that a transaction record is composed of three data fields: FIELD-A, FIELD-B, and FIELD-C. These are numeric fields that can be edited independently of the master record.

As shown in Figure 9-3, a separate module is defined for each of these input routines. The logic of these modules is similar to that of the master key input module. That is, the operator enters a value in response to the field prompt. The input value then is edited to verify that it is reasonable and within an expected range of values. If so, data entry continues with the next field. If, however, the input value does not pass the edit, an error message is flashed on the screen. The message can be displayed alongside the entered value. To highlight the error condition, the incorrect value may be displayed again in reverse video or blinking. The operator then must enter another value. The accept and edit routines are repeated until the entered value passes the edit.

Following the data entry session, control returns to the module UPDATE-MASTER-FILE with either a valid transaction record or a flag indicating the end of the session. If a valid record is present, the module PROCESS-VALID-TRANSACTION-RECORD is called; otherwise, a null, or blank, master key was input, an end-of-session flag has been set, and the program is terminated.

Because the input branch of the program is designed to pass only valid transaction records to the update branch, updating and logging are relatively straightforward processes. Figure 9-4 shows the structure chart expanded to include this further partitioning.

First, the image of the master record prior to updating is written to the log file. Next, the amount fields containing values are added

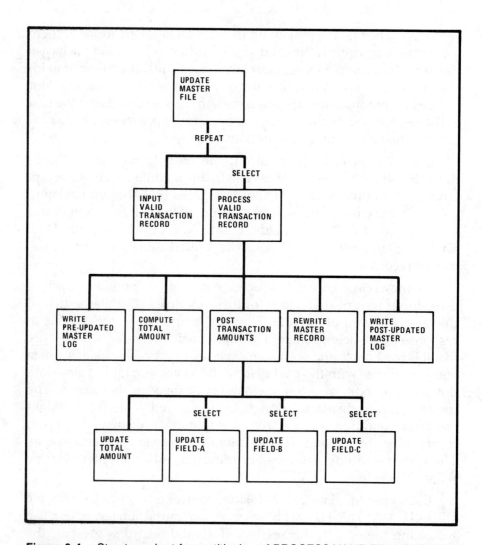

Figure 9-4. Structure chart for partitioning of PROCESS-VALID-TRANSACTION-RECORD procedure of master file update program.

to develop a transaction total. This total, as well as the individual amounts, will be added to the master record. Then, the transaction amounts are posted to the master record. The update total is posted to the file in all cases. The transaction fields, however, are posted only if they contain numeric values. Thus, each of the field-update modules is governed by a selection control structure. Following updating, the master record is rewritten to the file. Finally, processing is completed with the writing of a log record of the updated master record.

At this point, the program design identifies the basic set of modules necessary to perform processing. Their structure generally parallels the structure of business activities required to edit, post, and log transactions. Furthermore, a general computer processing model has been superimposed over the problem structure. Thus, functional decomposition attempts to model a program on the basis of the problem structure and, at the same time, recognizes that the design must facilitate computer processing. Integration of program control structures within the structural design of the software is a common approach.

The design, however, is still not in final form. For example, files must be opened and closed, flags must be initialized, and other pre- and post-processing activities must be added to the design to bring it closer to an operational model. Nevertheless, a formal, logical model of processing has been developed. The design is in a form that can be expanded easily to include these operational details.

Development of a general, logical structure for a software product, as well as its elaboration into an operational program, should be guided by the criteria of coupling and cohesion. At each level of partitioning, the impact of the design on these measures should be considered. Consider, for example, the need to add file opening and closing routines to the program illustrated above. One option would be to define an initialization module that would be activated one time at the beginning of processing. This module would call upon subordinate open routines for each of the files that were processed. Such a strategy normally would assume the existence of a corresponding termination module to close the files at the completion of processing. As pointed out in the last chapter, however, these modules would have temporal cohesion. The modules also would increase the coupling in the system because routines to open, input, output, and close files would appear throughout the program.

A second, and possibly preferable, option would be to apply the concept of information hiding. For instance, master file open and close routines, as well as the read and write functions, would be packaged together in the same module. This module would be called at the beginning of processing to open the file. The module would be called from the EDIT-MASTER-KEY module to input the master record, and then it would be called from the REWRITE-MASTER-RECORD module to output the master record. Also, it would be called at the completion of processing to close the file. The module would be designed as a subprogram with multiple entry points, with the main program passing flags to it to select desired processing. However, this need to pass control flags and any increased coupling that might result must be balanced against the increased coupling resulting from the first option.

Evaluation of Strategy

In general, the criteria used to guide partitioning are the business functions that are to be implemented. Even so, different designers may come up with vastly different designs depending on their perspectives on the business problem. Unfortunately, functional decomposition does not provide clear, exacting guidelines for how to go about partitioning. Thus, a lack of consistency in producing designs has been felt to be a weakness of this approach. There is no guarantee that two designers working on the same problem will develop the same or similar solutions. The design seems to depend on the experience and perspective of the individual designer.

A strength of the functional decomposition approach is that it is built upon sound principles. Use of top-down design techniques, procession through various levels of abstraction, identification of business functions, and use of the minimum set of program control structures to govern processing are all proven methods. However, the lack of formal guidelines for applying these techniques can make implementation of this strategy relatively inexact and inconsistent.

DATA FLOW STRATEGIES

The data flow approach to software design, also called *transform analysis*, has been derived, generally, from functional decomposition approaches. In addition, it provides methods for identifying and organizing program functions that model the structure of the problem. It is offered as an adjunct to functional decomposition for determining just how to decompose a problem structure and derive a parallel program structure. In addition to transform analysis, a special technique known as *transaction analysis* also is discussed.

Program structures are identified through data flow analysis. That is, the flow of data through a system and the transformations applied to those data are used in determining the overall relationships among problem functions. The structure of functions becomes the basis for the structure of software to implement those functions.

The data flow approach makes use of data flow diagrams to model the problem and to bring understanding about its structure. Design activity, then, translates that business model into a hierarchical organization of software modules that preserve, to the extent possible, the model of the original problem.

Transform Analysis

As a basis for software design, it may be possible to use data flow diagrams developed during the analysis and general design phase of the systems development life cycle. If the specifications prepared during analysis are not based on data flow diagrams, or if the diagrams are incomplete, the designer will have to develop his or her own low-level diagrams to model the system being implemented. The specifications for the new system, whatever their format, should be reviewed to make sure that all data streams are traced clearly and that all major data transformations are represented by clearly identifiable bubbles on the diagram.

As an illustration of the design process, first, an abstract example will be used. Figure 9-5 presents the data flow diagram that might have been drawn to model this abstract problem. Each bubble represents a successive transformation that is applied to the data as they flow through the system.

As in functional decomposition, design based on data flow begins with identification of a top-level module that represents the program function as a whole. Next, *first-level factoring* identifies the major processing functions that must be performed to accomplish this program function. At this point, the data flow diagram for the application is analyzed. A determination is made of the major processing branches of the system; for each major processing branch, a control module is defined subordinate to the top-level module. Analysis involves identification of three types of branches:

- *Afferent* branches encompass inputs to the program. An afferent, or input, branch is identified by those transformations that

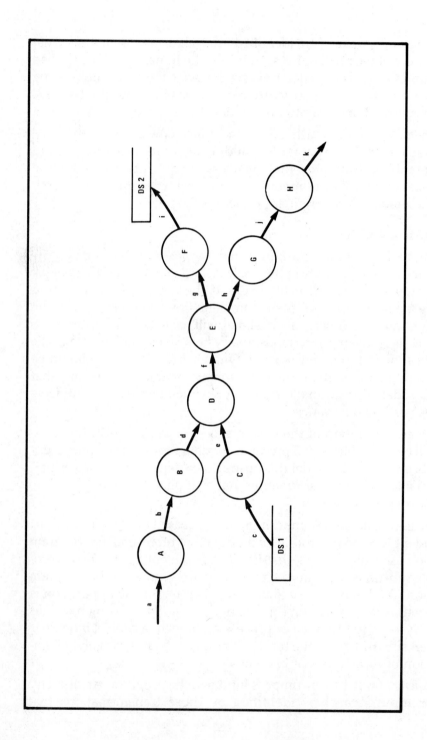

Figure 9-5. Absract data flow diagram modeling a portion of a system to be implemented as a computer program.

translate physical inputs into logical inputs that are ready for the main business processing functions. A physical input is a record or data item that resides in a file, that has been captured on some type of input medium, or that is being entered through a terminal keyboard or other type of direct-entry device. To be in a form acceptable for central processing functions, this physical input first must be stripped of its physical characteristics (through deblocking of files or construction of text from words, for example). Next, the input requires preparatory editing, validation, sorting, or other transformations to put it in proper format for processing. Thus, an afferent branch of a data flow diagram includes all transformations applied to inflowing data up to the point at which the data are available and in the proper format for processing. Figure 9-6A shows the data flow diagram for the above example as it might be partitioned to indicate the afferent branches of the system.

- *Efferent* branches of the data flow diagram encompass the movement from unstructured, unformatted output data produced by the system through the series of transformations that generate physical outputs. Thus, efferent data flows move outward from the main processing functions of the system. Efferent branches are identified by tracing backwards from final system output up to the points at which the data appear not to be flowing outward; that is, central business processing seems to be occurring. Depending on the particular application under study, there may be two or more efferent branches. For example, in Figure 9-6A, two such branches have been defined.

- The portions of the system that are not included in the afferent or efferent branches—everything that is left—are part of the *central transform* branch or branches. This portion of the system is responsible for the main business processing functions of the application and is the part in which input data streams are transformed to create output streams. The central transform in the master file updating system is shown in Figure 9-6A.

In factoring at the first level, a control module is defined for each afferent branch of the data flow diagram, for each efferent branch, and for the central transform or each logical group of transforms. The result is a high-level input-processing-output model of the system. As

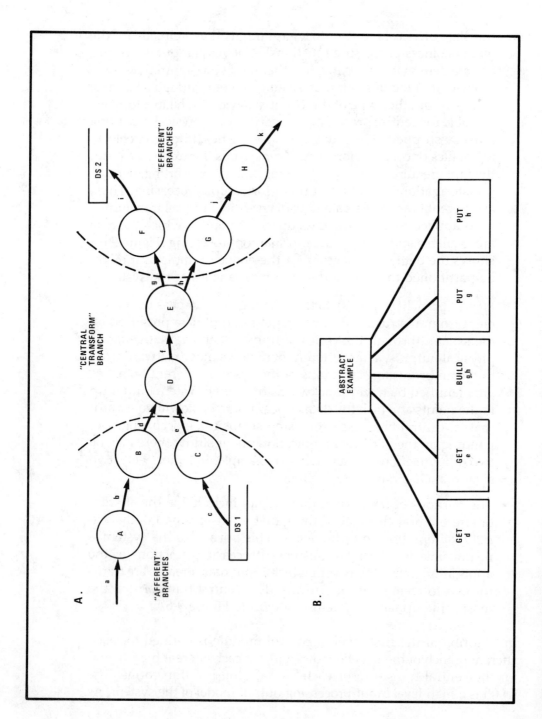

Figure 9-6. Partitioning of data flow into afferent, efferent, and central transform branches, and a resulting top-level structure chart.

illustrated in Figure 9-6B, each of these first-level control modules is responsible for carrying out the transformations required in each branch of the system. For example, the modules GET-d and GET-e are the control modules for the two afferent branches. The modules named PUT-g and PUT-h are defined to oversee processing in the two efferent branches. The central transform branch is controlled by the module BUILD-g,h.

Once the top-level structure of the system is derived, factoring proceeds iteratively, in top-down fashion, for each branch of the structure chart. As a general rule, separate procedures are followed for afferent and efferent branches and for transform branches. Typical procedures for afferent and efferent branches begin by tracing outward from the central transform. Each succeeding bubble of the data flow diagram becomes the basis for the next set of lower-level modules of the structure chart. Factoring then ends at the ends of respective branches, where physical inputs or outputs can be identified.

The process for factoring each afferent branch is shown in Figure 9-7. The module GET-d was identified in the first level factoring as a control module to provide data represented by the data flow "d" for use by the central transform. According to the data flow diagram in Figure 9-6A, the data flow "d" is produced by process B, and that process requires a "b." Thus the module GET-d can be factored into two processes as shown in Figure 9-7A: It receives a "d" by calling module B. But when calling B it must send "b." A new module GET-b is identified to control the process of providing occurrence of "b."

In a similar manner the GET-b module is factored as shown in Figure 9-7B. This results in a GET-a module that represents some type of physical input. Finally, Figure 9-7C represents a complete factoring of both afferent branches.

A related process is followed for factoring the efferent branches. These are illustrated in Figure 9-8. The module PUT-h was identified in the first level factoring as a control module to output data represented by the data flow "h" from the central transform. According to the data flow diagram in Figure 9-6A, the data flow "h" is sent along the output branch by process G, and that process produces output "j." Thus, the module PUT-h can be factored into two processes as shown in Figure 9-8A: PUT-h sends "h" to module G. Module G produces a "j" and returns it to the control module PUT-h.

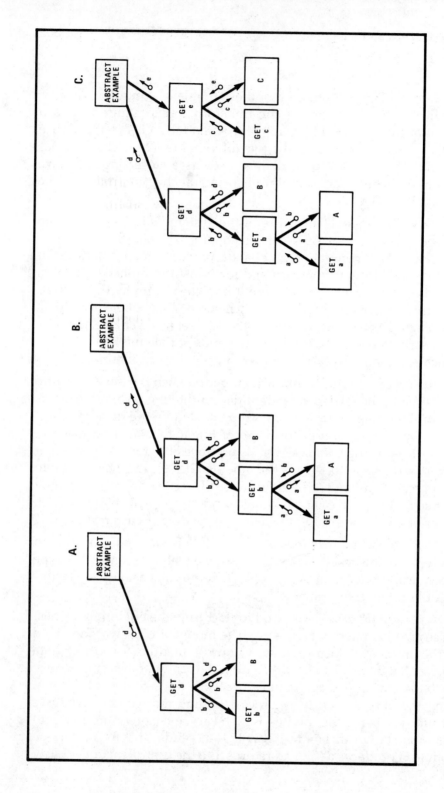

Figure 9-7. Factoring of afferent branches

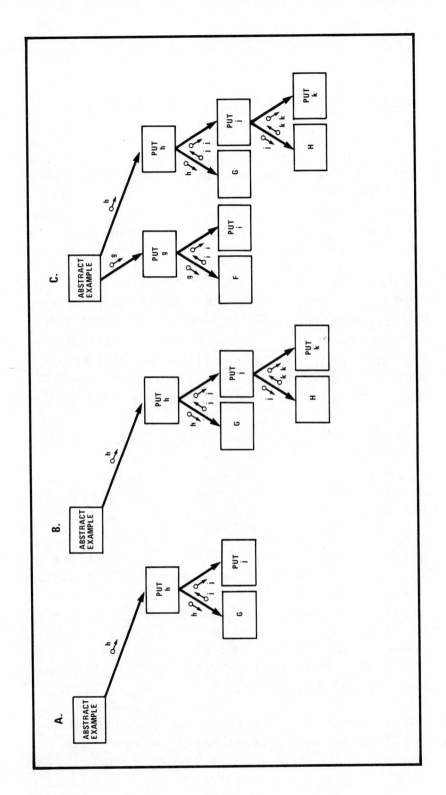

Figure 9-8. Factoring of efferent branches

A new control module PUT-j is identified to continue the process of factoring the efferent leg.

In a similar manner, the PUT-j module is factored as shown in Figure 9-8B. The result is a PUT-j module that represents some type of physical output. Finally, Figure 9-8C represents a complete factoring of both efferent branches.

For central transform branches, definition of a separate module for each bubble in a "middle-level" data flow diagram represents a starting point for partitioning. Beyond that, there are few specific guidelines. So, in many cases, central transform branches are factored following the same process as for functional decomposition. That is, the designer has to make a transition from a model that is problem oriented into program designs that are more closely machine related.

However, there are some rules of thumb that can be applied to determine when decomposition has gone far enough. Decomposition is complete when:

- There are no further identifiable subtasks to be added at lower levels.

- A standard, or library, routine can be applied to accomplish the functions for which a module has been identified.

- Modules are so small that they have reached a level at which coupling and cohesion guidelines would be violated by further partitioning.

Figure 9-9 represents the "first-cut" structure chart that results from applying the transform analysis process. In a real example, this structure chart then would be modified using some of the evaluation criteria discussed in Chapter 8.

Consider, now, the transform analysis process applied to a real example. The data flow diagram in Figure 9-10 models a loan quotation application. The task is to design a program to produce loan quotations. Figure 9-11 contains the relevant data dictionary entries.

The LOAN-REQUEST portion of the quote request specifies a desired loan type (auto or home, for example), the amount to be borrowed, the maximum payment the customer can afford, the frequency of payments and interest compounding, and a range of possible payback periods. For example, a home loan quote request might

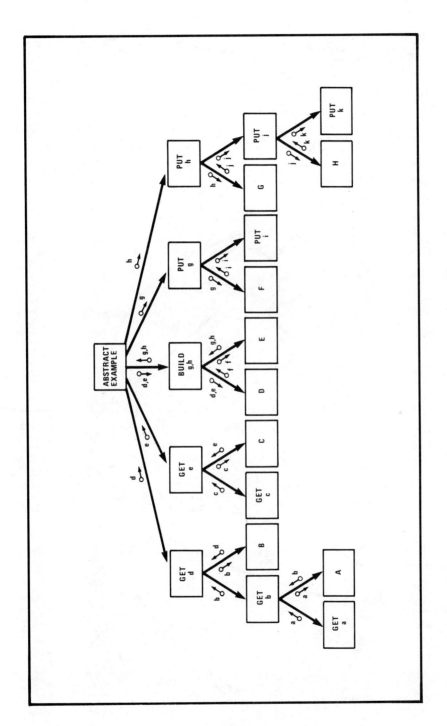

Figure 9-9. "First-cut" structure chart based on transform analysis.

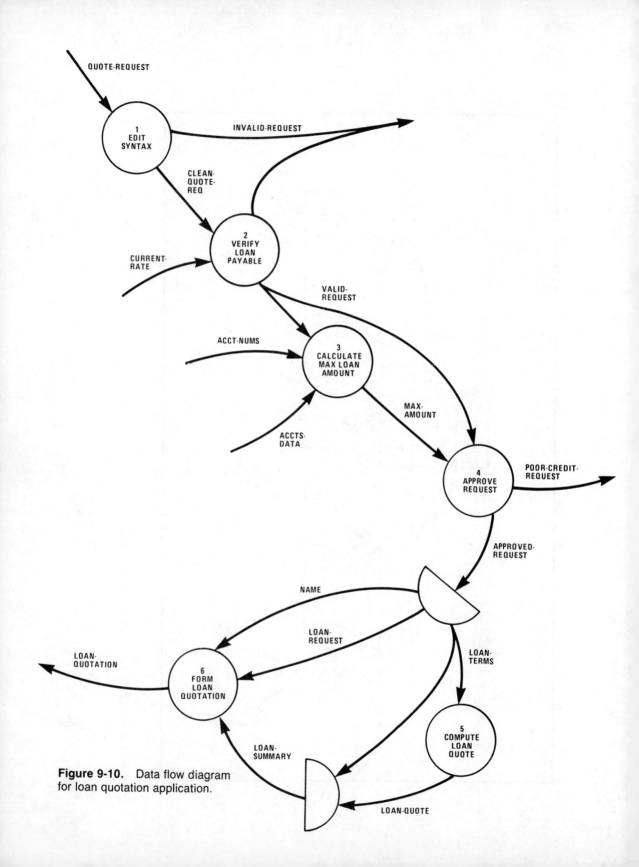

Figure 9-10. Data flow diagram for loan quotation application.

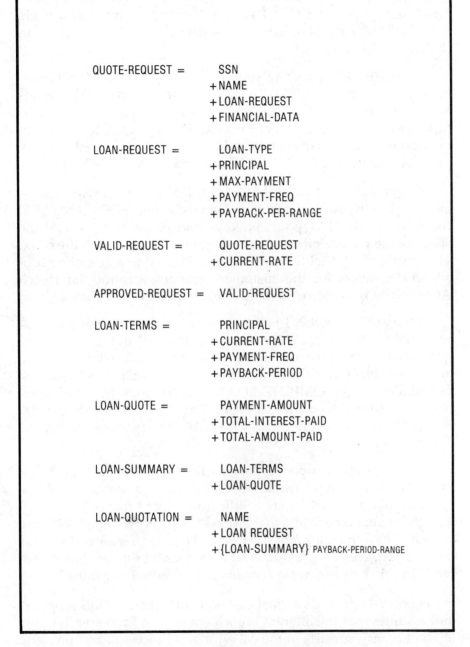

```
QUOTE-REQUEST =        SSN
                       + NAME
                       + LOAN-REQUEST
                       + FINANCIAL-DATA


LOAN-REQUEST =         LOAN-TYPE
                       + PRINCIPAL
                       + MAX-PAYMENT
                       + PAYMENT-FREQ
                       + PAYBACK-PER-RANGE


VALID-REQUEST =        QUOTE-REQUEST
                       + CURRENT-RATE


APPROVED-REQUEST =     VALID-REQUEST


LOAN-TERMS =           PRINCIPAL
                       + CURRENT-RATE
                       + PAYMENT-FREQ
                       + PAYBACK-PERIOD


LOAN-QUOTE =           PAYMENT-AMOUNT
                       + TOTAL-INTEREST-PAID
                       + TOTAL-AMOUNT-PAID


LOAN-SUMMARY =         LOAN-TERMS
                       + LOAN-QUOTE


LOAN-QUOTATION =       NAME
                       + LOAN REQUEST
                       + {LOAN-SUMMARY} PAYBACK-PERIOD-RANGE
```

Figure 9-11. Data dictionary entries for loan quotation application.

specify a range of payback periods from 25 to 30 years. In this case, the resulting LOAN-QUOTATION produced by the program actually would contain six sets of loan summaries: One each for payback in 25, 26, . . . 30 years.

Process 1: EDIT-SYNTAX simply does individual field and cross-field editing. The CURRENT-RATE, is read from a table and depends on the LOAN-TYPE, PRINCIPAL, and PAYMENT-FREQ. Calculations are done in Process 2: VERIFY-LOAN-PAYABLE to verify that the desired loan actually can be paid off in each requested payback period using the CURRENT-RATE and the MAX-PAYMENT.

Process 3: CALCULATE-MAX-LOAN-AMOUNT determines the amount that the bank is willing to loan, based on the LOAN-TYPE, the FINANCIAL-DATA submitted as part of the request, and the status of the various other accounts the customer has with the bank. The customer's Social Security number is used to access the set of account numbers for the customer's various accounts, and each ACCT-NUM is used to obtain the status information for that account.

If the MAX-AMOUNT is less than the requested principal, Process 4: APPROVE-REQUEST reduces the principal and accepts the request only for that portion of the payback period range for which the loan is payable. Using the resulting LOAN-TERMS for each payback period, Process 5: COMPUTE-LOAN-QUOTE does the calculations necessary to produce the corresponding LOAN-QUOTE details. Finally, Process 6: FORM-LOAN-QUOTATION assembles the several summaries into a single LOAN-QUOTATION.

The first step in applying the transform analysis process to this problem is to determine the afferent, efferent, and central transform branches of the model. There is little doubt that Process 1 is part of a single afferent branch and that Process 6 is part of the single efferent branch. The remaining questions are: How far forward does the afferent branch extend, and how far back does the efferent branch extend? In other words, what remains in the central transform?

Figure 9-12 presents a "first-cut" structure chart for this program and assumes that the afferent branch consists of Processes 1-4, the efferent branch consists of the collector and Process 6, and the central transform consists of Process 5. The router function is performed by the top module. The structure chart results from a mechanical application of the transform analysis process.

A closer look at the problem suggests changes in the structure chart. The input leg can be simplified by including the GET-CLEAN-QUOTE-REQ module in its parent and by letting the VERIFY-LOAN-PAYABLE module get the current interest rate for the particular loan type. In addition, superfluous data are being passed to GET-MAX-AMOUNT. The only data required are: SSN, FINANCIAL-DATA, and LOAN-TYPE. These changes are made in Figure 9-13.

When considering the two modules with heavy computation—VERIFY-LOAN-PAYABLE and COMPUTE-LOAN-QUOTE—a significant overlap is observed. The calculation COMPUTE-PAYMENT-AMOUNT has been factored out of these modules into a single computation module that, in turn, can be called by the other two modules. This change also is included in Figure 9-13.

Finally, the output leg in Figure 9-12 is awkward. For example, NAME and LOAN-REQUEST must be sent to PUT-LOAN-QUOTE on the first call for a customer, but not on subsequent calls. In Figure 9-13, the top module calls PUT-REPORT-HEADING to output the NAME and LOAN-REQUEST. The module then is responsible for collecting the individual LOAN-QUOTEs and forming the LOAN-SUMMARY.

It is helpful to understand clearly what the transform analysis process does—and does not—accomplish. The process separates the modules that deal with input and output from the main processing modules. Transform analysis further helps to organize the input and output legs. However, this technique does little for factoring the central transform, or main processing, of the program. As a first attempt, a separate module can be created for each process in the central transform. However, this approach clearly depends on the level of detail in the data flow diagram. If the diagram is too low-level, the result is too many small modules. On the other hand, a high-level data flow diagram probably will result in an incomplete factoring of the central transform. Thus, transform analysis does not provide a mechanical process that will result in optimum factoring of the central transform. However, basing the factoring on a ''middle-level'' data flow diagram does provide good initial guidance.

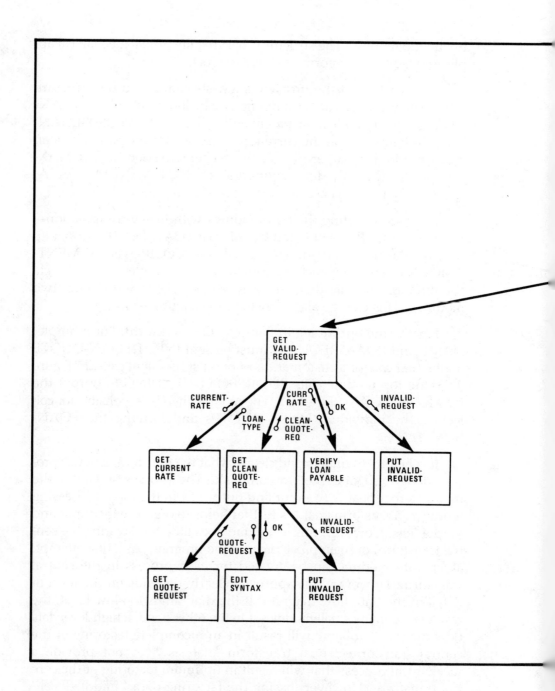

Figure 9-12. "First-cut" structure chart for the QUOTE LOANS program.

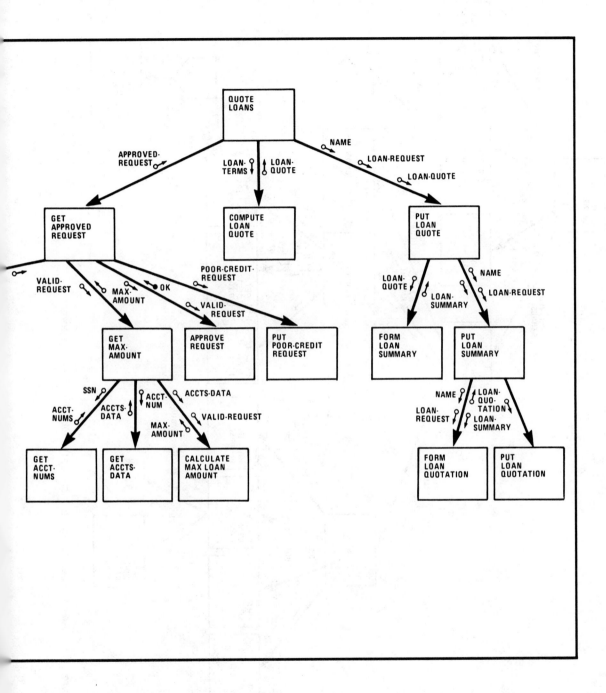

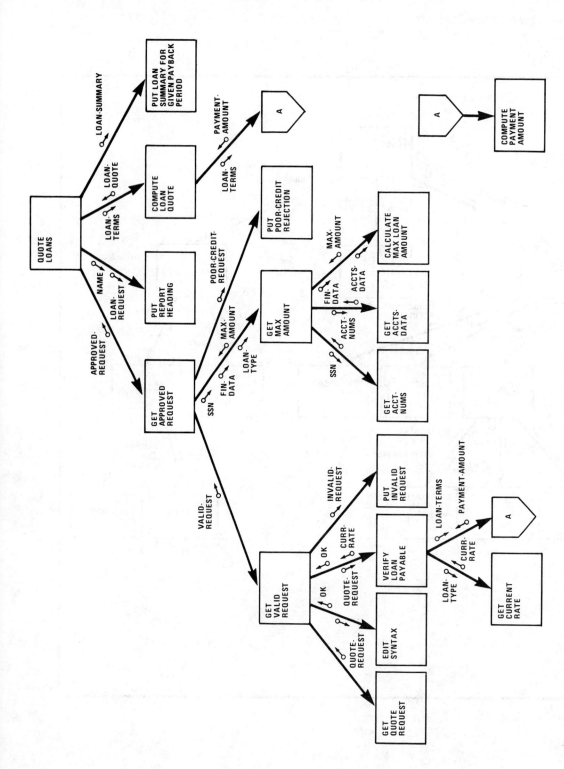

Figure 9-13. An improved structure chart for the QUOTE LOANS program.

Transaction Analysis

Often a program must process multiple types of transactions with extensive similarities in processing. A tempting design approach is to maximize the amount of common code by emphasizing these processing similarities and treating differences as exceptions. However, this approach can create maintenance problems because business changes tend to center around the transactions, not around the processing similarities. Thus, if a transaction is changed, large portions of the program also may have to be changed. Transaction analysis is a dataflow-oriented design process that organizes the design around the transactions in the higher-level modules and recaptures processing similarities using lower-level modules with high fan-in.

The general structure of a transaction-centered design is shown in the partial structure chart in Figure 9-14. The top module simply recognizes the transaction and routes it to the appropriate processing module. Each processing module knows the detail of that particular transaction—its data and the processing steps that must occur—but is unaware of the other transactions. The actual processing of the transactions is done by a third level of modules that process individual data structures. It is at this level that some benefit from processing similarities may start to occur. Thus, common detailed processes may be factored out of the third level modules to achieve even greater benefit from processing similarities.

The data flow diagram corresponding with a transaction center typically will show a data flow that branches where parallel processes handle each transaction type. Figure 9-15 shows a portion of a data flow diagram corresponding with the partial structure chart shown in Figure 9-14.

The process of deriving the structure chart from the data flow model is straightforward. Basically, the designer must be able to recognize situations in which multiple transactions are an inherent part of the logical or business-related processing. A transaction in this sense is any data structure that can assume one of a number of types. Several examples of common data processing transactions include:

- The most common type of transaction center occurs in a file maintenance program. An input transaction can be one of three types: an add transaction to insert a new record in the file, a

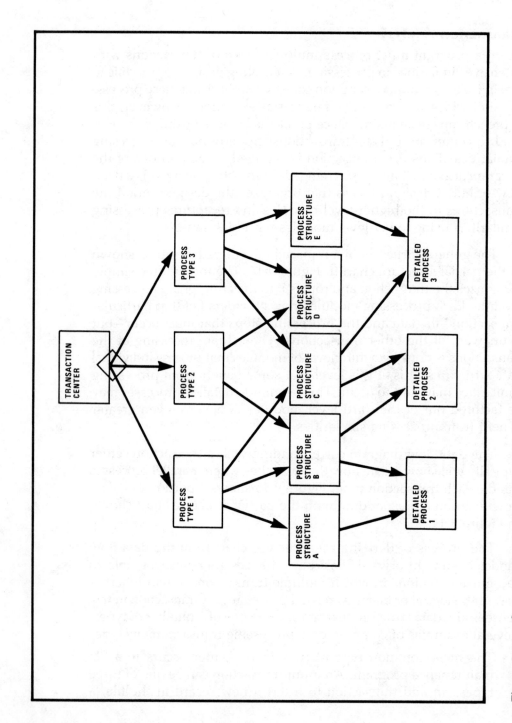

Figure 9-14. General structure for a transaction-centered design.

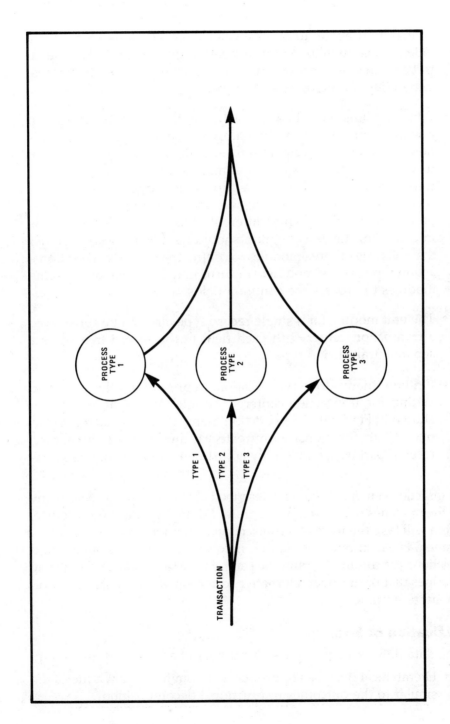

Figure 9-15. Partial data flow diagram corresponding with Figure 9-14.

change transaction to modify values in an existing record, or a delete transaction to remove a record from the file. The single input stream would be routed to one of three processing routines, depending on the transaction type.

- Edit programs typically process transactions of several types in a single input stream. An inventory system may process transactions representing receipts from vendors, parts usage, and inventory adjustments in a single stream. In a student records system, transactions for adding a class, dropping a class, and changing registration status between regular grading and pass-fail may be processed together. While different transactions will have fields in common, the edit program should have a transaction center that calls one of three modules to control the editing of each transaction type. These modules, in turn, may call some of the same modules to process the common fields.

- The edit module for a single record type also can be considered a transaction center with each field in the record treated as a separate transaction type.

- On-line, menu-driven applications represent another common setting for transaction-centered design. A typical example is shown in Figure 9-16. A function menu allows the user to select one of four functions, each treated as a transaction: inventory, receipt, part usage, inventory adjustment, and part status query.

Transaction centers of the type described above can appear within any of the branches of a data flow diagram. Editing routines, for instance, often will take the form of a transaction center located in the afferent branch. File maintenance routines involving additions, changes, and deletions are common within the central transform branch. Reporting routines that print under alternative formats usually would appear in an efferent branch.

Evaluation of Strategy

The data flow strategy presents a number of advantages, including:

- Hierarchical designs are produced through top-down methods, similar to the outcomes of functional decomposition.

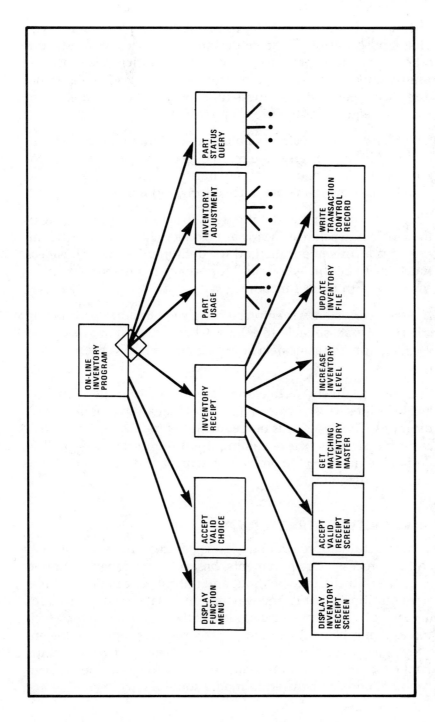

Figure 9-16. Structure chart for an on-line inventory program that presents a menu of transactions for the user.

- The methodology is sensitive to the design implications of coupling and cohesion. Because designs are based on an identification of business processing functions and their relationships, connections between program modules will have similar structures. Changes in the business functions can be traced easily to changes required in the program structure.

- Criteria for decomposition of afferent and efferent branches are included in the design strategy. In effect, input and output procedures are localized within separate branches of the system, thereby improving maintenance efforts and effects.

- There is assurance that the program structure corresponds with the problem structure as nearly as possible. The data flow diagram provides a convenient method for defining the major business functions to be carried out and for organizing the system in accordance with these functions.

- A design consistency is enforced because of uniform partitioning criteria. It is reasonable to assume that two or more designers working from the same data flow diagram will deliver comparable results.

- There are relatively reliable, mechanical means for deriving top-level designs of structure charts from analyses of data flow diagrams. The designer is not required to produce tentative, trial-and-error designs before coming upon the proper design. The design is inherent in the problem statement as documented in the data flow diagram.

DATA STRUCTURE STRATEGIES

The data structure approach to design is based on the principle that application processing requirements, basically, are dependent upon the structure of the data to be input and output. The data structures of an organization—that is, the way in which data are organized for input, storage, and output—are models of the organization for which the application is being developed. Data structures represent the entities the organization deals with as data items and relationships among those items. Thus, data structures serve to model the problem environment and can be used to model solutions.

The rationale behind the approach is that data structures have important similarities to processing structures. For example, data structures can be represented as simple sequences, repetitions, and selections organized hierarchically. A file is a repetition of records that, in turn, are sequences of fields. In some cases, records have two or more formats, in which case there is a selection structure. The same relationships, of course, pertain to program structures. Thus, it is possible to map a direct relationship between data structures and program structures.

Figure 9-17 illustrates this type of structuring of data within an input file. This same form of logical structuring occurs within system output. As shown in Figure 9-18, an output data structure such as a report can be represented as a hierarchy of data components related through sequences, repetitions, and selections. In this case, the report is shown to be comprised of a sequence of a report body and a set of report totals. The body of the report, in turn, is made up of repeated occurrences of detail lines, for which one of two different formats may appear. Each detail line and the report totals are composed of sequences of data fields.

Program structures can be derived from these data structures by considering the correspondence between input and output structures, as illustrated in Figure 9-19. At each level within the input and output data structures, program modules are defined to transform an input structure into its corresponding output structure. Thus, the hierarchy of the program is determined by the hierarchy of the data structures it processes.

Data Structure Technique

The data structure approach to software design follows a four-step process:

1. Define the logical structure of the input and output data.
2. List the processing operations that will have to appear in the program.
3. Devise a program structure based on the correspondence between the input and output data structures.
4. Allocate the processing operations to the appropriate modules in the program structure.

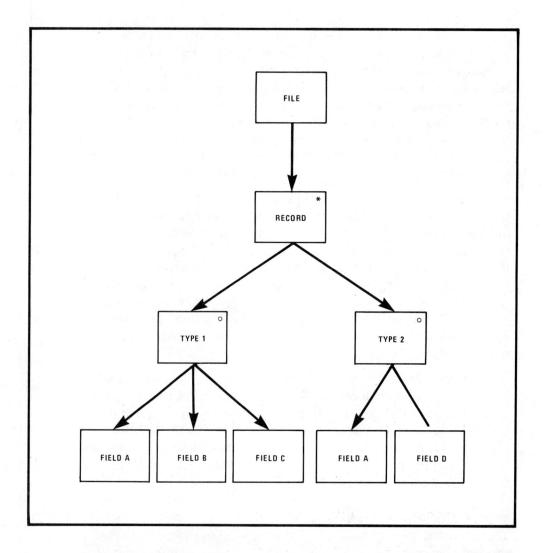

Figure 9-17. Input data structures such as transaction files often have hierarchical relationships between logical components (records and fields) that can be expressed as sequences, repetitions, and selections. Note the asterisk symbol (*) means an iteration of the structural component and the degree sign (°) indicates that one or the other component is present.

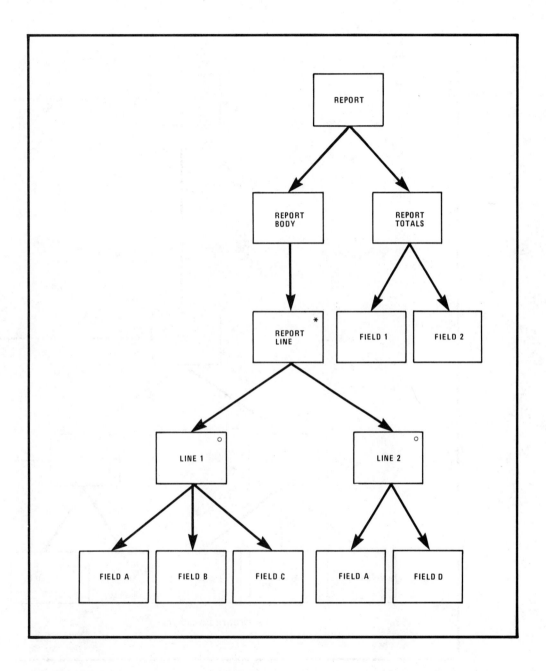

Figure 9-18. Output data structures such as reports can be represented as hierarchies of data components that are related as sequences, repetitions, and selections.

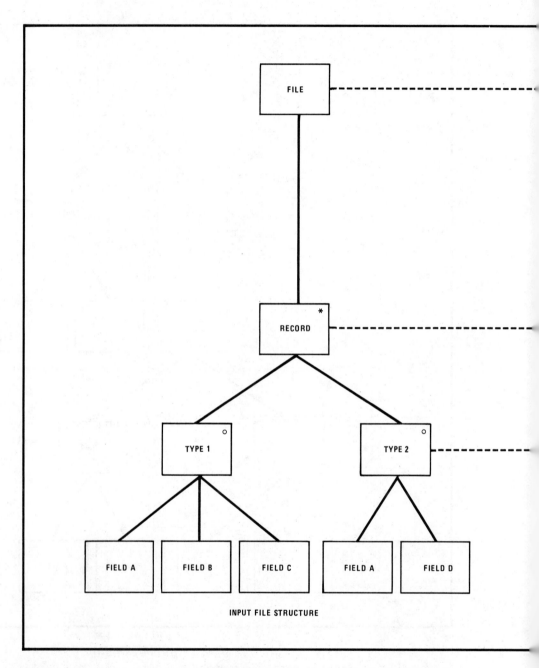

Figure 9-19. The data structure approach to design defines modules that produce output structures from corresponding input structures.

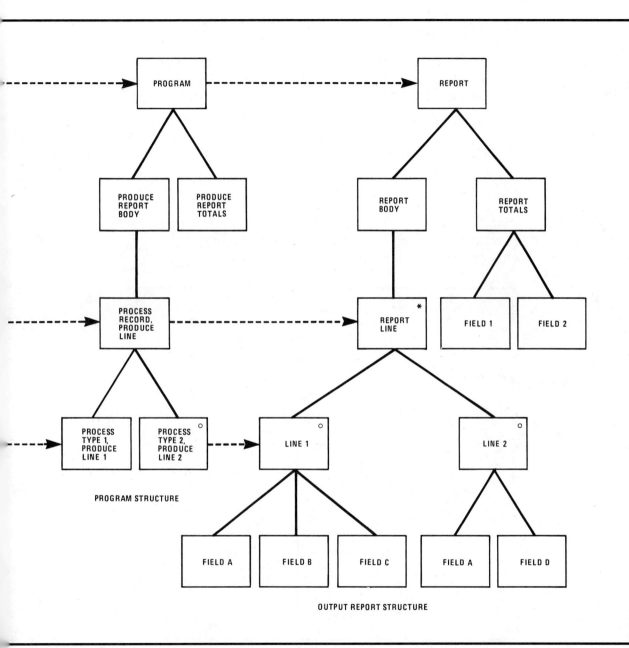

PROGRAM STRUCTURE

OUTPUT REPORT STRUCTURE

As an example to illustrate this strategy: one of the most common types of business applications that involve hierarchically structured input and output data is the *control break* report. The input file is composed of two or more levels of control keys and the output report has two or more levels of totals corresponding with the key groups.

A typical example of such an input file would be a sales file in which individual sales records are keyed to the salesperson number, the department number, and the division number within the company. Thus, hierarchically, the organization of the file is such that individual sales records are grouped by departments, which, in turn, are grouped by division. Logically, the input data structure is a repetition of salesperson records within a repetition of departments within a repetition of divisions. These relationships are illustrated in data structure notation in Figure 9-20.

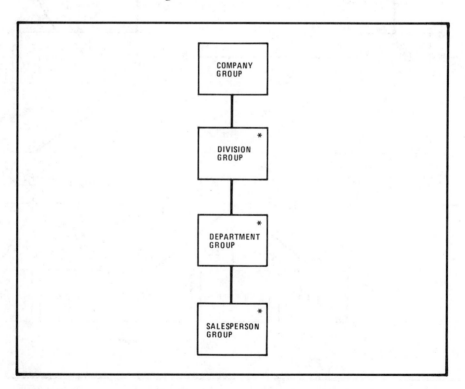

Figure 9-20. Structure of an input file with repeating groups organized by salesperson number, within department number, within division number.

The output record has a similar structure. Each detail line corresponds to an individual sales record. A collection, or repetition, of individual lines comprise a department report, which includes a departmental summary line. The collection of departmental reports comprises a division report with its accompanying summary total. The logical structure of this output report is shown in Figure 9-21.

After the structure of the data has been determined, the designer prepares a list of processing operations that will be included in the program. The following list applies to this program:

ONCE PER PROGRAM
    OPEN FILES
    PRINT DEPARTMENT HEADING
    INITIALIZE COMPANY TOTALS TO ZERO
    PRINT COMPANY TOTALS
    CLOSE FILES

ONCE PER DIVISION
    INITIALIZE DIVISION TOTALS TO ZERO
    PRINT DIVISION HEADING
    ACCUMULATE DEPARTMENT TOTALS INTO DIVISION TOTALS
    PRINT DEPARTMENT TOTALS

ONCE PER SALESPERSON
    INPUT SALESPERSON RECORD
    PRINT SALESPERSON LINE
    ACCUMULATE SALESPERSON TOTALS INTO DEPARTMENT TOTALS

The next step is to create a program structure that is modeled after the data structure. The designer looks for structural correspondence between the input and output structures and defines modules that will transform one structure into the other. The structure chart shown in Figure 9-22 would be produced for the control-break example. Note

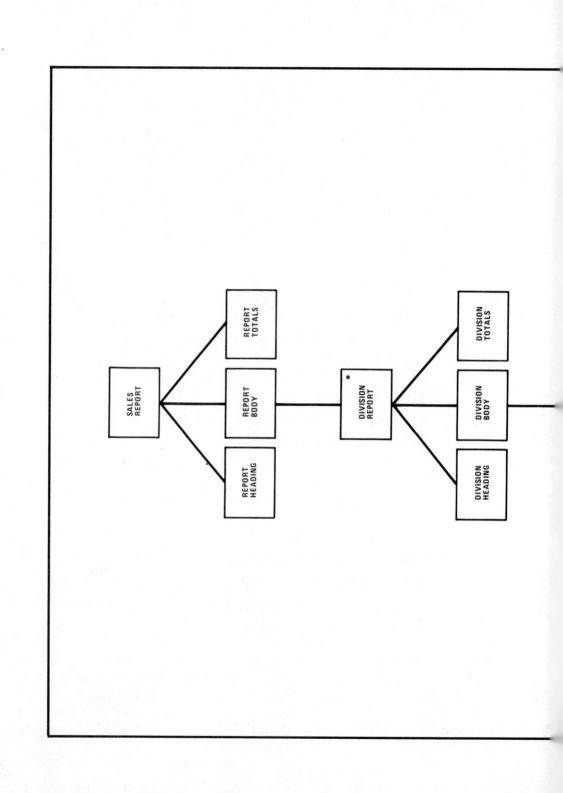

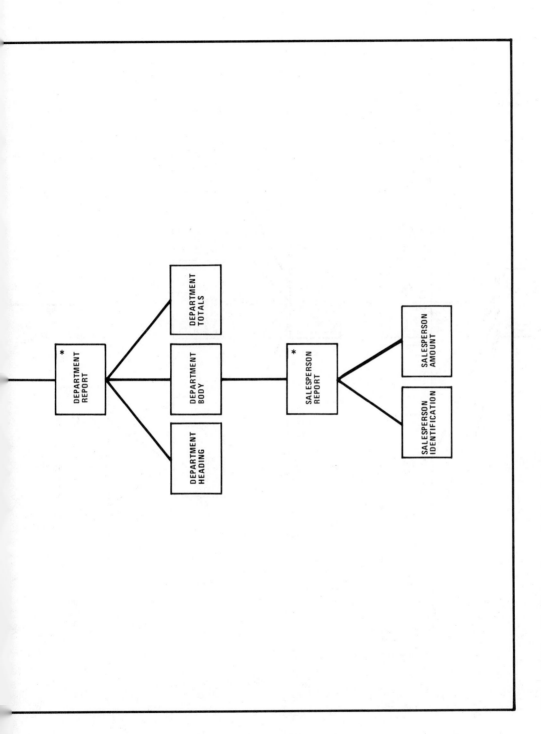

Figure 9-21. Structure of a control-break output report.

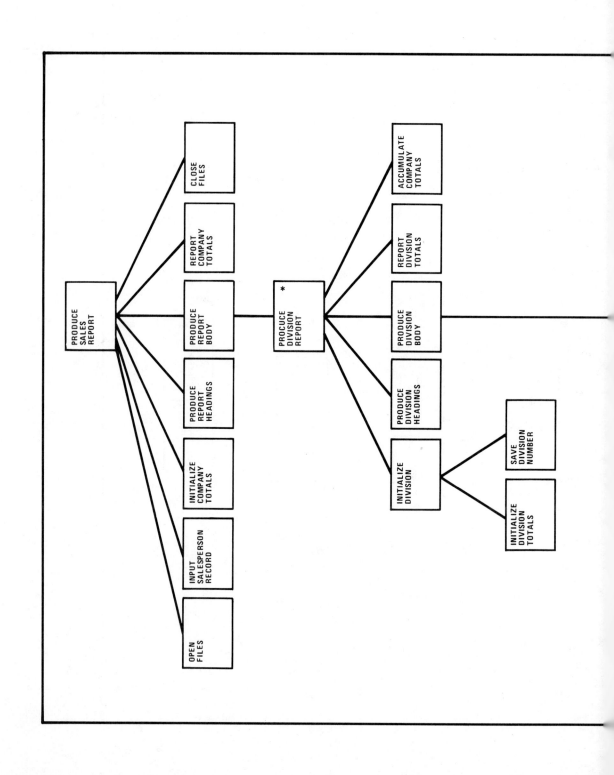

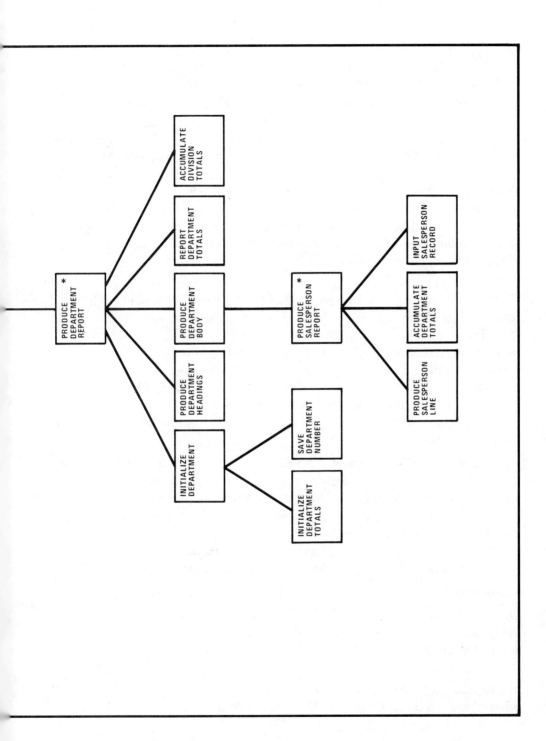

Figure 9-22. Structure of a control-break program designed through data structure method.

that each level in the structure chart contains modules that will transform the input structure for that level into the parallel output structure. In this case, there is a one-to-one correspondence among the input structure, the processing structure, and the output structure.

After the program structure has been derived, the processing operations defined earlier are assigned to the modules. This assignment is a relatively straightforward process; the modules at each level suggest the processing operations that should be included. Of course, there will be other computer operations that will have to be inserted into the program. These operations include the calls to the lower-level modules, specific formatting and printing routines, movement of keys into save/compare areas, and other such statements that make the program completely operational.

If the input and output data structures are similar, program structures are relatively easy to define. Problems can arise, however, if there are *structure clashes.* A structure clash, basically, is a noncorrespondence between input and output data structures. An example of a structure clash occurs regularly when a transaction file is created serially, in a time-oriented recording sequence. For processing against a master file, a sequential ordering of records is needed. Thus, the serial structure of the input file clashes with the sequential order of the output file. This clash must be resolved before a program structure can be designed.

Structure clashes are resolved by reordering the input file so that its structure matches that of the output file. This reordering may include writing a separate program to decompose the input data structure into elements that form an intermediate data structure that does not clash with the output structure. Reordering also could involve using a sort routine to reorder the file or using internal tables for placing the file in a structure that corresponds with the output file.

Evaluation

One of the appeals of the data structure approach is that software designs can be derived in a straightforward fashion from known entities. Procedures are relatively mechanical rather than judgmental.

The data structure approach appears appropriate for systems that are supported by common hierarchical data structures—particularly when there is a direct correspondence between input and output data structures. The correspondence of the data structures, in turn, provides a set of criteria for determining any given design. Conceptually, designs are relatively simple. That is, no software structure developed under this strategy is any more complex than the simplest kind of sequential batch processing program. In fact, the approach assumes that only sequential file processing techniques will be employed, even though file structures may exist as serial, sequential, direct, indexed, or database organizations.

A challenge in using the method lies in the need to resolve the structure clashes when they are encountered. Techniques to normalize the structures, or remove redundancies, can become much more complex than the original problem itself.

APPLYING THE DESIGN APPROACHES

It should be obvious from the preceding discussion that there is no best method for software design. There are alternative strategies that can be followed, each of which has particular advantages and disadvantages and through which different designs will be produced. These resulting designs will exhibit varying degrees of coupling and cohesion that make it difficult to compare the respective approaches. Naturally, the designer is looking for the best method to apply; however, it certainly is not practical to develop three separate designs using each of the strategies and then compare them.

For lack of definitive criteria for choosing a design strategy, possibly the most practical solution is the following:

- Apply the data flow strategy for initial program partitioning. Develop a data flow diagram describing, in business processing terms, the major procedures and the data flows into and out of those procedures. Carry data flow diagramming to the level at which it becomes difficult to document the processing without considering computer processing techniques or logical program control structures. Where it is appropriate, produce data flow diagrams describing transaction centers. Then, use the resulting diagrams for first-cut factoring of the system into major processing routines.
- Factor the program on the basis of data flow and functional decomposition techniques. Consider, for each processing routine,

specific computer techniques for implementing the function and the logic of sequences, repetitions, and selections that must be employed to overlay an execution structure on the functional structure of the system. For each design decision, consider the effects upon coupling and cohesion.

- Throughout the design process, look for occasions to utilize data structure approaches. In particular, modules that access hierarchically structured input data and that produce hierarchically structured output data become candidates for data structure design techniques.

- Once a design has been developed, apply cohesion and coupling criteria to the entire structure. Pay particular attention to input and output interfaces. These interfaces offer possibilities for employing information hiding techniques to buffer the program from changes in the external environment.

Application software design, in summary, can be viewed as a skill, or craft, that is still under intensive development. The goal of study in this area is eventually to replace intuitive, creative, and individualistic skill with standardized, predictable, and mechanical methods for developing software. Ultimately, these results will be achieved. However, today, in most shops, mental effort is the primary tool of the software designer. The best available mechanisms for guiding the development process guide the thinking of the designer rather than prescribe results.

The primary motivation for the development of the design strategies reviewed above is the alleviation of maintenance problems. It has been estimated that up to 80 percent of programming effort in the typical computer department is devoted to the modification of existing software. Consequently, new applications average 24 to 36 months behind schedule. The solution to this dilemma, it is reasoned, is to design software that is easier to modify and maintain, thereby releasing human effort to development of new applications. Maintainability has been a continuing theme throughout this discussion and is evident in much of the rationale presented for the design approaches.

The magnitude of the maintenance problem is considerable. The problem is so large that, realistically, the impact of any design strategy is marginal at best. What will be required are order-of-magnitude changes in approaches to software development. Such changes will come as the need disappears to develop coded programs using traditional higher-level languages. Changes will be brought on by the practicality of using database software and a much higher-level instruction set for defining and organizing processing functions. Until that time, however, the strategies and techniques reviewed here will serve to bring some order and logic to a most difficult process. Although these design approaches may not completely solve the maintenance problem, they still represent systematic methods for developing effective, quality software.

Summary

The objective of software design is to produce designs that effectively package statements into modules and into programs. Design decisions must minimize coupling and maximize cohesion between modules.

Approaches for developing software designs include functional decomposition, data flow, and data structure.

Functional decomposition focuses on the identification of processing functions through a process of top-down partitioning. Abstract functions are decomposed into more detailed functions, proceeding successively down the program hierarchy. Data flow approaches identify functional components by focusing on the transformations applied to data as they flow through the system. A data structure approach considers the logical structure of the data to be processed by the program as the framework for deriving a parallel program structure.

In functional decomposition, program modules represent processing tasks defined at several levels of abstraction. At the top levels, decomposition is carried out in light of the business functions that are to be performed. At lower levels, these business functions must be translated into computer processing techniques. There is a transformation of the design from basically a problem-related structure to a machine-related structure.

In applying functional decomposition, the criteria used to guide partitioning are the business functions that are to be implemented. However, no formal guidelines exist and the resulting design may be based on the individual judgment of the programmer.

The data flow approach, or transform analysis, is an adjunct to functional decomposition for determining just how to decompose a problem structure and derive a parallel program structure. The flow of data through a system and the transformations applied to those data are used in determining the overall relationships among problem functions. Data flow diagrams are used to model the problem. Design activity translates that business model into a hierarchical organization of software modules that preserve the model of the original problem.

Design based on data flow begins with definition of a top-level module. Then, first-level factoring identifies the major processing functions. Further analysis involves identification of afferent branches (program inputs), efferent branches (data flows to physical outputs), and central transform branches (remaining portions of the diagram).

Decomposition is complete when there are no further identifiable subtasks to be added at lower levels, when a standard, or library, routine can be applied to accomplish the functions for which a module has been identified, or when modules are so small that they have reached a level at which coupling and cohesion guidelines would be violated by further partitioning.

The use of data flow diagrams is limited to determining the overall logical structure of a program rather than determining its execution structure. Beyond establishing this logical structure, the designer must apply other techniques such as functional decomposition to factor the design to the lowest levels.

The data structure approach to design is based on the principle that application processing requirements are dependent upon the structure of the data to be input and output.

In applying the data structure approach, the programmer must: define the logical structure of the input and output data, list the processing operations that will have to appear in the program, devise a program structure based on the correspondence between the input and output data structures, and allocate the processing operations to the appropriate modules in the program structure.

The data structure approach appears to be appropriate for systems that are supported by common hierarchical data structures. This is particularly true when there is a direct correspondence between the input and output data structures.

A coordinated design approach would be to apply the data flow strategy for initial program partitioning, factor the program on the basis of functional decomposition techniques, look for occasions to utilize data structure approaches, and then apply cohesion and coupling criteria to the entire structure.

Key Terms

1. functional decomposition
2. data flow approach
3. data structure approach
4. stepwise refinement
5. level of abstraction

6. transform analysis
7. transaction analysis
8. first-level factoring
9. afferent
10. efferent
11. central transform
12. control break
13. structure clash

Review/Discussion Questions

1. What is the objective of software design?
2. What are coupling and cohesion, and how may these criteria guide the partitioning process?
3. What are three approaches used for developing software designs?
4. What is the process of functional decomposition?
5. What are the strengths and weaknesses of functional decomposition?
6. What is the data flow approach to software design?
7. In applying a data flow approach, what is involved in the identification of afferent, efferent, and central transform branches of a data flow diagram?

8. Why is a data flow approach applicable only to high-level program modules?

9. What is the data structure approach to software design?

10. When dealing with a data structure approach, what is a structure clash?

11. For what types of design problems is a data structure approach most appropriate?

12. How may the approaches of functional decomposition, data flow, and data structure be coordinated in developing a software design?

Project Assignments

The Appendix presents three case studies that can be used in support of project assignments. All three scenarios provide functional system specifications that would result from systems analysis. Included are data flow diagrams, data dictionaries, and process narratives that would result from systems analysis. Continuing the work begun in Chapters 6–8, assignments that relate to the content of this chapter will be found in *PART 3: Detailed Systems Design*.

TEST SPECIFICATIONS AND PLANNING 10

LEARNING OBJECTIVES

On completing reading and other learning assignments for this chapter, you should be able to:

☐ Describe the types of testing for which specifications are developed in this activity.

☐ State the objectives of the test specifications and planning activity.

☐ Describe the scope of this activity.

☐ Give the end products of this activity.

☐ Tell what personnel are involved in preparing test specifications and planning tests.

☐ Tell how testing relates to top-down design strategy.

ACTIVITY DESCRIPTION

The preparation of test specifications is a planning-oriented activity. Therefore, its position within the systems development life cycle overlaps Activity 7: Technical Design, and Activity 9: Programming and Testing. Thus, at the start of this activity, a technical, or implementation-oriented, design has been developed for the new system. Before programming can begin, standards must be set for acceptability of the programs that will become the technical implementation of the user specification. It should be stressed that the specifications and plans developed during this activity go beyond program tests. All of the manual procedures, the forms, and the verification of value for the results produced are encompassed in the specifications and the test procedures that are evolved. The specifications and procedures for tests that are developed in this activity are applied at several levels:

- *Unit testing,* or *module testing,* is applied to individual program modules. Tests determine whether the modules are logically and functionally sound. Unit testing is done by using the modules to process test data and examining the outputs to determine that results are as expected.

- *Integration testing* is applied to interfaces between modules. This type of testing is done in parallel with unit testing. That is, as individual units are tested and accepted, the modules then are run together to determine that the interfaces between them are workable and produce expected results. Integration testing tests the transfer of both data and control among modules for a program.

- *Function testing* seeks to identify any variances between the results of program processing and the specifications for the programs agreed to and approved by users. Function testing concentrates upon the results produced by complete programs to be sure that results meet user expectations and requirements.

- *System testing* deals with the integration of a system, or integrated group, of application programs with system software, hardware, peripherals, manual procedures, and any other system components. This category of testing extends beyond the computer system to encompass all related procedures and processing. The idea of this kind of testing is to try out the system as an operational and functional entity. Thus, the tests check to see whether the training and reference manuals are adequate to cover instructions and operational problems that might arise. A system test simulates actual operation of the entire system prior to its conversion and continuing use. In effect, the system test is a preview of how the overall system will work. This testing is for throughput, capacity, timing, and backup and recovery procedures and is applied under control of CIS operations personnel just before the system actually is turned over to users.

- *Acceptance testing* is performed by the user. Such testing occurs immediately prior to operational status. In effect, an acceptance test is a dress rehearsal to be sure that the users are ready to bring the new system into operation and to use it for ongoing production.

The first three of the tests identified above deal with application software. The specifications and the data to be used for these tests are incorporated in software design activities, as discussed in subsequent chapters. The first of the tests, unit testing, deals with the lowest level of program entity, the module. To carry out tests at this level, data are selected that apply and test the workability of the programming logic for the instructions within each module. The second level of testing, then, deals with the capabilities for pairs and groups of modules to interact with one another. The dimensions tested deal with the passing of data and control among modules. Function testing deals with the level of programs at which identifiable end products are produced. A function, in this sense, is a program that accepts input and produces usable output. The programs involved can be complete systems or parts of larger systems of programs. The standards applied in measuring the results of tests are user expectations.

The last two levels of testing described above deal with multiple programs that comprise complete processing systems. The philosophy represented by the system test is that CIS professionals should satisfy themselves that the systems of programs are performing to expectations before the systems are turned over to users for acceptance tests under conditions that simulate actual use. Even though system tests are performed by CIS professionals, the criteria for evaluating the results come from user-approved documentation and specifications.

OBJECTIVES

Three important objectives should be met through completion of the tasks in this activity:

- Define all conditions that must be tested at the program, subsystem, and system levels.

- Prepare all necessary test data and create test files to be applied during Activity 9: Programming and Testing.

- Prepare a system test plan for coordinating the tasks of Activity 11: System Test.

SCOPE

This activity accepts as input the technical specifications produced during Activity 7. These specifications are reviewed. The individual programs and program modules are identified as work units for which test specifications and test data must be assembled.

The activity then encompasses the identification of all products to be tested during Activity 9 and the preparation of data for the testing of each.

At the conclusion of this activity, a set of documentation should exist that includes specification and test data. As appropriate, the test data may be incorporated in input files that can be used in actual test processing. These tests should cover the complete application system from the lowest module level through to complete implementation of computerized and associated procedures.

END PRODUCTS

Separate sets of end products are produced to incorporate specifications and plans. The specification products include the following:

- *Program test specifications* include extensive, detailed lists of all conditions that must be tested. For each specification, there is a related reference to the test files to be used.

- *Test data files* are prepared for each program. Emphasis is upon the quality of data in terms of the specific properties of each module or program to be tested.

- *Job stream test specifications* list the conditions under which each job stream is tested. These specifications assume successful completion of tests for individual modules and programs. Again, corresponding test files are referenced in these specifications.

- *System and subsystem test specifications,* as well as any other required testing procedures, are documented, with accompanying cross-references to test files.

In a separate set of documents, a *test plan* outlines the steps to be followed in testing an application. Components of a test plan include:

- The steps, or phases, of a test plan are spelled out and objectives are stated for each phase.

- Completion criteria, or acceptance criteria, are established for each testing task.

- Complete schedules are developed for testing tasks.

- Responsibilities for performance of testing tasks and completion of tests and their covering documentation are prepared.

- Provision is made for the computer time needed to perform the tests.

- An overview document encompasses plans for integrating all of the tasks associated with program and system testing.

THE PROCESS

In effect, the process followed in testing programs and systems is in the inverse order of the sequence of work performed as part of the systems development life cycle. These relationships are represented graphically in Figure 10-1. As shown, the progression of work in systems development moves from systems analysis, to systems design, to software design, and to the design of individual modules. The preparation of specifications and the performance of program tests, by contrast, moves in the other direction. That is, the first testing activity is unit testing. Unit testing is applied to the results of module design, the last of the systems development procedures. This inverse relationship follows through to the final testing activity, the acceptance test, which is applied to the product of systems analysis.

This relationship stems from the nature of systems development itself. That is, as each product is developed during the systems development life cycle, the criteria for evaluation and testing that product also are prepared. This correspondence between development and quality assurance is inherent in the process of systems analysis, systems design, and the overall development cycle.

Implementing the Project Plan

The underlying quality assurance that is part of the entire systems development life cycle also is built into the documentation that forms

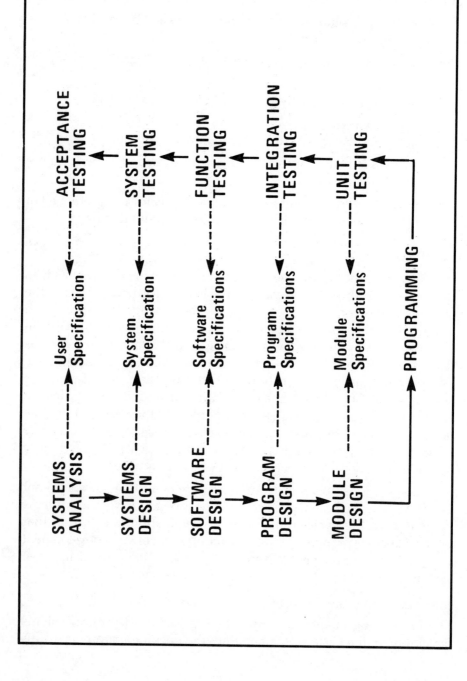

Figure 10-1. Relationships between systems development and testing activities.

the test plan. Note that there is a hierarchy inherent in the basic approach to testing. Systems and systems of programs are designed on a top-down basis. Testing, however, typically proceeds on a smallest-to-largest basis, following, in effect, a bottom-up approach to testing that is the inverse of the design strategy. Thus, modules are tested individually. Then, the interfaces between modules are tested, followed by tests of programs, and then the entire system. Finally, an acceptance test is done under conditions that closely resemble actual use.

At each stage in this testing plan, there are objectives to meet and criteria to satisfy. These criteria are established by a philosophy that tests the full range or scope of each module, group of modules, program, and system. Accordingly, the criteria for acceptance are inherent in the methodologies followed in the development of specifications and test data, as described in subsequent chapters.

The philosophy for the acceptance of test results is noteworthy in its practicality. That is, it is a fair assumption in any set of logical procedures as complex as a system of programs that it will be impossible either to test a system totally or to find all potential errors that may exist within the system. However, recognizing this limitation, it is essential that some sort of reasonable assurance of performance be designed into the test plan. So, the approach taken is to determine the parameters, or boundaries, of each module, of groups of modules, of programs, and of the system as a whole. Testing is at the boundaries, rather than over a full range of all possible combinations of data or processing conditions. That is, it is assumed that, if all programs are tested to the outer limits of their performance, there is a reasonable assurance of reliability.

In applying tests to systems and their components, it is inevitable that errors will be encountered. Programs and procedures do not anticipate all logical paths or occurrences. It would be impossible to predefine the total conditions that may be faced by any complex system of programs. Thus, testing becomes a matter of identifying and correcting errors as they occur. Following correction, retesting takes place.

As a test program moves through its cycle of finding and correcting errors, each correction, in turn, can cause situations that create new errors. Thus, for some period of time, the incidence of errors encountered usually will increase. There is no way to say, for any given program or program element, how long this process will continue. However, it is a typical pattern that the incidence of errors grows as a testing program is launched. Then, at some point in time that seems to vary for different programs, the incidence of errors begins to decrease. An important part of testing philosophy and policy lies in knowing when the incidence of errors has decreased to a level that makes a system or component of a system acceptable. Theoretically, it could be possible to keep testing until no more errors occurred. However, to follow a test program through to this level would be impractical and exorbitantly costly. If testing were done to this level, it is literally possible that a system would never be implemented. Rather, some time after the incidence of errors begins to decrease significantly, a decision must be made about the existing level of confidence. When the level of confidence in the system reaches a point of acceptability, testing can be terminated. This level of confidence may be reached a relatively short time after the decrease in error incidences is noted.

Thus, the level of errors and the pattern of errors should dictate the cycle for testing of any given program component or system. This is the only measure that should be followed. Attempts are sometimes made to set time limits on testing programs. Arbitrarily, for example, it may be decided that a given program can be tested in two weeks. At the end of this time, the plan would call for discontinuing the testing. It is not practical to predict or commit to the establishment of quality on a time basis only.

When a testing program reaches the point where the entire system is to be exercised as a single, coherent entity—system test—a special objective viewpoint should be brought to bear. Although the system is not ready, at this point, to be put into user hands, some outside objectivity is desired. A sound, often-adopted practice is to have system test procedures performed by key users and CIS professionals who were not part of the project team. Occasionally, outside consulting organizations are asked to perform system test functions.

From a process standpoint, the acceptance test can be a critical juncture in the activities within a systems development life cycle. This testing activity involves broad, active user participation, with the users beginning to assume control over the system. Thus, in planning for acceptance testing, provision should be made for full briefings and preliminary training of user personnel who will be involved.

PERSONNEL INVOLVED

For the most part, the same persons who complete the work in Activity 7: Technical Specifications are responsible for completing the tasks of this activity. Ideally, an independent group should perform most of the test specifications and planning. Persons who do the design and programming work approach these tasks with the expectation that their final products will work correctly. There may be reluctance to test vigorously in an attempt to verify the incorrectness of the system. Independent testers would be less hesitant about proving problems with the system. As the activity draws to a close, members of the user group should be involved in establishing the specifications and data to be used in acceptance testing.

CUMULATIVE PROJECT FILE

Existing documentation already in place in the cumulative project file is not affected directly by this activity. However, parts of the cumulative documentation file, specifically those statements that spell out performance criteria for the new system, become the guidelines for development of test specifications and plans. Thus, the documentation created during this activity—specific test specifications, a testing plan, and files of test data—enhance and build upon the cumulative project file.

Summary

The specifications and procedures for tests in this activity involve unit testing (or module testing), integration testing, function testing, system testing, and acceptance testing.

Objectives of this activity are to define all conditions that must be tested at the program, subsystem, and system levels; to prepare all necessary test data and create test files; and to prepare a system test plan for coordinating the tasks of the final system test.

The scope of this activity includes review of the specifications produced during Activity 7, then the individual programs and program modules are identified as work units, and test specifications and test data are assembled.

End products include program test specifications, test data files, job stream test specifications, and system and subsystem test specifications. A separate set of documents includes a test plan.

Testing typically proceeds on a smallest-to-largest basis, following a bottom-up approach that is the inverse of the design strategy. Modules are tested individually. Then, the interfaces between modules are tested, followed by tests of programs, and then the entire system. Finally, an acceptance test is done under conditions that closely resemble actual use.

For the most part, the same persons who complete the work in Activity 7: Technical Specifications are responsible for completing the tasks of this activity with eventual involvement of users in acceptance testing.

The documentation created during this activity—specific test specifications, a testing plan, and files of test data—enhance and build upon the cumulative project file.

Key Terms

1. unit testing
2. module testing
3. integration testing
4. function testing
5. system testing
6. acceptance testing
7. program test specification
8. test data file
9. job stream test specification
10. system test specification
11. subsystem test specification
12. test plan

Review/Discussion Questions

1. What types of testing are involved in the specifications and procedures produced by this activity?

2. Which three types of testing are applied primarily to application software?

3. What two levels of testing deal with multiple programs that comprise complete processing systems?

4. What are the objectives of the test specifications and planning activity?

5. What is the scope of this activity?

6. What are the end products of test specifications and planning?

7. What does a test plan include?

8. In what sequence are modules tested?

9. How does the process of testing compare with the sequence in which a system is designed?

10. What personnel are involved in this activity?

11. What types of documentation generated by this activity are incorporated in the cumulative project file?

Project Assignments

The Appendix presents three case studies that can be used in support of project assignments. All three scenarios provide functional system specifications that would result from systems analysis. Included are data flow diagrams, data dictionaries, and process narratives that would result from systems analysis. Assignments that relate to the content of this chapter will be found in *PART 4: Software Design and Test Specifications*.

11 SOFTWARE TESTING STRATEGIES

LEARNING OBJECTIVES

On completing reading and other learning assignments for this chapter, you should be able to:

- [ ] Describe the nature of testing and the rationale for conducting software tests.
- [ ] Tell how the quality assurance steps of conducting walkthroughs and performing desk checking can supplement testing.
- [ ] Describe the processes of black-box and white-box testing and explain how each may be applied in testing modules.
- [ ] Tell what types of errors are sought, what principles are applied, and what types of data are employed in unit testing.
- [ ] Describe incremental and nonincremental integration testing and describe how incremental integration testing may be conducted in either a top-down or a bottom-up sequence.
- [ ] State the criteria applied in function testing.
- [ ] Tell what tools may be used in debugging.

NATURE OF TESTING

Software testing is a process for executing a program with the intent to cause and discover errors. This purpose contrasts with the common misconception that the goal of testing is to prove that a program works correctly. Just the opposite is true: The goal of testing is to force a program to work incorrectly and then to discover the causes of these errors and to fix them.

The rationale of testing is that, when designing software, it is virtually impossible to consider every kind of processing condition that possibly can arise. Unforeseen conditions are bound to occur. The purpose of testing, then, is to subject a program to a wide variety of stresses, both probable and improbable. The objective is failure, not proof that a program works. A program can work perfectly under ideal circumstances and still produce incorrect results under unforeseen conditions.

Putting it another way, testing is a destructive process. A successful test destroys the image of invincibility that people may have had for a program. Having discovered conditions under which programs can fail, however, the project team can devise constructive solutions to improve software quality.

At the same time, a rule of reasonableness should prevail. Just as it is impossible to design a piece of software to meet all contingencies, foreseen and unforeseen, so also is it impossible to design tests that can cause all of the unforeseen processing problems that a program can encounter during its entire useful life. Therefore, part of the challenge of testing design lies in establishing practical limits on the amount of testing to be done and the costs to be borne for this function. Consequently, to make testing as productive as possible, strategies are needed to ensure identification of the errors most likely to occur.

ROLE OF TESTING

Testing provides a final measure of quality assurance for a software product during the detailed design and implementation phases of the systems development life cycle. As a quality assurance measure, testing, in turn, is a final step in a series of checkpoints applied to assure the quality of software development.

For example, during the activities of the analysis and general design phase, systems specifications and any programming specifications that evolve from them are reviewed carefully. The purpose of these reviews is to ascertain that the logical and physical requirements placed upon programs are correct, will be realistic, and can be executed to produce quality products that meet user needs.

During design-related activities of the systems development life cycle, the structured methodologies described earlier in this book are applied as quality assurance techniques. That is, by following structured approaches to software design, there is some built-in checking and validation that the designs are sound and will result in quality programs. In addition, design activities should include walkthroughs or manually stepping through documentation end products that ultimately will be used as a basis for software development.

During the implementation phase of a project, there is continued use of structured methodologies and walkthroughs as quality assurance techniques. For example, structured programming adds a new dimension of quality assurance, as does the *desk checking*, of program code. Finally, software testing involves the actual running of programs on computers with test data. This chapter concentrates on software testing techniques and on the selection of data to be used for these operational tests. During installation of the new system, further quality assurance is applied through the monitoring of results of initial operations upon live, or real-world, data.

TYPES OF SOFTWARE TESTING

Tests of application software are applied at several levels:

- Unit testing, or module testing
- Integration testing
- Function testing
- System testing
- Acceptance testing.

The discussions that follow in this chapter deal with unit, integration, and function testing. Later chapters deal with system and acceptance testing.

Testing Strategies

Depending upon the individual unit of software and the conditions under which that unit will be applied, an individualized testing strategy is devised and carried out. Strategies, in turn, are formed through application of two approaches, or techniques. It should be stressed that these methods are not separate and exclusive. Rather, both are available and are used in some sort of tailor-made mix, depending on the nature of individual programs and the objectives to be met. The two methods are:

- Black-box testing
- White-box testing.

Black-box testing. *Black-box testing* is used to review module or program functions from the outside. As previously discussed, a black box is a processor about which an observer needs to know nothing of its internal workings in order to use it. For testing purposes, inputs and outputs are monitored and judgments are reached on the basis of results. There is no concern in black-box testing with the internal logic of a module or program. Rather, the testing is applied to determine whether inputs are accepted as planned and whether outputs meet expectations.

As a general rule in black-box testing, there is no attempt to uncover all errors. The testing that would have to be applied simply would be too exhaustive because, given no knowledge of the logic of a program, it would be necessary to devise massive inputs to test for all possible logical processing paths. Then, it would be necessary to review massive outputs to compare results with expectations. Instead, it is more effective to combine black-box testing with a review of program logic.

White-box testing. *White-box testing* considers the logic of modules as if the processor were clear, or transparent. Included are examinations of procedural details, tests covering the execution of all statements in the module, and tracing of all decision paths within the logic of an individual module. Note that white-box testing is carried out at the module level. It would be virtually impossible to apply white-box testing to an entire program at one time. Even in a relatively simple application program, there could be literally millions of potential logic paths open to data being processed. Therefore, it would be impractical to attempt complete, exhaustive program testing at a white-box level. Instead, logic testing is done at the module level and then is combined with black-box testing to establish satisfaction with executions of modules and with interfaces among modules.

General Strategy

The typical approach to the design of software testing, then, is to apply black-box tests to evaluate program and module functions. Further, white-box testing is applied selectively to evaluate program and module logic. In general, white-box testing will supplement black-box testing. Because exhaustive testing of all functions and logic paths is not feasible, the designer must decide what subset of all possible tests will uncover the greatest number of errors. As with other areas of systems development, test strategy development becomes largely a cost-benefit trade-off.

UNIT (MODULE) TESTING

Unit testing seeks to find errors within the logic and function of individual modules. The types of errors sought include:

- Interface
- Input/output
- Data structure
- Arithmetic
- Comparison
- Control logic.

Interface errors.   This type of test is applied to program segments in which control and data are passed from one module to another. A typical example involves the transfer of processing control from a module to a subroutine or subprogram. The object of the test is to determine that the arguments passed to the subroutine correspond with the parameters received. Tests are applied to be sure of correspondence in the number of data fields, the attributes (type and size) of the data fields, and the order of transmission and receipt.

Input/output errors.   Where external files are to be read into or written out of a program, this type of test is applied to make sure that all records are transmitted and received as expected. Tests are applied to record attributes, including numbers of fields, sizes of fields, and types of data contained within fields. In addition, errors are sought in record formatting, in the organization of files, and in the use of keys. The idea is to make sure that keys are used properly within records and that files are structured and referenced correctly through use of keys.

Other input/output errors for which checking is done include proper procedures for the opening and closing of files and for the handling of errors identified on input or output. Further, tests are applied to look for errors in the flagging of end-of-file conditions and in the proper processing of null, or empty, files. With direct access of files, tests are applied for the errors that would occur if a record with a matching key is found, or not found, as appropriate.

The ultimate purpose of tests for input-output errors is to find errors in system outputs. Test procedures should print or display outputs as appropriate. These outputs then should be checked against specifications for the modules under test to determine that accurate processing was performed upon test data and that expectations are being met—relative to the content as well as the formatting of outputs.

Data structure errors. These tests seek errors in the handling or building of data elements that are defined and generated within processing modules. Examples would include program-generated tables or interim records used in transforming data. Tests are applied for correctness of formats and attributes. Other tests may deal with the correctness of table definitions, including subscripting procedures and table sizes. Other checks would be made for consistency in the use of names, for proper initialization in the use accumulators and counters, and for completeness of specifications for data items.

Arithmetic errors. The results of calculation instructions within modules are tested to find errors involving failure to define properly all data items included in arithmetic instructions. For each data item, tests are applied to make sure that the data are in the proper mode for execution of the instructions, that the sizes of intermediate and final result fields are large enough to accommodate results of computations (eliminating the potential for errors through truncation), that computations are executed in the proper order to produce specified results, and that the value zero is not used as a divisor. (If division by zero is encountered, the program should have provision for handling this condition).

Comparison errors.   These tests look for errors involving presentation of data items of different modes or data types for comparison functions. The goal is to ensure that a program or module will not permit comparisons of alphabetic fields with numeric fields or among numeric fields of mixed modes. As with other error testing functions, the idea is to present such conditions to the system and to be sure that it can handle them and recover from any consequences. Another comparison test deals with the order of evaluation of relational operators in data comparisons. It can be difficult to be sure that such functions as multiple-nested comparisons are performed in the order intended. Tests should look for errors; a lack of errors would indicate that the computer is performing the comparisons in the same order intended by the programmer.

Control logic errors.   A computer is a sequential machine. Thus, no special tests are needed for sequential control functions. However, specific tests should be applied to selections and repetitions, which, in effect, are situations in which the natural processing sequence that the computer applies is altered.

Tests governing selections determine whether valid execution paths are established for all conditions to be tested and for all values of the data elements that are tested. Selection tests also ascertain that all branch points in selection functions are properly labeled and that there are exit points from all open paths.

For repetition control structures, one of the tests applied is to be sure that the loop index or subscript is initialized properly and that it is incremented (and incremented properly) with each iteration of the loop. Tests also ascertain that end-of-loop flags are set, tested, and implemented properly to avoid closed-loop situations in which programs are ''hung up'' indefinitely. All subscript values within repetition structures should have subscript values within anticipated ranges.

The table in Figure 11-1 summarizes the types of error tests made during unit testing.

DEVELOPING UNIT TEST DATA

Most tests at the module level are applied to check the logic of the design and the logic paths within the coding. Thus, most testing of

1. INTERFACE
   A. Correspondence in number of data elements passed and received
   B. Correspondence in attributes of data elements passed and received
   C. Correspondence in order of data elements passed and received

2. INPUT / OUTPUT
   A. Proper file opening and closing routines
   B. Proper definitions of files, records, and fields
   C. Proper use of keys for accessing and writing records
   D. Proper handling of end-of-file conditions
   E. Proper handling of input/output errors

3. DATA STRUCTURES
   A. Proper derfinitions of table sizes and attributes
   B. Correct definitions and uses of search subscripts and indexes
   C. Consistent use of data names
   D. Proper initialization of constants
   E. Proper initialization of accumulators and counters
   F. Proper format and attribute definitions of data items

4. ARITHMETIC
   A. Proper size, type, and precision characteristics of intermediate-result fields
   B. Proper size, type and editing characteristics of final-result fields
   C. Proper modes of data items involved in computations
   D. Proper sequencing of arithmetic operations
   E. Ability to handle division by zero

5. COMPARISON
   A. Correspondence in attributes of data items to be compared, or tested
   B. Proper order of comparisons in multiple AND and/or OR relations

6. CONTROL LOGIC
   A. Provision for branching paths for all selection results
   B. Provision for common exit point from all selected paths
   C. Proper initialization and incrementation of loop subscripts
   D. Provision for exits from loops

Figure 11-1. Types of error tests applied during unit testing.

program modules is done through white-box techniques. Logical checking, as described above, is accomplished through white-box testing.

However, program modules do not exist as stand-alone software elements. Modules are interdependent; there must be processing interfaces among modules. These interfaces, in turn, are tested most effectively through black-box techniques because emphasis on the interface focuses on inputs and outputs. Thus, at the module testing level, it is a common practice to apply white-box testing supplemented with black-box methods at the entry and exit points of individual modules. The principles applied to the testing of modules include:

- Test all statements at least once.
- Test all possible combinations of execution or logic paths.
- Test all repetitions over the full ranges of their indexes or subscripts.
- Test all entry and exit points for each module.

The challenge of unit testing, then, is to derive sets of test data that apply the required tests to individual modules. That is, not all tests must be applied to every module. Thus, in devising a testing plan, the idea is to determine which of the tests listed above are appropriate for a specific module, then to devise data that apply these tests.

Deriving Test Data for Modules

Test data to be processed within modules must be selected and organized to exercise and determine the validity of the logic under which the data will be processed. Four types of data are accumulated and applied to newly developed software modules:

- Range of values
- Categories of values
- Discrete values
- Ordered sets of values (for testing sequential files and tables).

These four kinds of test data, when selected carefully for testing purposes, will represent any data that can be presented to the module for processing. In other words, these four types of data cover all contingencies; the data take into account any data items that may be presented within any module, for any purpose, with any type of value.

Range of values. Ranges of values are a consequence of most arithmetic functions applied by programs. For example, if an instruction calls for multiplying two three-digit numbers by each other, it can be predicted that the range of values for the derived answer will be anywhere from zero through 999,999. For such an instruction, the ranges of data values can apply to both multipliers, as well as to the answer.

Thus, the procedure in testing the range of values is to look at each data item within a module to identify whether its values are expected to vary within some range, then to devise data that test the ability of the program to handle these ranges. This selection should be done systematically rather than at random and does not require testing for all possible values. Instead, only the boundaries of the ranges that can result should be identified, then tested. For any range test, four boundaries can be identified:

- Value at the lower boundary of the range
- Value at the upper boundary of the range
- Value that is less than the lower boundary
- Value that is greater than the upper boundary.

Note that emphasis is on the boundaries of the range, rather than at the midpoints. The boundaries, in this case, are reasonable expectations for the data to be handled rather than absolute mathematical limits.

As a simple example, consider a gross pay calculation for an hourly payroll. Ranges would be set for expected values of the lowest number of hours that a person could work in any given week and the highest number of hours. To illustrate, these values might be 8 and 80. Then, values would be determined for the lowest and highest expected hourly pay rate—possibly $4.75 at the low side and $21.50 at the high side. Values then would be derived for the test data that were both at and beyond these upper and lower boundaries of hours worked and pay ranges. For example, for hours worked, the values

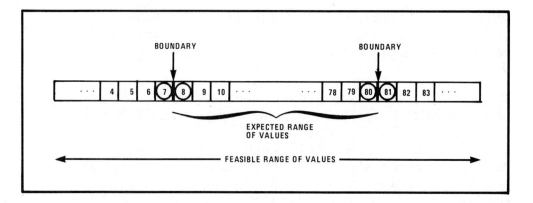

Figure 11-2. In generating test data items that can take on a range of values, testing values are selected from the values at the boundaries of the expected range.

presented in test data might be 7, 8, 80, and 81 as shown in Figure 11-2. Each of these values then would be multiplied by four pay rates that are at and beyond the lower and upper boundaries. These data would test the ability of the gross pay calculation module to handle both anticipated values and exceptions.

Similar procedures can be applied when data to be processed are extracted from tables. The values tested in this case are subscript values. That is, the range of values for the table encompass a value of 1 through the length of the table. Thus, if you had a table with 25 elements, you would derive test subscript values of 0, 1, 25, and 26.

A similar practice could be applied to loops. Values entered would be for the number of iterations anticipated. The idea is that the subscript controlling the loop would represent a range of values to be tested. At the same time, tests for both valid and invalid conditions for terminating the loop can be made.

Categories of values. Data for testing categories of values are applied within modules that use specific types, or attributes, of values within fields. In such situations, data presented can be either one of the expected types of values or one of the unexpected types.

To illustrate, consider a module in which an on-line user must enter an alphabetic field in response to a request for a customer name. For testing purposes, the actual name entered is not important.

Rather, the test is applied to determine whether the module can handle both alphabetic and nonalphabetic data. A representative value would be any combination of valid alphabetic characters within the size parameters of the field. A nonrepresentative value would be any other combination of characters. Thus, in testing a module that incorporates data with certain attributes, it is sufficient to present a single representative test value and a single nonrepresentative value.

Discrete values.   Test data also must be generated for data items that will take on selected, discrete values. In these cases, it is necessary to supply test values that match identically all expected execution values and also to provide a test value that is unexpected.

For example, consider an update module that accepts transaction records identified for addition, change, and deletion. In this instance, all transaction codes should be supplied for testing along with a code value that is invalid. On other words, all valid data values and a single invalid value comprise the test data.

Tests for discrete values are applied in checking the branching results in selection control structures. They provide an indication that processing paths exist for all expected values and that the module can handle unexpected values.

Ordered sets.   Over and above ranges, categories, and discrete values, the ordering of data within files or tables also may be critically important to the execution of a program. For example, in a sequential master file, it is important that records be processed in key order. The same may be true within tables. For modules that include the handling of sequential files or table data, it is necessary to test the ordering of records. Tests are applied to:

- Select the first element in an ordered set
- Select the last element in an ordered set
- Select an item known to be missing
- Select excessive cases, both high and low.

The principle is the same as for a boundary test except that the tests are applied to the sequencing, or ordering, of records rather than to the values of the records.

For example, if a module executes a table search, test data would be devised that cause the instructions to search for the first and last values in that table, to search for an item known to be missing from the table, and also to process subscripts that are both lower and higher than the range of subscript values for that table.

In devising actual test data, it would be a good practice to prepare a checklist of the types of errors that can be made for the data in a given module. This checklist, in effect, can become a matrix for determining which of the four types of test data described above can be used to search for the errors that can exist within program modules.

It should be noted that it is not necessary to perform tests within all modules that process this data. Usually, edit modules perform many of the testing functions to ensure that only valid data are passed to processing routines. Therefore, unit tests can assume the existence of valid data within reasonable ranges and categories. Edit modules, on the other hand, should employ full tests of both valid and invalid data.

INTEGRATION TESTING

Program integration encompasses the procedures followed for connecting modules to form programs. *Integration testing* encompasses the exercise of these program connections to determine their soundness and workability. Thus, integration testing is applied to the interfaces among modules within any given application program. Two approaches can be taken to integration testing:

- Modules can be added to one another for testing on an individual basis, possibly as new modules are written. This procedure is called *incremental* testing.
- All modules within a program can be developed first, then joined and tested as an entity. That is, all interfaces among all modules are tested at one time, for the entire program. This procedure is known as *nonincremental* testing.

Nonincremental testing can be extremely difficult and, therefore, is not recommended. If an interface problem does arise and multiple interfaces are tested in a single operation, it can be extremely difficult and tedious to pinpoint the exact location of the problem. By contrast, incremental testing has the effect of building quality into a program as it goes together. If any problems do develop, they can be identified and pinpointed specifically to known sets or groups of modules. To illustrate, if a group of modules that already has been tested is known to work satisfactorily and a new module is added, any execution problems must exist either in the new module or in the interfaces between the new module and the previously tested modules.

Thus, incremental testing is the only method recommended and described in the presentations that follow. Two methods can be used in applying incremental testing:

- Top-down
- Bottom-up.

There also can be combinations of the top-down and bottom-up approaches.

Top-Down Testing

Top-down testing is applied to modules on a top-down basis, or proceeding from higher-level modules down through detailed modules.

That is, interfaces among modules are tracked from top to bottom, following paths derived from the structure chart for the program. To illustrate top-down integration testing, consider the structure chart in Figure 11-3. This structure chart will be used as a model to demonstrate the course of top-down methods and also to illustrate the strategy involved.

Figure 11-4, based on the same program structure, shows how the top-down approach progresses. Figure 11-4A indicates that testing begins with the topmost module (A). Connections deriving from this module are to modules at lower levels that have not yet been implemented. Therefore, the portions of the modules that will handle the interfaces are prepared and used in executing test data applied to the connections between Module A and its subordinate modules. These testing routines are called program *stubs*. Such routines simulate the processing that will take place when the modules actually are coded. During testing, stubs provide targets for subprogram calls so that superordinate modules can exercise their control functions. Stubs also may pass data to superordinates to test interfaces. Usually, the passed data are generated as literal values within the stubs, since the stubs

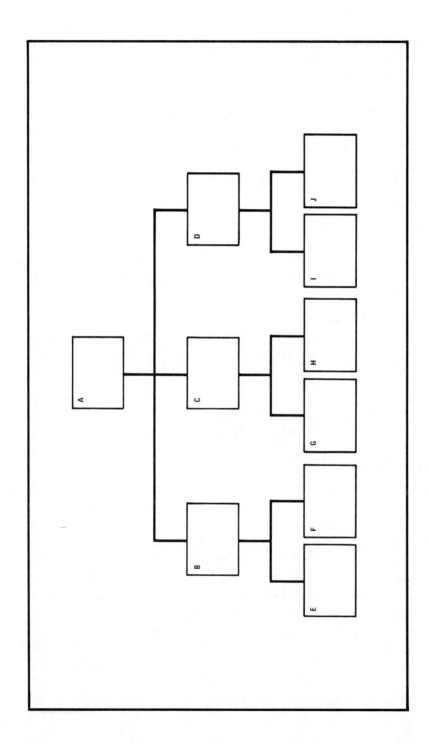

Figure 11-3. Structure chart for program to be tested incrementally.

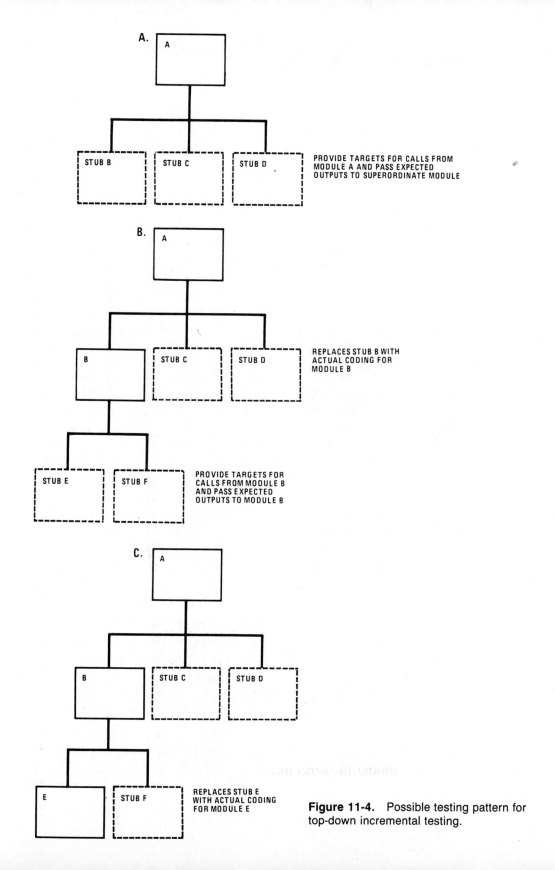

Figure 11-4. Possible testing pattern for top-down incremental testing.

have not yet been coded as complete modules. Use of stubs is a common practice in program testing.

Figures 11-4B and 11-4C show how testing progresses to succeeding, lower levels through use of stubs to process data that cross the interfaces to be tested.

Although it is not apparent from the illustration, the practice followed is to test the input branch of a program first, wherever possible. Then, the output branch is tested as soon as possible thereafter, leaving central processing modules for last. This practice has evolved because it is easier to provide test cases for processing modules once routines for handling inputs and outputs have been established. Exceptions would exist for programs having individual processing modules that are critical to the capabilities of the entire program. In such cases, the critical module should be tested as early as possible.

Evaluation of the top-down approach.   Testing with a top-down approach provides early verification of the major interfaces and also of the overall control logic, which must emanate from the top of the program's structure. Also, by starting at the top and working down, it becomes possible to demonstrate the program functions for the complete, overall program at an early point in the testing procedure. These overall functions cannot be demonstrated if modules at lower levels are tested independently. For the project team, this approach also can serve as a useful public relations tool with the user, who can see some tangible results. When combined with a phased implementation strategy, such usable portions of the system can be completed and tested while other, detailed pieces remain incomplete.

However, the top-down approach also presents an inconvenience. It can be difficult to provide test cases when testing begins at the highest level. The lower-level input and output modules have not yet been completed and tested. Thus, it becomes necessary to develop stubs to represent the lower-level modules in test situations. Eventually, these stubs will be replaced by complete modules. At this point, however, testing may require multiple versions of the same stub, providing multiple test data to the higher-level modules.

Another possible disadvantage is that testing of modules in input and output branches of a program is often carried out before the intervening processing modules are ready. It may be difficult, occasionally impossible, to provide processing stubs that present all of the output configurations that should be tested.

Bottom-Up Testing

The strategy of *bottom-up* testing is to begin at the lowest level in a program structure and move progressively higher as modules are tested and integrated. The progress of bottom-up testing for the program shown in Figure 11-3 is diagrammed in Figure 11-5. Figure 11-5A indicates that testing begins at the module appearing at the lowest and leftmost position in a structure chart. The same principle applies here as for top-down testing. That is, it is desirable, though not absolutely essential, to begin with the input branch, then go to the output branch as the overall testing strategy. However, depending on individual situations, variations can exist.

The illustration in Figure 11-5A shows that a *driver* has been developed for the tests of a low-level module. A driver plays a role similar to that of a stub, from an inverse position. That is, a driver is a test module, or program routine, that generates calls to lower-level modules and passes data across an interface. A driver simulates the actions of the group of yet-to-be-developed modules superordinate to the tested module.

In Figure 11-5B, the same driver is used to test a second low-level module. Eventually, then, the driver is replaced by a finished module and a new driver is used at the succeeding, higher level. This situation is illustrated in Figure 11-5C.

Evaluation of bottom-up testing approaches.   Advantages for the bottom-up approach to testing include early identification of any detailed processing flaws that might exist across low-level interfaces and early exercising of input and output test cases. A disadvantage is that this approach puts off the ability to form an overall, skeletal program until all modules have been tested and are in place.

Combined Testing Methods

There are no hard-and-fast guidelines about when top-down and bottom-up approaches to testing are combined most effectively. However, it is important to be aware that these types of situations may arise. Individual work schedules or needs of individual programs may create situations in which it is profitable to start not only at the top

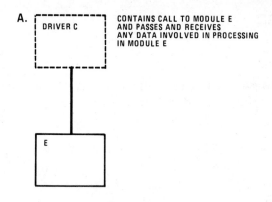

A. DRIVER C   CONTAINS CALL TO MODULE E
    AND PASSES AND RECEIVES
    ANY DATA INVOLVED IN PROCESSING
    IN MODULE E

E

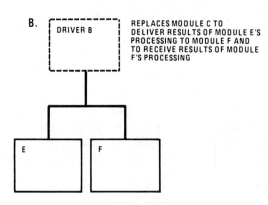

B. DRIVER B   REPLACES MODULE C TO
    DELIVER RESULTS OF MODULE E'S
    PROCESSING TO MODULE F AND
    TO RECEIVE RESULTS OF MODULE
    F'S PROCESSING

E   F

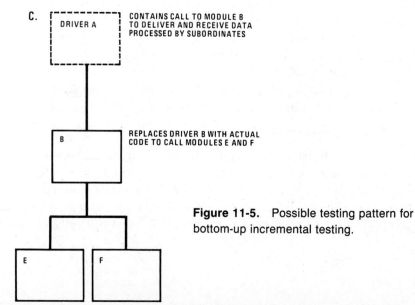

C. DRIVER A   CONTAINS CALL TO MODULE B
    TO DELIVER AND RECEIVE DATA
    PROCESSED BY SUBORDINATES

B   REPLACES DRIVER B WITH ACTUAL
    CODE TO CALL MODULES E AND F

E   F

Figure 11-5. Possible testing pattern for bottom-up incremental testing.

but also at the bottom and work toward the middle. Conversely, in specific situations, it may be best to start in the middle and work toward both the top and the bottom.

All that can be said for certain is that the same methodologies apply on a module-by-module basis. That is, it is necessary to devise a plan for incremental integration testing of the connections and interfaces among all modules in a system. (It should be pointed out that it is not always necessary to add just one module at a time. For example, modules with no intended impact on one another or in completely different parts of the structure may be combined in a testing step, provided that care is exercised.)

FUNCTION TESTING

Function testing applies black-box techniques in looking for errors or failures within a complete software package. Chronologically, function testing follows module testing and integration testing. It is impossible to test a complete system until all of the modules have been tested and necessary adjustments made. Then, the modules have to be integrated individually or in small groups. Ultimately, the testing process builds up to the point at which a complete program is operational. By this time, white-box methods for detailed examination of individual functions or procedures are no longer practical. The amount of detail involved simply would be unmanageable.

The special role of function testing is to exercise the entire program to be sure that external user specifications for inputs and outputs are met. Test cases applied are those developed during systems analysis and general design—activities in which the users participate actively. Criteria applied in function testing include measures of compliance with:

- Input formats
- Output formats
- File organization
- File access
- Human/machine interfaces.

Function testing does just what the name implies. Program functions are exercised to be sure that the software is operating as designed. However, function testing still falls short of performance or systems tests. At this point, there is no attempt to exercise the system at full production volumes; neither is there any attempt to overload the system with work to see how it reacts to unforeseen strains. The intent is to look at and evaluate the soundness of the software. Thus, function testing is still within the domain of CIS professionals rather than users.

Input records presented for a function test would include both valid and invalid data fields. Input records would include situations in which alphanumeric data were too long for the fields designated to hold them, in which the data within fields were positioned incorrectly, in which numeric data were entered into alphanumeric fields, and so on. In other words, the same principles appropriate to the preparation of test data at the module level also apply at the function level. Input and output parameters are examined and both valid and invalid data elements are presented as a means of determining the ability of the program to handle such data.

Particular attention should be paid to the points at which people and machines interact. In these instances, function tests deal with the documentation and with the reactions of the system to stimuli from users. Ideally, for example, a person with no existing knowledge of the system would be asked to read the manual and perform input functions. Separately, a more experienced person would input erroneous or invalid data. In each case, part of the testing lies in evaluating the clarity of messages to the user generated by the system. These messages would include both error descriptions and prompts about what should be done next.

At all levels of testing—module, integration, and function—the underlying purpose is to cause the system to fail, to generate and deal with errors.

DEBUGGING

Debugging is a result of testing. The purpose of testing is to cause errors. Debugging, then, takes over to locate and correct those errors.

Several tools or approaches are available to help identify and locate errors. These include:

- Memory dumps
- Execution traces
- Program desk checking
- Hypothesis testing.

These techniques are well established. The first three items have been standard practices in the CIS environment for many years.

Memory dumps are simply printouts showing status of memory at a given moment, usually at the point at which an error has been detected. Memory dumps make it possible to search for specific data items that may have been processed erroneously.

Execution traces cause the computer to print out a sequential log of program execution. A typical trace will list the names of modules as they are executed. An execution trace serves as a tool to determine the processing sequence of the program. If a program is aborted, or "blows up," because of an error, it becomes possible to find the last module that was executed. This point, then, is where troubleshooting begins.

Some programming languages have built-in tracing features that can be used in the same way as utilities are used within some programs. In other instances, programmers may have to insert PRINT or DISPLAY instructions to cause the program to trace processing sequences.

Another debugging method is desk checking a program. Desk checking is done through a detailed examination of source code that, in effect, executes the logic in the mind of the reviewer rather than in the circuits of the computer. An experienced programmer can track the logic and processing of data through a program by reviewing its source code in detail.

Hypothesis testing looks at the program through a more analytical method. In effect, the programmer applies troubleshooting or problem-solving techniques. Once an error is identified, analysis is performed to determine what kinds of programming errors might have caused this specific result. A hypothesis is formed about where the error probably exists within the program and what kind of mistake caused the detected error. Data then are developed specifically to test these hypotheses and to pinpoint the place in the program where modification is needed.

In effect, the first three of these error detection methods are actions to be taken. The final method represents a thought process that can be applied. In the hands of a knowledgeable person, hypothesis testing can identify and correct causes of problems more readily and effectively. Further, hypothesis testing minimizes the need to resort to trial-and-error methods involving extensive use of the computer itself to find program errors.

Error Correction

The correcting of program errors involves trade-offs between immediacy or expediency and long-term effectiveness.

The expedient way simply is to locate the problem and patch the program. In effect, the result is to program around the error rather than to deal with any logical or functional problems. Patching can get the job done; however, the patch may not necessarily fix the problem. The problem may be buried and may cause other problems downstream. Thus, patching may be justifiable as a temporary measure or in an extreme emergency but should not be a standard practice. In addition to degrading, potentially, the quality of a program, a patch usually is not documented up to established standards in a given installation. Therefore, when a program requires modification or maintenance at a future time, there may be no adequate documentation to work with.

The far more preferable approach is to reexamine the program and produce a quality product right at the outset. Some redesign may be involved, as well as the writing of new code and some additional testing. However, in the long run, increased maintainability is worth the effort.

If these procedures are carried out properly, the results of testing should be far more valuable than the sum of the individual parts of the testing plan. That is, there are identified types and levels of tests that can be applied. Cumulatively, the results should produce more than a series of checks on somebody's list. The idea of testing is to produce quality software. Thus, elements of concern and pride should go into the testing. After all, testing is the procedure by which CIS professionals get their product ready to present to users, to the customers who have demonstrated a need for a quality product. So, in a sense, the philosophy that goes into a testing program is as important as the series of events that constitute a test plan.

Summary

Software testing is a process for executing a program with the intent of causing and discovering errors. The rationale of testing is that it is virtually impossible to design a piece of software that will deal with every kind of processing condition that possibly can arise.

Quality assurance steps that supplement testing include walkthroughs and desk checking.

Testing strategies include black-box testing and white-box testing. Black-box testing looks at modules in terms of the inputs applied and the outputs produced. White-box testing examines the internal logic of modules. In general, white-box testing will supplement black-box testing.

Types of errors sought in unit testing include interface, input/output, data structure, arithmetic, comparison, and control logic.

The principles applied to the testing of modules include testing all statements at least once, testing all possible combinations of execution or logic paths, testing all repetitions over the full ranges of their indexes or subscripts, and testing all entry and exit points for each module.

Types of data accumulated and applied to newly developed software modules include ranges of values, categories of values, discrete values, and ordered sets of values (for testing sequential files and tables).

Integration testing encompasses the exercise of program connections to determine their soundness and workability. Integration testing may be incremental or nonincremental. Incremental testing may be conducted with either a top-down or a bottom-up approach.

Testing with a top-down approach provides early verification of the major interfaces and also of the overall control logic, which must emanate from the top of the program's structure. Also, by starting at the top and working down, it becomes possible to demonstrate the program functions for the complete, overall program at an early point in the testing procedure. However, it can be difficult to provide test cases when testing begins at the highest level.

Advantages for the bottom-up approach to testing include early identification of any detailed processing flaws that might exist across low-level interfaces and early exercising of input and output test cases. A disadvantage is that this approach puts off the ability to form an overall, skeletal program until all modules have been tested and are in place.

Criteria applied in function testing include measures of compliance with input formats, output formats, file organization, file access, and human/machine interfaces.

Tools that aid in the process of debugging include memory dumps, execution traces, program desk checking, and hypothesis testing.

The correcting of program errors involves trade-offs between immediacy or expediency and long-term effectiveness.

Key Terms

1. software testing
2. desk checking
3. black-box testing
4. white-box testing
5. program integration
6. integration testing
7. incremental
8. nonincremental
9. top-down testing
10. stub
11. bottom-up testing
12. driver
13. function testing
14. debugging
15. memory dump
16. execution trace
17. hypothesis testing

Review/Discussion Questions

1. What is the goal of software testing?

2. What two quality assurance tests supplement testing?

3. What is involved in black-box testing?

4. What is involved in white-box testing?

5. What types of errors are sought in unit testing?

6. What principles may be applied in unit testing?

7. What types of data are applied in unit testing?

8. What are two approaches to integration testing, and what is involved in each?

9. What are two approaches to incremental testing, and what is the process of each?

10. What criteria are applied in function testing?

11. What tools aid in the process of debugging?

12. What trade-off is encountered in the correction of program errors?

Project Assignments

The Appendix presents three case studies that can be used in support of project assignments. All three scenarios provide functional system specifications that would result from systems analysis. Included are data flow diagrams, data dictionaries, and process narratives that would result from systems analysis. Continuing the work begun in Chapter 10, assignments that relate to the content of this chapter will be found in *PART 4: Software Design and Test Specifications*.

ACTIVITY 9:

12 PROGRAMMING AND TESTING

LEARNING OBJECTIVES

On completing reading and other learning assignments for this chapter, you should be able to:

☐ Describe the work done during the programming and testing activity.

☐ Tell why the preparation of documentation is an important part of this activity.

☐ State the objectives of the programming and testing activity.

☐ Describe the scope of this activity, including the delegation of programming tasks and the setting of checkpoints.

☐ Describe the end products of programming and testing.

☐ Identify and describe the steps in the process of program development.

☐ Tell how programs are cataloged and documented.

☐ Tell what personnel are involved in this activity.

☐ Tell what documentation is added to the cumulative project file by this activity, including the use of program folders.

ACTIVITY DESCRIPTION

If programs are being developed from scratch for an application, the programming and testing activity can be both the most extensive and the most expensive in an entire systems development project. To develop an original set of programs, members of the project team, chiefly programmers and their supervisors, go through the entire process of program development involving the multiple steps to build an understanding: develop a solution for the problem or need; design, write, and test programs; and perform all of the debugging activities needed to assure program reliability.

If proper emphasis is placed on the programming activity, it also can involve a highly challenging and creative series of tasks. Programs that are designed properly run properly. Programs that run properly impart a feeling of great achievement to the members of the project development team.

Administratively, this activity involves intensive involvement in documentation. Once programs are up and running, they are only as good as the supporting documentation. Programs are dynamic; continuing revision and maintenance activities are needed. These ongoing activities depend almost entirely on existing documentation.

The documentation and task completion emphasis during the programming activity should place heavy emphasis on design documentation. If the programming effort is managed and executed properly, considerably more time will go into design functions than into the actual coding. There can be a temptation, particularly if the writing of code is to be done on-line, to de-emphasize documentation. Remember that the important documentation lies in developing the structures, the data dictionary, the systems flowcharts, program structure charts, and the other design bases on which the coding will be built. It may be true that source code listings can be generated and updated by computers. But, the physical production of listings is almost incidental by comparison with the broader goals of programming documentation.

OBJECTIVES

The objective of this activity is to produce a complete, documented, and tested system of programs that are ready for testing with manual and other external functions as a complete system. The testing within this activity is limited to unit, integration, and function testing. All of the programs should function to specification and on their own.

However, the testing portion of this activity does not encompass complete system or acceptance testing.

SCOPE

The programming and testing activity begins with a review of program specifications prepared during technical design. Specification documents include structure charts, data dictionaries, form layouts, record layouts, and samples of any interactive dialogue or menu requirements that may be appropriate. In addition, there will be systems flowcharts that served as a basis for development of the structure charts. Such systems flowcharts can be important references in establishing final parameters for program modules.

On most medium- or large-scale systems, there will be a number of programmers assigned to this activity. Thus, an important consideration in the scope of work lies in the delegation of assignments to programmers with appropriate experience levels and backgrounds. Work assignments should include professional development plans and opportunities for individuals and for the programming group as a whole. The scope of the activity should involve a continuing series of checkpoints at which module designs and code are desk-checked by the people who wrote them and then are examined by supervisors or peers in walkthrough sessions.

As each program module is written, it is tested. Modules go together in building-block fashion until the programming effort is completed.

After an interval that can range from weeks to several months, all of the modules are written and tested at the job-stream level.

END PRODUCTS

End products are function-tested programs that are documented fully, tested, cataloged in the program library, and supported with operating instructions for personnel in the computer operations center.

THE PROCESS

Programs are developed in an orderly process, typically with five steps, as illustrated in Figure 12-1. These steps, briefly, encompass:

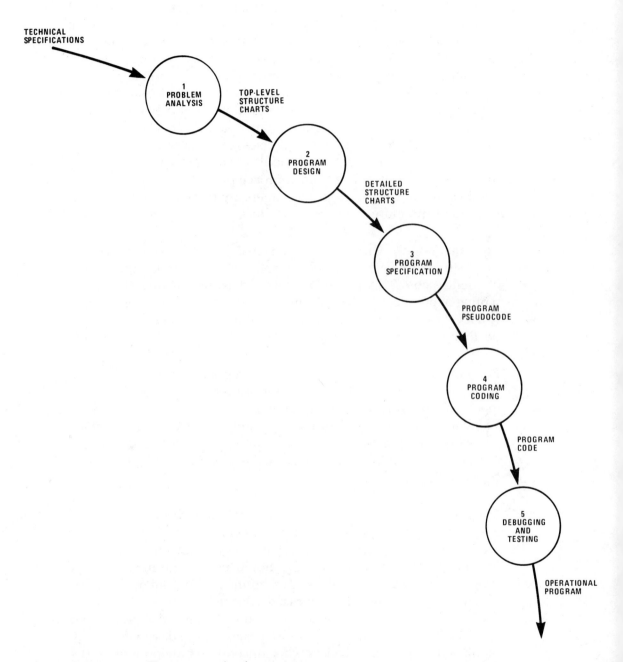

Figure 12-1. The program development process.

Problem analysis. In Step 1, the programmer reviews technical specifications and builds his or her understanding of the nature of the problem to be solved or need to be met. In some instances, complete structure charts for the programs will exist. In other instances, the programmer will have to develop structure charts. This initial phase of the program development process may involve meetings with other programmers to assure an understanding of the boundaries of modules to be developed by different people. Also discussed will be the intermodule communication interfaces for passing data. Programmers may work individually on modules or be part of teams assigned to groups of modules. The team approach prevails on larger systems, and in situations in which programmers are receiving on-the-job instruction from more senior professionals.

Program design. This design step, Step 2, involves elaboration of existing structure charts to a level of detail needed to support the writing of pseudocode. This phase, however, does not involve the actual writing of pseudocode. Rather, the structure of the program is detailed to a level needed to support pseudocoding, which takes place in the next step.

Program specification. In Step 3, the programmer who will do the actual writing of instructions prepares pseudocode for each assigned module. This pseudocode designs the processing logic for the module. Pseudocode should be reviewed thoroughly in walkthrough sessions and with a programming supervisor. This review serves both to check the individual module and to oversee the needed module-to-module compatibility.

Program coding. If the process has been followed effectively, Step 4 is a relatively straightforward translation of existing specifications into instructions written in a selected programming language. Bear in mind that the writing of code should not be the major part of program development. Rather, if planning and design are handled effectively, programming becomes a routine procedure. In some cases, coding is done by the same person who prepared the design and specification documentation. In other instances, the design documentation is turned over to coders who concentrate only on writing detailed instructions.

Program debugging and testing. In Step 5, for each module, bugs identified by the compiler may be corrected immediately. Logical errors are identified and corrected through testing involving the processing of data. Known data are processed under the program and results are compared with predetermined specifications. Variances indicate logical errors, which must be tracked through reviews of the logical structure of tested modules or groups of modules. As coding corrections are made, retesting takes place. As modules are written, unit, integration, and function testing are performed.

Throughout the program development process, extensive use is made of walkthroughs. At each step, programming teams meet for reviews of designs and code. Critiques are aimed at assuring quality products before effort continues to the next programming step.

Cataloging

Once programs have been developed and tested, they are made part of the program library. Depending upon local standards, cataloging may involve setting up what amounts to three separate libraries:

- Source program libraries containing the original source code
- Object program libraries containing compiled versions of programs and/or subprograms
- Core-image libraries containing executable versions of the programs.

Operations Documentation

Documentation should be prepared for all responsibilities and activities required of computer center personnel. For on-line programs, emphasis will be on maintenance or troubleshooting activities. For batch programs, documentation will be more extensive, involving complete setup, run, and follow-up procedures. Setup procedures deal with program loading and peripheral device preparation. Run procedures detail the console messages that can be expected—and reactions that should be taken—while the program actually is being executed. Follow-up would include instructions for handling backup, logging, or new generation files that would be created, as well as disposition for the existing media. In addition, instructions must be provided for the handling of outputs and the return of program media to the library.

PERSONNEL INVOLVED

Personnel who complete this activity are entirely within the programming group of the CIS function. In large shops, and for large jobs, programmers may work in teams. If so, there will be a team leader responsible for all of the programs associated with the application under development.

CUMULATIVE PROJECT FILE

A program folder is prepared for each separate program. Folders also are prepared for each special program required in conversion of the new system. These special programs may be required to convert files from existing formats to those required by the new system. Use of such programs will be one-time only. However, full documentation support still is needed. Each program folder contains:

- The latest source code listing or reference to the listing in the source code library
- The specifications for the program
- The test log for each program, including notations on each test run, the date, the test file version used, the results, and notations on any modifications made.

This documentation, of course, is added to the existing documentation for earlier activities in the project. Personnel who will perform succeeding activities for the training of users and the testing of complete systems, may require access to earlier documentation, as well as to these end products.

Summary

To develop an original set of programs, members of the project team go through the entire process of program development involving the multiple steps needed to build an understanding: developing a solution for the problem or need; designing, writing, and testing programs; and performing all of the debugging activities needed to assure program reliability.

The objective of this activity is to produce a complete, documented, and tested system of programs that are ready for testing with manual and other external functions as a complete system. The testing within this activity is limited to unit, integration, and function testing.

End products are function-tested programs that are documented fully, tested, cataloged in the program library, and supported with operating instructions for personnel in the computer operations center.

Programs are developed in an orderly process that encompasses problem analysis, program design, program specification, program coding, program debugging and testing.

Program cataloging may involve setting up source program libraries containing the original source code, object program libraries containing compiled versions of programs and/or subprograms, and core-image libraries containing executable versions of the programs.

Personnel in activity include the programming group, which may be divided into teams headed by team leaders.

A program folder is prepared for each separate program. Each program folder contains the latest source code listing; the specifications for the program; and the test log for each program, including notations on each test run, the date, the test file version used, the results, and notations on any modifications made.

This documentation, of course, is added to the existing documentation for earlier activities in the project.

Review/Discussion Questions

1. Why is documentation important in the programming and testing activity?

2. What is the objective of the programming and testing activity?

3. What is the scope of this activity?

4. What steps are involved in program development?

5. What types of testing are applied during this activity?

6. What are the end products of the programming and testing activity?

7. What is involved in program cataloging?

8. What personnel are involved in the programming and testing activity?

9. What information does a program folder contain?

10. What might operations documentation include?

Project Assignments

The Appendix presents three case studies that can be used in support of project assignments. All three scenarios provide functional system specifications that would result from systems analysis. Included are data flow diagrams, data dictionaries, and process narratives that would result from systems analysis. Continuing the work begun in Chapters 10 and 11, assignments that relate to the content of this chapter will be found in PART 4: Software Design and Test Specifications.

USER TRAINING 13

LEARNING OBJECTIVES

On completing reading and other learning assignments for this chapter, you should be able to:

☐ Describe the purpose of the user training activity.

☐ Explain the role of CIS professionals in training user personnel who, in turn, conduct the training of user operations personnel.

☐ State the objectives and scope of user training.

☐ Tell what personnel are involved in user training, including the advisory role of CIS professionals.

☐ Describe the preparation and use of reference manuals and training manuals in this activity.

☐ Tell what documentation from this activity is incorporated into the cumulative project file.

ACTIVITY DESCRIPTION

The activity of user training is preparatory in nature. Note that it is positioned in the systems development life cycle between activities that cover programming and testing and the activity for full system test. Thus, the purpose of this activity is to prepare users for the acceptance test that they will perform and then for the ongoing use of the system. Because of this stepped nature of the responsibilities that will follow, user training may take place in a series of steps. There may be separate training for different user groups—for managers who will evaluate outputs, for input operators, for clerical personnel—and also separate training within these groups for persons who will be involved in the acceptance tests and for those who then will implement the system on an ongoing basis.

Physically, the activity will involve CIS professionals as trainers of trainers. That is, initial groups of users will be instructed on the operation of the system so that they, in turn, can instruct their peers. It is extremely important that the training of users be a user responsibility. Users must work closely with CIS personnel to determine what training is needed. Also, users must ascertain when they feel members of their group have had enough training. They must look critically at both training programs and acceptance testing activities to determine whether there are loopholes that must be closed within the training curriculum or methodologies.

User training will be a mix of instructional sessions and hands-on practice to provide simulated experience in the use of the system that is coming to life. The objectives should be to build an understanding of the needs being met by the new system, of the philosophies followed in meeting those needs, and also of the specific techniques being applied. In the process of these training sessions, one of the elements being tested is the adequacy and appropriateness of the user training documentation. There should be two categories of documentation ready by the time training of users begins. One category includes the manuals used for ongoing operation of the system. This form of documentation typically comprises *reference manuals*, or procedure manuals. The second category includes separate *training manuals*. Training manuals, in effect, provide cross-reference for and guide the use of reference manuals. Reference manuals, in turn, are structured so that people who need guidance on the job can look up and be directed specifically to the areas that will help.

OBJECTIVES

The objectives of this activity include:

- On completion of the tasks in this activity, user managers should have a clear understanding of how the new system will operate.

- The user operating staff should be trained fully in the tasks they will perform and the responsibilities they will assume under the new system.

- All user personnel should acquire sufficient documentation to guide them in day-to-day operations and problem solving.

- A supervisor or manager within the user group should be designated as coordinator or liaison person for the resolution of future training needs or the securing of answers to questions that might arise.

SCOPE

Depending upon the nature and procedural content of any given system, this activity may begin before programming has concluded. For example, if there are extensive manual procedures that already have been developed and implemented through to the availability of manual forms and records, training could begin on that aspect of the system while program writing is proceeding. Bear in mind that many user procedures may not involve using a computer directly. For example, it is entirely possible for the personnel who will prepare and use new source documents to learn these procedures—and possibly put these aspects of a system into use—before the computer programs are implemented. The more prepared user personnel are, the easier the transition will be into the computerized portions of the new system. Further, if microcomputers are being used in conjunction with a new system, this portion of the application may be implemented through the availability of existing software packages. If so, user training can and should proceed at the earliest possible moment. Again, the goal is to have the user group as comfortable as possible with both the concepts and the mechanics of the new system and to do this at the earliest possible time.

The breadth of the training effort should encompass all personnel who will be affected by the new system in any way. Training should be tailored to the needs of individuals. Those who will be involved actively in new procedures should receive in-depth, detailed training. Those who simply must be aware of the existence of the new system without having to change the content of their jobs should be briefed on what is happening.

At least a select group of user personnel should be trained and ready to go by the time the new system comes into its acceptance testing. As soon as possible after the completion of tests, training should move forward and encompass all affected personnel.

There should be feedback throughout the training activity. Documents should be modified and revised instructions should be administered as needed or appropriate.

The training effort should remain intact at least until the new system has been operated through one full, functional cycle. Thus, the experience and expertise of trained user personnel will be available to deal with any personnel problems that may result from conversion to the new system.

END PRODUCTS

Procedures manuals.   Manuals, as the term is used here, are documents that direct people in performing manual procedures within a computer-based system. In effect, manuals do for people what programs do for computers. However, people and computers are different. Manuals should reflect these differences.

The guiding principle in developing a manual should be that people are in the system because they are able to apply judgment. Any functions that can be automated completely would be done by a computer. Thus, emphasis in the procedures manual should be on those points in the system at which people assure quality or apply judgment. Care should be taken to explain the reasons for doing things in the ways that have been specified.

This type of presentation helps to convince people that the jobs they are doing are worthwhile. A well-executed procedures manual should have the effect of selling the person doing the work on the value and importance of his or her job. Unfortunately, many procedures manuals give the impression of talking down to the people who actually do the work—of emphasizing the steps taken rather than the importance of the results. Such manuals, rather than guiding and helping people, encourage feelings of boredom and futility. Thus, they defeat their own purpose, contributing to a lack of quality rather than assuring that standards are met.

Some content items within procedures manuals that can help build human understanding and interest are:

- Explain the purpose and value of the overall system of which the individual is a part.

- Identify the customer, or user, of the outputs produced by each task.

- Describe, specifically, what successful performance will look like and what will be expected from the person handling each task.

- Describe any and all quality standards that should be met within the context of the job description itself.

- In describing procedures, cover each step to be taken in sequence. Be sure to identify the starting and completion points for each step, as well as the overall continuity between steps.

- Whenever a judgment or decision is to be made by a human operator, emphasize the value of this judgment and its contribution to the success of the system. Follow the decision-making model in identifying what is to be decided, what alternatives are available, and the conditions under which each alternative should be selected.

- Encourage people to apply judgment. That is why they are part of the system. Include instructions on how individuals can make suggestions to improve the system or to streamline the work flow.

The same guidelines apply in developing procedures manuals for computer console operators. The more a manual can do to help make an operator feel important because of his or her ability to apply judgment, the more effective that manual will be. Conversely, the more a manual tends to treat a person as an attendant waiting upon a machine, the less effective that manual will be.

Training manuals. The job of training operators and users for installation and use of a new system should be approached with some humility. When it comes to using computer information systems, experience is still the best teacher. There is no way that a trainer, no matter how skilled, can impart all of the knowledge and experience needed for smooth, continuous operation of a computer information system. This kind of skill and experience can only be built on the job.

Therefore, materials and presentation programs for training sessions should be prepared with the full knowledge that it will be almost impossible to complete the job of training personnel during the brief classroom sessions that are made available. Recognizing this, the training program should concentrate on teaching people to meet needs or solve problems on the job. This is a more practical approach than undertaking a probably impossible task of teaching all of the operations, functions, and skills that will be needed on a relatively complex job.

Training materials have a different purpose than procedures manuals. Training manuals should be designed as easy-to-use references. Thus, for example, it is perfectly acceptable to have a reference in a training manual that simply tells an operator what page of the procedures manual to turn to for instructions on a given job. The training manual can then offer hints aimed at helping the operator to master the functions described in the procedures manual. It is not necessary to duplicate all of the procedures manual content in a training manual. Rather, the idea is to help the operator feel comfortable with the procedures manual so that it can be used as a job aid.

In a CIS environment, there are many opportunities to use the computer itself as a training aid. This is particularly true in the training of operators working at video display terminals. Many "user-friendly" systems build in options in which operators can ask for prompting or help from the computer itself. Under one option, for example, the operator simply enters a question mark at the beginning of a line on the terminal, then presses the return key. The computer is directed to display a menu of assistance routines that the operator can call up.

Another common technique, used in data entry systems, is to display blocks of data at the top of the screen. These identify the codes or formats to be used by the operator. As the operator learns the job, this display on the screen can be eliminated.

Above all, effective training programs *teach operators to learn*. A training effort should never downgrade people to the level of machines by attempting to "program" them to make the correct responses.

The end products of this activity will be the two documents referred to earlier—training manuals and reference manuals. Following conversion to the new system, the reference manual will be the ongoing documentation needed by user operating personnel. However, because it may be necessary to train new groups of user personnel periodically, the user's manual, or training manual, should be kept up-to-date, and additional training sessions should be planned as required.

Either intangibly or specifically, an end product of this activity should include designation of personnel within the user group who will be specialists in training new employees or solving the problems for the existing staff.

CUMULATIVE PROJECT FILE

Procedures and training manuals are added to the cumulative project file. No additional documentation is required to support this activity. Content of the procedures and training manuals will vary according to methods followed in individual organizations and according to the needs of the systems. However, the following is a general outline of the elements of the procedures manual:

- Table of contents
- Narrative overview describing what the system does and where it fits within the user organization
- Individual job descriptions and procedures
- Input and source data forms, including any necessary explanations, preparation instructions, and manual processing procedure descriptions
- Data entry instructions, including procedures, controls, error correction instructions, and procedures for handling exceptions
- File maintenance descriptions, including explanations of any updating responsibilities of users, error correction, and backup procedures (Such descriptions are provided for both master files and for any tables that need updating.)
- Output documents or reports, accompanied by explanations of content, lists of distribution, and objectives for intended use
- Any needed policy statements
- Procedures for updating the manual.

The training manual includes:

- Copies of any special handouts needed for training sessions should be included in the manual. Examples would include completed forms, and so on.

- Copies of the operating procedures should be included in the training manual. Operating procedures should be placed in chronological sequence, rather than in reference order. Thus, users should be able to follow the training manual from beginning to end. By contrast, the reference manual is segmented according to function, facilitating on-the-job reference.

The procedure manual becomes part of the cumulative project file documenting the system. The training materials used entirely for training sessions need not be included in the permanent project file.

THE PROCESS

The training of users in the implementation of a new system can present an extremely delicate challenge. People who are earning a living at specific jobs—and, therefore, who are professionals in their respective fields—are being asked, in effect, to go back to school. This very requirement may pose a psychological threat to some people. Thus, one of the challenges in structuring a training program lies in putting people at their ease and establishing a feeling of mutual growth for all persons involved. To do this, the presentation techniques should be structured for credibility. Credibility exists by pre-training persons who can address the future trainees within the user group as peers. The idea, basically, is that the trainers can say, convincingly, that the new procedures are not difficult, based on personal experience. The trainers also can say that they have found the new procedures helpful in job performance and capable of enhancing the smoothness and efficiency of operation in the department. A CIS professional would not have the credibility to make these statements. Therefore, CIS professionals should work closely with and indoctrinate selected users. These users, in turn, should train their peers.

In actual training, there should be a careful mix between the building of background understanding and the establishment of confidence in the actual procedures. If appropriate, terminals should be available at the training site. Trainees should perform enough functional cycles of their respective activities so that they feel comfortable, have confidence in their ability to remember the procedures, and know they can perform efficiently.

If it is necessary to train people in the use of on-line procedures before programming actually is completed, it may be necessary to develop and implement prototype programs that simulate the functions of the actual system. Similarly, it may be necessary to prepare pilot output reports that can be used for reference in simulated job situations. It isn't necessary that the programs or simulated documents be total and complete. Rather, their scope should be great enough to represent the job conditions for which people are being trained.

PERSONNEL INVOLVED

Except for standby, resource personnel from the CIS group, training is entirely a user activity. Personnel involved include the users appointed as trainers, user management, supervisors within the user group, and all operating personnel affected by the new system. Any CIS personnel involved should maintain as low a profile as possible and should be available for consultation with user trainers. However, the users are beginning to take control of the system and must be encouraged in every way possible.

Summary

The purpose of the user training activity is to prepare users for the acceptance test that they will perform and then for the ongoing use of the system.

Physically, the activity involves CIS professionals as trainers of trainers. User training is a mix of instructional sessions and hands-on practice to provide simulated experience in the use of the system.

Documentation prepared includes reference manuals for use by operators and training manuals for instructional use.

Objectives of this activity involve understanding of the new system by user managers, training of user operating staff, receipt of needed documentation by all user personnel, and designation of a liaison person within the user group for handling future questions.

The scope of the training effort should encompass all personnel who will be affected by the new system in any way. Those who will be involved actively in new procedures should receive in-depth, detailed training. Those who simply must be aware of the existence of the new system without having to change the content of their jobs should be briefed on what is happening.

The process of this activity involves the simulation of actual system use with user personnel to build confidence and competence.

With the exception of some CIS professionals in an advisory capacity, this activity is conducted primarily by users.

The reference manual includes a table of contents, a narrative overview describing what the system does and where it fits within the user organization, individual job descriptions and procedures, input and source data forms, data entry instructions, file maintenance descriptions, output documents or reports, any needed policy statements, and procedures for updating the manual.

The training manual includes copies of any special handouts needed for training sessions and also copies of the operating procedures placed in chronological sequence.

The procedure manual becomes part of the cumulative project file documenting the system. The training materials used entirely for training sessions need not be included in the permanent project file.

Key Terms

1. reference manual
2. training manual

Review/Discussion Questions

1. What is the purpose of the user training activity?

2. From the standpoint of CIS professionals, what is the meaning of user training?

3. What documentation is prepared or completed during this activity, and how is this documentation used?

4. What are the objectives of the user training activity?

5. What is the scope of this activity?

6. What is the process of user training?

7. What personnel are involved in the user training activity?

8. What does a reference manual contain?

9. What does a training manual contain?

10. What documentation from this activity becomes a part of the cumulative project file?

Project Assignments

The Appendix presents three case studies that can be used in support of project assignments. All three scenarios provide functional system specifications that would result from systems analysis. Included are data flow diagrams, data dictionaries, and process narratives that would result from systems analysis. Continuing the work begun in Chapters 10–12, assignments that relate to the content of this chapter will be found in *PART 4: Software Design and Test Specifications*.

ACTIVITY 11:

14 SYSTEM TEST

LEARNING OBJECTIVES

On completing reading and other learning assignments for this chapter, you should be able to:

☐ Describe the activity of system testing, encompassing system tests and acceptance test functions.

☐ State the objectives and scope of the system test activity.

☐ Give the end products of the system test activity.

☐ Tell what personnel are involved in system and acceptance testing.

☐ Tell what documentation is updated during this activity and how preparation is made for the subsequent review phase.

ACTIVITY DESCRIPTION

The system test activity encompasses both system tests and acceptance test functions. That is, at the beginning of the activity, CIS professionals perform all of the procedures and use all of the equipment associated with the new system and perform testing of all programs, manual procedures, off-line equipment, and so on. When the CIS group is satisfied, users are phased into the activity and perform all testing for ensuring the competence and proficiency of personnel who ultimately will be responsible for operating the system.

Within this framework, two types of tests are performed. The first type of test is a *functional test*. Procedures are designed to make sure that all of the operations covered in the specifications actually can be performed and that expectations for results actually can be met.

The second type of test is *performance testing*. These procedures test the quality of performance rather than straightforward functions. For example, one of the performance tests that should be applied is to overload the system with volumes of transactions—up to the point at which the processing volume cannot be handled. The purpose is to make sure that the system can respond to such exceptional situations. Other performance tests deal with response times, delivery times, the usability of outputs, security (from an access standpoint), backup procedures that assume temporary system failure, and recovery from system failure.

These test procedures should be complete before installation of the new system begins.

OBJECTIVES

Prior to the system test activity, it has been ascertained that workable programs exist. The objective of this activity, then, is to test the system as an entity. Testing includes evaluating the ability of the system to function in coordination—integrating manual, off-line, machine, and computer processing to be sure that all procedures are compatible and can be integrated.

A related concern and objective lies in testing the adequacy of the user reference manual to be sure that it is both appropriate and adequate for guiding users in their day-to-day operation of the new system.

SCOPE

This activity begins only after all programs have been written and have been tested at the unit, integration, and function levels. Thus, when this activity begins, programs have been tested individually. But, systems of programs have not yet been tested as integrated entities.

The activity encompasses an initial set of tests applied by CIS professionals. Then, the professionals back away, retaining only minimal, advisory involvement as a previously trained group of users takes over and operates the system under conditions that are as realistic as possible.

At the conclusion of this activity, the detailed design and implementation phase comes to a close. The system is ready for file conversion and installation for ongoing use.

END PRODUCTS

End products of this activity include:

- A proven system ready to be put into operation
- People who have used the system and who have built both the competence and the confidence necessary to put it into operation
- Reference manuals that have been tested under life-like conditions and have been proven adequate
- Run books and any other needed manuals for the computer operations staff
- Outputs or responses that have met expectations.

THE PROCESS

At this level of testing, one of the objectives is to determine how well the system can be operated by people who are not familiar with it. Therefore, one of the procedures to be applied should be inclusion of either CIS personnel who have not been associated with the project previously or the hiring of outside organizations, possibly consultants, to perform system test functions.

Units or functions tested should reflect, as closely as possible, actual inputs, processing, and outputs to be delivered by the system. Full replication of actual conditions may not always be possible because of the availability of computer time, delivery of special forms, and so forth. Also, because existing procedures already occupy allocated slices of computer time, there may not be time to exercise the new system fully during normal business hours. Such situations should be considered one of the realities of systems development. Many system tests are conducted in the evening hours or over weekends for this reason.

At some point during system tests, the computer should be loaded with transaction and processing volumes that far exceed the specified limitations of the system. The object is to find out whether and how well the system can deal with overloads. A concerted attempt should be made to force the system to a level at which it cannot handle the work load that has been presented. Recall that the purpose of testing is to expose the conditions under which the system will fail. Accordingly, system test is, in effect, a stress test. In forcing failure, the idea is to determine whether recovery can be implemented as planned.

Another system capability to be tested is response or delivery time. For on-line systems, there should be specifications about the time required for a system to receive a transaction, complete processing, and return a response to the user. On batch systems, there should be schedules that cover the elapsed time from receipt of source documents to the delivery of outputs. This time interval is referred to as *turnaround time.* Volume and timing requirements of system operation should be tested to discover the point at which timing standards cannot be met.

System testing is the last opportunity to reaffirm that users actually want the outputs that they have specified and that these results will be of practical use. The measure of usability also extends to determining that the reference manuals provided are adequate for implementing the new system.

Although outside the realm of routine operation, procedures for backup and recovery should be considered as important portions of the system to be tested. A system that cannot provide backup service or recover from disruption is, effectively, worthless. Therefore, all backup and recovery procedures should be used. One of the important results of this phase of testing is that users build the confidence to know that they can stay in business if the computer goes down.

Security measures as applied by hardware and software should be tested as well. System tests need not be concerned with physical measures within a computer installation itself because these measures are part of the overall operations function. However, strenuous attempts should be made to access system files and procedures illicitly. If the system itself is sufficiently sensitive, a consultant who specializes in breaking hardware and software access codes should be engaged to find out just how secure the system is and how well it will protect the privacy of the files it processes.

It is worth repeating and stressing that, when users are going through acceptance testing, the CIS staff should stay as far away from the system as possible. Project team members should be designated to field questions and provide assistance when requested to do so. This support should be provided as realistically as possible, simulating conditions that will exist when problems come up after the system is in regular use.

Problems will be discovered during system tests. For each problem, an evaluation should be made of its importance and potential impact on actual operations. Only corrections necessary for reliable use of the new system should be made prior to file conversion and installation. Other required changes should be scheduled as maintenance projects.

PERSONNEL INVOLVED

During system tests, the project team leader and members of the technical design staff monitor results. However, other experienced CIS professionals, either from the same shop or from an outside organization, should perform the actual test procedures.

During acceptance testing, users should perform all functions, interacting only with designated CIS resource persons.

CUMULATIVE PROJECT FILES

During this project, reference manuals to be used in conversion and operation of the new system should be updated as necessary, reproduced, and distributed to authorized users. The same should apply to operations manuals, which should become part of the operations center library.

Any required changes in programs or system documentation should be incorporated in library copies as well. Procedures should be followed for identifying and dating all changes to existing documents.

The accumulated documentation from the earlier phases and activities should be assembled and prepared for examination during the review phase.

Summary

The system test activity encompasses both system tests and acceptance test functions. Two types of tests are performed, functional testing and performance testing.

The objective of this activity is to test the system as an entity. A related objective is assuring the adequacy of the user reference manual.

The scope of this activity encompasses an initial set of tests applied by CIS professionals. Then, the professionals back away, retaining only minimal, advisory involvement as a previously trained group of users takes over and operates the system under conditions that are as realistic as possible.

At the conclusion of this activity, the detailed design and implementation phase comes to a close. The system is ready for file conversion and installation for ongoing use.

End products of this activity include a proven system ready to be put into operation, trained users, proven reference manuals, run books and any other needed manuals for the computer operations staff, and outputs or responses that have met expectations.

At some point during system tests, the computer should be loaded with transaction and processing volumes that far exceed the specified limitations of the system. The object is to find out whether and how well the system can deal with overloads and to determine whether recovery can be implemented as planned. Another system capability to be tested is response or delivery time.

Security measures as applied by hardware and software should be tested as well.

During system tests, the project team leader and members of the technical design staff monitor results and other, objective CIS professionals perform the actual test procedures. Users perform acceptance testing.

Reference and operations manuals, programs, and system documentation are updated as required during this activity. The accumulated documentation from earlier phases and activities should be assembled and prepared for examination during the review phase.

Review/Discussion Questions

1. What tests are performed during the system test activity?
2. What is the objective of the system test activity?
3. What is the scope of this activity?
4. What are the end products of the system test activity?
5. Beyond functional system testing, what specific system capabilities must be tested during this activity?
6. What personnel are involved in system testing?
7. What documentation is updated during this activity?
8. For what purpose is the accumulated documentation from earlier phases and activities assembled during this activity?
9. In testing a system as an entity, what system elements may have to be integrated?
10. Why might it not be possible to replicate fully actual conditions for purposes of testing the system?

Project Assignments

The Appendix presents three case studies that can be used in support of project assignments. All three of the scenarios provide functional system specifications that would result from systems analysis. Included are the data flow diagrams, data dictionaries, and process narratives that would result from systems analysis. Continuing the work begun in Chapters 10–13, assignments that relate to the content of this chapter will be found in *PART 4: Software Design and Test Specifications.*

INSTALLATION AND REVIEW PHASES III

The two phases of the systems development life cycle covered in this part are critical to the success of any CIS project. These phases mark the points at which the project effort delivers its promised payoff, then is measured to determine whether expectations match promises.

During the Installation Phase, the existing system or ad hoc methods that were formerly in use are replaced. The new system takes over and becomes one of the resources of its organization. For users, the new system rapidly becomes part of a way of business life. For CIS professionals, the system becomes part of the background applied to the challenges of new projects.

The final life cycle phase, Review, is dedicated first to evaluating the work done on the now-active system. In addition, review procedures provide opportunities to improve systems development practices within the organization, to evaluate performance of individual team members, and to establish a list of any required modifications that should be implemented for the new system.

The actual skills applied in the final two phases of a project structure go beyond the scope of a classroom-oriented course. These practical areas of systems development skill can be honed in future project assignments within the CIS curriculum. Thus, this text emphasizes the content from which you can derive the greatest benefits at this stage of your experience.

15 INSTALLATION

LEARNING OBJECTIVES

On completing reading and other learning assignments for this chapter, you should be able to:

☐ Identify the scope and major achievements associated with system installation.

☐ Describe the four basic methods of system installation and explain the advantages and disadvantages of each.

☐ Identify the steps involved in file conversion.

☐ Discuss the transitions to user ownership and system maintenance.

PHASE DESCRIPTION

Installation is the fourth phase in the systems development life cycle. The position within the life cycle of this phase and the two activities that comprise it are illustrated in the flowchart in Figure 15-1.

This phase is critically important for two reasons. First, it marks the culmination of development efforts and the realization of the proposed new system. Second, it is a critical transition time for the users. Actual, realized—as opposed to projected—benefits will depend on how the user group learns to adapt to the system during the installation phase.

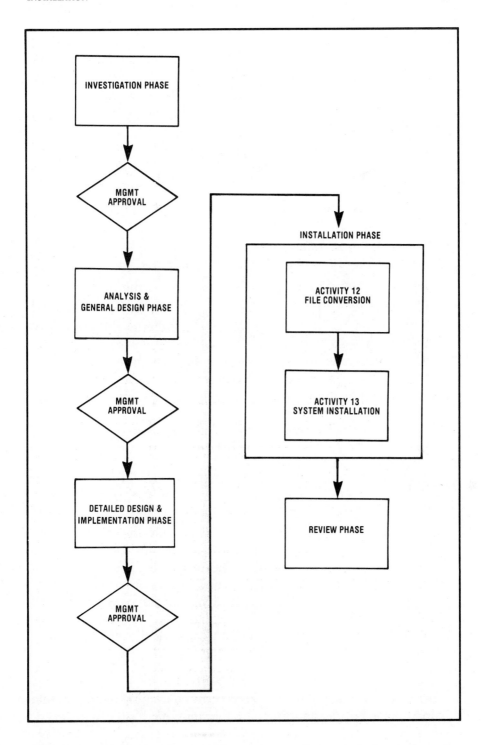

Figure 15-1. Diagram shows activities of the Installation phase in relation to the overall systems development life cycle.

During this phase, users actually take over ownership of the new system. That is, development is complete. The new system is operational. The new files exist and are in day-to-day use. A relationship has begun between users and CIS operations, with systems analysts fading gradually from the picture.

The actual method of installation for the new system will depend upon its design, upon the needs and preferences of user managers, and upon the risks that can be tolerated. Options include:

- Cutover can be abrupt, with the old system simply discontinued and the new one begun at the same instant.

- The old and new systems can be operated in parallel for some time while results of the two systems are compared.

- There can be a parallel conversion in which the old system is gradually phased out while the new one takes over.

- *Version installation* techniques can be used. Under this approach, the system is divided into a series of functional areas, or incremental steps, called *versions*. These versions can be installed in any of the three ways described above. However, the entire system will not be fully implemented until all of the versions are in place.

These options, and the trade-offs involved, are discussed later in this chapter.

OBJECTIVES

This phase has two principal objectives:

- First, the existing system is replaced with the new, tested, documented system. This replacement assumes user acceptance and ongoing user responsibility for the system. The project team is disbanded, and analysts and programmers are removed from routine involvement with system operation. In connection with system implementation, all files utilized by the old system are converted to the new system, and use of the old system is discontinued.

- Second, to maximize the potential benefits of the new system, the user must be taken beyond simple how-to training to an intimate and detailed understanding of the system that has been installed.

SCOPE

This phase begins with the existence of a complete system—or, if appropriate, a version—that has been completely tested under realistic conditions and is ready for installation. In addition, any needed file conversion programs have been written and tested.

The phase ends with a new system, or version, implemented and in day-to-day operation, with no further intervention or supervision by members of the project team. The conclusion of this phase is also marked by discontinuance of the old system.

END PRODUCTS

No major new end products are produced during this phase. Rather, previously designed and developed products are implemented. New files are created or converted and put into regular use in support of the new system.

In addition, all previously prepared documentation is updated and placed in maintenance status. Copies are distributed to persons who need them, and arrangements are made to update all copies as necessary to reflect modification or maintenance of the system. Processing schedules or calendars to be followed by both users and the CIS operations group are established and put into regular use.

THE PROCESS

This overview of the installation process focuses on three main concerns:

- Strategies for file conversion
- Basic system installation alternatives
- Personnel transitions that must occur as the responsibility for the system moves from the development team to the user organization.

File Conversion

File conversion strategies vary with the complexity of the system and with the installation method used. In some cases, file conversion can be completed quickly and with few complications. In other situations, however, the conversion of files can be relatively complex.

Problems, when they arise, center around conversion from the old system to the new one and the possible need to support both systems or parts of both systems concurrently. There is apt to be some delay, or lag time, during which the old system still needs its files while the new one also requires access to files that have already been converted. Thus, one system or the other may be operating with files that are not completely current.

These problems can be avoided if the nature of the system lends itself to an abrupt conversion. For example, changes in general ledger accounting systems are typically made at year-end just to avoid this type of problem. At the end of the fiscal year, the old system begins its closeout routines. The new system then starts up with all balances at zero. If payments or bills are received that should be accounted for in the prior year, these can be processed under the old system. Transactions dated after the first of the year are run through the new system. There are no processing conflicts or file conversion problems because a clean break has been made.

However, this kind of transition is not always possible. For example, suppose a conversion is being made in an accounts receivable system. The existing files represent all unpaid bills owed to the organization. As the new system is implemented, all of the data from the existing files must be captured into the new ones. In the interim while the old system is still in operation, there could be trouble in finding such information as current customer account balances for the purposes of credit authorization. To avoid such problems, arrangements are usually made to maintain and access both files during some interim period while files are being converted. In general, file conversion involves the following basic procedures:

- Prepare existing computer files for conversion. This means that all master files should be brought up to date. Accuracy should be verified. Errors should be identified and corrected.

- Prepare existing manual files for conversion. Post manual data to new system maintenance input forms, and do data entry of manual data.

- Build new files and validate them as they are created.

- Begin maintenance on the new files. Input data continue to update old files until after implementation, but the converted files must also be updated. The basic procedure is: First, establish a cutoff date for each file to be converted. Then, any input documents that represent transactions after the conversion data are batched and used for periodic updating of the converted file until installation of the new system.

- Make a final check of accuracy, or balancing, between the new files and the old ones.

Installation Alternatives

The methods used in converting files will depend at least in part upon the alternative selected for installation of the new system. The installation technique selected depends chiefly upon the nature of the new system and the trade-offs involved in the various installation alternatives. The basic alternatives are:

- Abrupt cutover
- Parallel operation with a single cutover point
- Parallel operation with a gradual shift from the old system to the new
- Version installation.

Abrupt cutover.   As illustrated earlier, an abrupt cutover involves a simultaneous dismantling of the old system and start-up of the new one. It's just that simple. At a predetermined time, the old system no longer exists. The new one handles all transactions.

One advantage of this approach, in situations where it can be used, is that costs are minimized. There are no transition costs because there is no transition.

In some cases, an abrupt cutover may be the most natural, if not the only, way to solve a problem. In addition to the year-end conversion of accounting systems, consider situations in which a new system changes the way a company does business. Consider, for example, what happens when a supermarket installs a checkout system using the universal product code. Under the old system, all prices had to be entered into keyboards by checkers. Checkers needed extensive, manual reference files to look up prices if packages weren't marked clearly. Under the universal product code, store personnel can stop marking individual packages and simply place prices on the shelves where the products are displayed. Pricing is done by an on-line computer. A computer price file replaces the manual reference file. Abrupt cutover represents a natural way to convert to the new system.

The major disadvantage of this approach is that it can carry a high risk. In an abrupt cutover, the old system is stopped. If a major problem develops with the new system, it may be very difficult—perhaps impossible—to return to the old system. Depending on the system and its role in the organization, the ability to carry on the business could be curtailed.

Parallel operation, single cutover. Under this approach, both systems are operated concurrently for some period of time. Often this parallel operation period coincides with business processing cycles, such as weeks or months. During this interim period, all input transactions are used to update the files that support both the old and the new systems. A balancing between results of the two systems is performed regularly.

An advantage of this approach is that risks are relatively low if problems arise in the start-up of the new system. The corresponding disadvantage is the cost of operating both systems concurrently.

A typical use of this approach occurs when a computerized system is replacing manual procedures. Users already trained in the manual procedures simply carry on for a while, phasing out after the new system has proved itself.

Parallel operation, gradual cutover. Again, both systems are operated concurrently. However, rather than having a single cutover point between systems, the old system is discontinued gradually. Discontinuance of the old system can be according to geographic location, type of business, or other criteria.

Advantages of this approach, once again, include minimizing the risk associated with any problems that may arise with the new system. Costs are more moderate with a gradual cutover than they are if the old system is continued for a predetermined period of time. With a gradual cutover, the old system can be discontinued as quickly or slowly as management feels comfortable with the new one.

A disadvantage lies in the possible confusion that can result if people are unsure about which system to use.

To illustrate how this approach might work, an order entry system might discontinue the old system for one sales district at a time. Another possible approach would be for cutover to occur according to the order processing points. For example, on a given day, one warehouse would put all of its orders through the new system, discontinuing the old. In the water billing system introduced earlier in this book, the cutover could be done by billing cycles. That is, the new system could be applied to one cycle of customers at a time.

Version installation.   Under this approach, a basic set of capabilities is implemented within the first version of the system, then additional capabilities are added in subsequent versions. Each version goes through a complete implementation and installation cycle of its own. Some time after the first version is operational, procedures, programs, and files are implemented for the second version, which then goes through the installation process. This procedure is repeated for each version. In effect, version installation involves breaking the proposed system into a series of incremental steps, or versions, at the end of the analysis and general design phase and then implementing and installing the system one step at a time.

To illustrate, consider the example of the supermarket that installs a universal product code checkout system. In the first version, files, programs, and procedures would be set up for identifying and pricing products at the checkout counters through use of the computerized system. In the second version, inventory levels could be added to the product file. This would make possible stock control at the individual store level. Then, in a third version, stock replenishment, or requisitioning of merchandise, could be added based on sales histories that are accumulated within the computer files. Thus, each version adds capabilities to a basic system rather than requiring development of an entirely new system each time.

Version installation represents an option that can modify the steps followed in the later phases of the systems development life cycle. However, the basic project structure and methodologies remain intact. If a version installation option is used, the complete system should still be analyzed as an entity and carried through, again in its entirety, at least to the end of the analysis and general design phase. The system can then be partitioned into versions for performance of the remaining activities of the detailed design and implementation phase and the installation phase.

Some definite advantages accrue from the practice of designing a system as an entity even though it may be implemented and installed in different versions. One of these is that the database can be designed with the total system needs in mind. Thus, the complete system will be supported as each version is implemented and installed. This is because the interrelationships among versions are understood from the outset. Similarly, some of the application programs in the total system may be shared by different versions. If the entire system has been designed in advance, the finished programs will be more appropriate for the final jobs they will be expected to do.

If version installation is used, there will be some modification in the structure of the systems development life cycle. Steering committee decisions at the conclusion of the second and third phases of the project will be limited to one version at a time. This means that the resources allocated in individual decisions will be smaller. On the other hand, financial feasibility may be difficult to justify for the first version because costs may be relatively high and benefits limited. Moreover, changes in cost pictures over time could result in decisions not to implement successive versions after the first one has been installed.

Another possible advantage of implementing the system one version at a time, rather than completing the entire system at once, is the potential for greater responsiveness. Because a part of a system can be implemented in less time, at lower cost, it becomes possible to demonstrate results more quickly and to build credibility for the system as it is unfolding. At the same time, this approach makes it possible to realize some of the benefits of the system at an earlier point than would be possible if the entire system were being implemented at once. This partial completion can be particularly attractive if tight deadlines are involved. For example, it may be possible to implement the version that meets regulatory needs immediately, leaving enhancements to the system for later implementation.

Problems to avoid in connection with version installation involve inconsistencies in overall design that can come with the breaking down of a system into separate versions. Further, version implementation may diminish control over overall system costs.

In summary, version installation can be a potentially valuable option under the right circumstances. However, this methodology requires careful thought and planning to identify proper versions and to provide for tight management control over resources committed to the project over time.

Personnel Transitions

Up to this point in the project, users and systems analysts have literally formed a team. Close relationships and understandings may have evolved, and friendships may have been built.

Whatever the personal relationships that may exist at this point, installation marks a transition. Once the system is installed and in regular operation, it belongs to the users. Users are the owners of operational systems. Systems analysts have completed their mission; it is time for them to move on to other projects. At this point, it is part of their job to disengage as expeditiously as they can.

Another systems analysis responsibility associated with installation is to avoid making any system changes that are not absolutely necessary at this point. A list of maintenance changes to be made following implementation is initiated in the previous phase and continued through installation. What must be avoided is a situation in which a new system remains incomplete—or is permitted to overrun costs exorbitantly—because of a flurry of last-minute modifications and changes. A management nightmare in the systems development field is the system that is 99 percent complete—indefinitely.

One way to be sure that the old system is discontinued is to terminate use of its documentation and programs. The documentation and programs for the old system should be relegated to the archives as part of the installation procedure.

Two special concerns are worthy of additional notice at this point: assuring that the user has the understanding to make the best possible use of the system, and establishing a procedure for moving into the maintenance phase of the system life cycle.

Building user understanding. Despite the user training activity during the detailed design and implementation phase, the installation of a new system is seldom routine. The problems usually involve the user. Systems analysts can predict ahead of time what the computerized portion of the system will do, but it is very difficult to predict what the user will do. Often users have a much bigger problem adapting to the new system than the developers had building it.

A good user training program is only the first step in building user understanding of the new system. Also, training programs tend to be most effective at the clerical or operational level. This type of program should be followed, during and after installation, by a series of discussions with users at all levels. At the start, these discussions can center around perceived problems with the system. Later, for situations in which several users interact with the system in much the same way, these sessions can emphasize insights gained by individual users in making full use of the system. The point is to encourage the user to go beyond a merely mechanical use of the system to a deeper understanding of its capabilities. The user should learn to exploit the system. It is not at all uncommon for a user who understands how a system works to successfully apply parts of the system in ways that would not have occurred to the original project team.

There are two main requirements for developing user understanding of the new system. The first is that the system work effectively—that it be reliable and easy to use. A system that is straightforward, with clean input requirements and clearly understandable outputs, is far superior to a system with numerous functions, some of which may not always perform reliably, and more complex, unnatural rules for describing inputs and outputs. The second requirement is an alert user management—one that provides the motivation and education required to make effective use of the system.

Transition to maintenance.   Ongoing maintenance of a new system is considered to begin from the time the installation phase ends. To the extent possible, maintenance projects should be held for consideration until after the post-implementation review phase, when it becomes more feasible to consider the results of the system and to put the need for and role of maintenance in perspective.

However, there may be requirements for maintenance that just can't wait until formal reviews have been performed. When maintenance is needed, it should be done following in-place maintenance procedures, rather than as an extension of the development project.

Within each CIS operation, there will usually be one or more staff analysts with maintenance responsibility. Maintenance requests should be routed through these regular channels. Normally, early maintenance requests will involve either correction of errors considered to be important or minor procedural changes that are easy and quick to make. Major enhancements of the system, unless they are mandated by regulatory agencies or changes in corporate policy, should wait until after the review phase is completed.

Whenever maintenance begins, it is important that procedures for updating documentation and keeping all document files current be instituted at the same time. Standards for current, accurate documentation should never be compromised for any kind of maintenance project.

PERSONNEL INVOLVED

File conversion work during this phase is handled by programmers and analysts. Installation responsibilities are coordinated among analysts, key users, and CIS operations personnel.

Summary

The installation phase marks the culmination of development efforts and the realization of the proposed new system. It is also a critical transition time for users, as they take over ownership of the new system.

The method of installation depends on the type of system, the needs and preferences of user managers, and the risks that can be tolerated. Options include an abrupt cutover, parallel operation with a single cutover, parallel operation with a gradual cutover, and version installation.

This phase has two principal objectives. First, the new system must be placed in full day-to-day operation, and the old system discontinued. Second, the user must gain an intimate and detailed understanding of the new system.

The general procedure for file conversion involves the following steps: Prepare existing computer files for conversion by updating, verifying accuracy, and correcting errors. Prepare existing manual files for conversion. Build and validate new files as they are created. Begin maintenance on the new files. Make a final check of accuracy, or balancing, between the new files and the old ones.

Once the system is installed and in regular operation, it belongs to the users. Systems analysts have completed their mission and should disengage as expeditiously as they can. Above all, systems analysts must avoid making any system changes that are not absolutely necessary at this point.

The user training program should be followed up, during and after installation, by a series of discussions with users at all levels.

Ongoing maintenance of a new system is considered to begin from the time that the installation phase ends. To the extent possible, maintenance projects should wait until after the post-implementation review phase has been completed. If maintenance is required, however, it should be handled through normal maintenance channels. Updating of documentation must accompany any maintenance project.

Key Terms

1. version installation
2. version

Review/Discussion Questions

1. What are the most important results of system installation?

2. What are the four basic alternative approaches to system installation?

3. When is system installation considered complete?

4. Under what circumstances are file conversion problems most likely to arise? Why?

5. What are the basic steps involved in file conversion?

6. What are the principal trade-offs between abrupt cutover and parallel operation with a single cutover? Under what circumstances would you be likely to choose one approach over the other?

7. Describe a situation in which parallel operation with a gradual cutover might be the most appropriate installation method. Why?

8. Explain how a version installation might be combined with each of the other three installation methods.

9. What are the main responsibilities of the systems analyst as the installation phase draws to a close?

Project Assignments

The Appendix presents three case studies that can be used in support of project assignments. All three of the scenarios provide functional system specifications that would result from systems analysis. Included are data flow diagrams, data dictionaries, and process narratives that would result from systems analysis. Continuing the work begun in Chapters 10–14, assignments that relate to the content of this chapter will be found in *PART 4: Software Design and Test Specifications*.

16 REVIEW

LEARNING OBJECTIVES

On completing reading and other learning assignments for this chapter, you should be able to:

☐ Identify the scope and objectives of system development recaps and post-implementation reviews following systems development projects.

☐ Describe the end products and other results of the post-implementation reviews.

PHASE DESCRIPTION

The position of the review phase within the systems development life cycle—as well as the component activities that make up this phase—are shown graphically in the flowchart in Figure 16-1. A major characteristic of this phase is that it is completed in a short time span that is devoted to intensive study and analysis of project results. The phase begins with Activity 14: Developmental Recap. This activity is devoted to an in-depth study of the developmental activities that have just been completed. The purpose of the recap is to prepare specific suggestions aimed at:

• Helping individual team members to perform more effectively on future project assignments

• Sharpening management skills for the organization as a whole and for the project team leader in particular

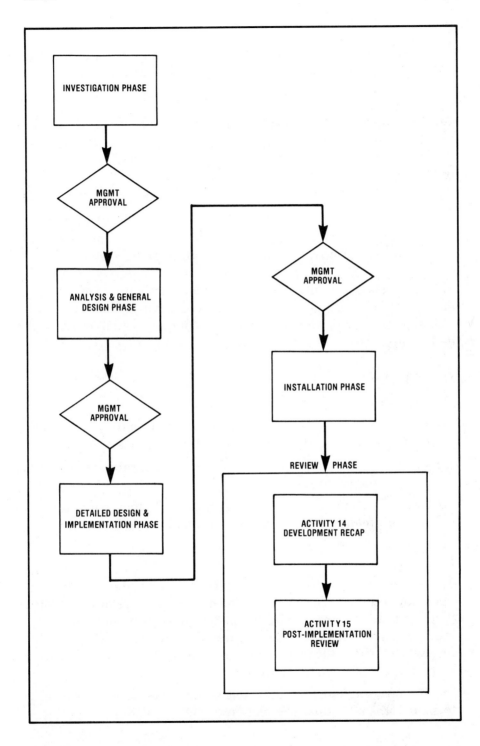

Figure 16-1. Diagram shows activities of the Review phase in relation to the overall systems development life cycle.

- Finding approaches that might enhance or improve the organization's skills and methods in systems development.

Activity 15: Post-Implementation Review is conducted after the new system has been in operation for some time. This time lapse permits the new system to become a regular part of day-to-day operations and the people involved to gain a measure of detachment. The purpose of this review is to:

- Evaluate how well the system has performed in meeting original expectations and projections for cost/benefit improvements.
- Identify any maintenance projects that should be undertaken to enhance or improve the implemented, ongoing system.

This second activity within the phase is particularly useful as a review of projects for which personnel or other cost savings were projected. This activity provides an opportunity to compare actual results with earlier projections.

OBJECTIVES

Objectives for this phase are:

- Review systems development results in terms of the effectiveness of the life cycle and the management techniques applied.
- Review the new system to determine whether projected benefits have actually been realized.
- Review the new system to determine whether enhancement through maintenance projects is desirable and justifiable.

SCOPE

The development recap should begin immediately after the system is operating routinely. Even if some lingering tasks or details associated with installation remain to be completed, there is no point in delaying the recap activity. On the contrary, it is better to conduct this review while project team members are still available—and while their memories are still relatively fresh.

The post-implementation review is usually performed some four to six months after final completion of the installation phase.

END PRODUCTS

This phase produces two end products:

- The systems development recap report
- The post-implementation review report.

Systems Development Recap Report

This document is prepared for CIS management. Its contents, of course, reflect the nature of the individual project. However, certain basic items should be included:

- Development costs should be analyzed. The presentation should compare the projected budget with actual costs, breaking the figures down by cost category within each activity. Any significant variances should be analyzed and explained.

- Working time on the project should be reported and analyzed. Comparisons should be made between budgeted and actual working hours for each activity. Variances should be analyzed to determine cause, including rework required by user changes; rework required to meet outside mandates, such as rulings from regulatory agencies; rework caused by design errors; rework caused by programming errors; overruns due to failure of development team members to complete work as scheduled; and overruns due to errors in estimation.

- Any design errors identified during the review should be described and classified according to their nature and should be related to any required rework.

- Programming errors should be similarly reported and classified.

- Suggested revisions in the systems development methodology should be described and evaluated.

- Any other suggestions or insights should be described.

Post-Implementation Review Report

This report is prepared for review by CIS and user departments. It may also be delivered to the steering committee. The following elements should be covered:

- The original requirements and objectives that led to the systems development project should be listed. Accompanying this list should be an evaluation of the extent to which the original requirements and objectives are being met by the installed system.

- The costs of developing and operating the new system should be reviewed and compared with original cost estimates.
- The originally projected benefits should be compared with the benefits actually realized.
- The new system should be reviewed as a functional entity to determine, first, whether any steps can be taken to realize more of the original or additional benefits, and, second, whether any modifications are needed in the near future.

THE PROCESS

The process approach for each of the activities in this phase is fairly straightforward.

Development Recap

Systems development is difficult to structure and do well—largely because it is so people-oriented. Because of these special challenges, a great deal of interest and effort is devoted to developing new techniques and methodologies that can make the development process more effective. But for an organization to incorporate these new ideas and approaches into its own systems development methodology, it is necessary periodically to pause, reflect on past experiences, and suggest modifications based both on these past experiences and on new techniques that have been developed.

The purpose here is to give the project team, and the organization, an opportunity to reflect on the project that has just been completed and to draw lessons and recommendations for improvement from the experience. As a starting point for this activity, the project leader prepares statistical reports that recap the development effort. These include comparisons between projected and actual expenditures, in money and in working time, for each activity. Causes should be assigned to any variances reported. Causes of variances should be readily supportable from statistics gathered during the development project. These may include specification changes that resulted in rework, identifiable errors that led to rework or overruns, inaccurate original projections, or performance by team members that was different from what was expected.

Team participation.   The development recap activity offers professional growth opportunities to each member of the project team. To realize these benefits, a series of meetings should be held to deal with the activities or phases of the project. Persons who were active during each of the phases or activities covered by a meeting should be present. With this level of participation, persons attending the meeting can understand and participate in the reviews and critiques of the work done. Based on their experience, participants can take part in brainstorming sessions aimed at improving project development and administration methods. Active participation in sessions of this type should enhance the professionalism of each of the individual team members.

Skill is needed to keep meetings of this type on track and productive. They must be approached with a positive attitude. The emphasis must be on making positive recommendations for future development work—not on retributions for past mistakes. Otherwise, the activity can degenerate into a forum for finger pointing and excuse making. The meetings themselves should be relatively brief. For example, a recommended schedule might include two one-hour meetings on each of the three key phases of the project.

Importance of the development recap.   Many systems development life cycles do not list a development recap as a separately identified activity. This is understandable. First, there is the pressure to "move on." While the recap may be seen as "nice to do," there is normally a backlog of development projects awaiting action, and management sees greater payoff in beginning them with no further delay. Second, without the proper approach and management backing, the recap may be seen as a threat to members of the project team. They may view it only as an exercise in covering past errors and failures.

However, as stated above, without this recap activity it is very difficult for an organization to break out of its old approaches to problems and take advantage of advances being made in the systems development area. With a separately identified development recap activity, and with positive management support and expectations, the stage is set for growth in the ability of an organization to respond to systems development needs.

Post-Implementation Review

This is an actual review of the new system after it has been installed and operating for four to six months. Depending on the size of the system, this review may require the efforts of one or more analysts. These analysts may or may not have been members of the original project team.

Standard systems analysis techniques are used, including interviews with users and operations personnel. Data are collected on processing volumes and operating costs as a basis for analysis and comparison with the projections made during the feasibility study and updated at the end of the analysis and general design phase.

The job of the systems analysts completing this work is, in part, to determine whether user objectives are being met. The results of the new system are compared with the stated objectives. These objectives are contained in the user specification and also in the new system design specification—documents produced during the analysis and general design phase. In addition, the analysts are charged with determining whether projected benefits for the new system are being realized. A comparison is made between existing costs and benefits and those projected during the analysis and general design phase of the development project.

Any problems noted should be analyzed and described. If appropriate, recommendations for corrective action should be submitted.

TRANSITION INTO MAINTENANCE

At several points in discussions covering activities within the systems development life cycle, this book indicates that the scope of the project should be retained within boundaries approved by the management-level steering committee. If possible, extensive modifications of or additions to a system under development should be held until after installation, considered during the review phase, and scheduled for maintenance projects. Thus, almost from the day a system becomes operational, it becomes a candidate for enhancement, updating, or revision through maintenance projects. Maintenance projects stem from four major sources:

- Mandates. Government regulations or industry requirements may dictate the inclusion of new or modified features within an existing system.

- Management policies or strategies. As management develops the strategic plans for a company, new requirements emerge in all of the systems that constitute that company. Information systems are an integral part of these corporate systems and must be changed to support the directions established by management.

- User perceptions. Users become progressively more sophisticated as systems are developed and applied. The more experienced they become, the more opportunities users are apt to discover for enhancing the systems they have. Most systems development requests originate with identification of user needs. The same is true for systems maintenance. An alert group of users will generate a continuing stream of opportunities to enhance existing systems.

- Technological advances. Computers are at the hub of one of the most dynamic fields of endeavor in the world. Whole new generations of concepts and equipment are brought into use every five or seven years. In between, there are many additional, significant improvements in hardware and software. At any given time, new equipment or new methodologies may make it economically feasible or competitively necessary to modify existing systems to accommodate advanced capabilities.

As the CIS field itself has matured, maintenance has become an increasingly important segment of the activities of systems analysts, systems designers, and programmers. Current estimates are that, in ''mature'' CIS shops, maintenance projects now occupy as much as 60 to 70 percent or more of the time of systems and programming personnel. Thus, cumulatively, maintenance projects can be greater factors than systems development projects in terms of working days of systems and programming time consumed.

As the working environment in the CIS field assumes these dimensions, it becomes opportune to notice and comment upon the close similarities between maintenance and systems development projects. Make no mistake about it; systems maintenance is a project-oriented kind of undertaking. In maintenance, just as in new systems development, it is essential to go through a series of steps that involve understanding the problem, analyzing needs and opportunities, then devising solutions and designing them technically before going ahead with implementation. In other words, maintenance is not a patchwork undertaking. If maintenance is treated on a "quick and dirty" basis, there is a great danger that the systems being maintained will, in fact, be destroyed by those who want to enhance them. All of the principles about sound design and maintainability apply in maintenance projects just as they do in systems development projects. Therefore, it follows that all of the skills acquired in connection with the study of systems development are equally appropriate in maintenance projects. The principles of sound systems maintenance are essentially the same as for systems development. Only the scale tends to be different. Quality is quality, wherever needs are encountered.

Summary

After a new system is operational, there should be two reviews of results—a recap shortly after implementation and a post-implementation review four to six months later.

The recap should begin as soon as the system is operating routinely. Its purpose is to help individual team members perform more effectively on future project assignments, sharpen management skills, and find approaches that might enhance or improve the organization's skills and methods in future systems development projects.

The systems development recap report prepared for CIS management should compare projected costs, both in money and in working time, against actual costs. Figures should be broken down by category within each activity, and any significant variances should be analyzed and explained. Any design errors or programming errors that may have necessitated reworking should be reported and classified. Finally, any suggested revisions in the systems development methodology should be described and evaluated.

The purpose of the post-implementation review is to evaluate how well the system is meeting original expectations and projections, and to identify any maintenance projects that should be implemented to enhance or improve the system.

The post-implementation review report is prepared for review by CIS and user departments, and may also be delivered to the Steering Committee. This report should include a list of the original requirements and objectives and an evaluation of the extent to which these have been met. Developmental and operational costs of the new system should be reviewed and compared with original cost estimates, and the originally projected benefits should be compared with the benefits actually realized. Finally, the new system should be reviewed as a functional entity to determine, first, whether any steps can be taken to realize more of the original or additional benefits, and, second, whether any modifications are needed in the near future.

Review/Discussion Questions

1. What reviews should be conducted following implementation of a new system?

2. What information should be gathered and reported during the recap activity that takes place shortly after implementation?

3. What information should be gathered and reported during the post-implementation review activity that takes place four to six months after implementation?

4. What can users gain from review activities that follow implementation of a new system?

5. What can CIS management gain from review activities that follow implementation of a new system?

6. What can top management of a company gain from review activities that follow implementation of a new system?

7. How can procedures for systems development be employed for systems maintenance?

Project Assignments

The Appendix presents three case studies that can be used in support of project assignments. All three of the scenarios provide functional system specifications that would result from systems analysis. Included are data flow diagrams, data dictionaries, and process narratives that would result from systems analysis. Continuing the work begun in Chapters 10–15, assignments that relate to the content of this chapter will be found in *PART 4: Software Design and Test Specifications.*

APPENDIX:
SYSTEMS DESIGN
PROJECTS

INTRODUCTION

This appendix presents case scenarios that can be used as the basis for project assignments that parallel the material covered in the text. Assignments call for student completion of activities related to the transition from systems analysis into systems design, to system design activities, and to software design. Although programming assignments are not included within these projects, system analysis, system design, and software design will have proceeded to a level of detail sufficient for assignment to a programming staff.

The assignments presented here are tailored to class-project conditions. Some systems development tasks typically followed in industry settings have been modified or deleted to fit assignments within realistic educational boundaries. However, the fundamentals of design as a problem-solving task have been retained. The basic problem-solving strategies, techniques, and tools of design are applied within the context of the case projects.

The cases have been divided into four major parts, corresponding generally with the life cycle activities that:

- Produce a complete, detailed system specification

- Provide a transition between the products of systems analysis and the development of a general system design

- Produce a system design that documents the software products, data files, and other hardware and software resources required to implement the system

- Develop a software design as the basis for programming the system.

Each part of the assignments calls for materials to be produced and submitted for evaluation. These milestone events will be necessary for keeping projects on track and for providing feedback to students.

The four major parts of the assignments require students to complete designs for basic, functional systems. These systems implement the business procedures described in the case scenarios. With the scenarios as starting points, further design efforts are needed before the systems designs can be implemented. For example, the systems may require file creation programs to build master files; maintenance programs for additions, changes, and deletions involving transaction and reference files; backup programs for protection of master files, and other, added operational procedures required for system integrity and continuity. The case scenarios describe the logical business systems. However, the designer may have to develop additional processing routines to configure viable systems. In some cases, the existing documentation outputs from analysis will have to be modified to fit solutions within hardware and software constraints.

Each of the case problems presents the opportunity for supplemental design projects. The instructor may choose to require these projects as part of the basic cases or may assign them as options for additional credit. If students are working in project teams, the supplemental projects should be considered as part of the normal case requirements.

The cases can be completed by students individually or in project teams, as directed by the instructor. This latter arrangement will give students an excellent sense of the pressures, responsibilities, and other interpersonal experiences encountered by systems development professionals. The ability to work as contributing members of project teams can be valuable and, if possible, should be emphasized throughout the case projects.

It is assumed that complete systems analysis documentation exists as the starting point for the cases. That is, the projects assume existence of a set of data flow diagrams, data dictionary entries, and process narratives that describe a new system under development. There will be occasion for the student to expand and modify these documents as design realities are faced. However, in most cases, the documents represent system specifications that should not be changed. System users have signed off on the specifications and expect systems that deliver these outcomes. Nonetheless, the case scenarios are not formulated to constrain the designer. Rather, the cases are general enough to allow alternate solutions to each of the problems and to encourage creativity and problem-solving insight from the student.

The process narratives included in the case scenarios provide more implementation details than would be included in standard process descriptions. In fact, the narratives could be considered as general designs rather than as statements of business policies. The inclusion of this amount of detail is intentional. The narratives provide examples of feasible alternatives to general designs that may be developed by the student. Thus, these narratives can serve as starting points for detailed designs based on the examples. The designs also can be used as models for the kinds of decisions that must be made by students working through their own analysis projects. For this reason, some guidance is offered for proceeding into the design phases of the project. However, the student should not feel constrained by the amount of detail provided.

The case scenarios presented in this appendix should be considered as alternatives to analysis cases that the student may have completed already. Those students who have worked through the systems analysis projects that are described in the companion text, *Computer Information Systems Development: Analysis and Design*, have a ready-made project for study. If possible, these projects should be used as the basis for design. Doing so will give the student a sense of the continuity of systems development projects. It also may point out to the student the importance of having a complete, exacting set of system specifications as the foundation for design, especially if the student has to work with his or her own less-than-adequate specifications.

The case projects present general scenarios that can be expanded or contracted to fit educational requirements. The suggestions listed below should not be viewed as limitations on other valuable activities that could be included to simulate realistic systems development projects. Instructors should devise project structures and sets of activities that best serve their educational goals and requirements. By the same token, care should be taken to keep the projects within reasonable boundaries as student projects. There may be a temptation to expand the systems into fully operational, commercial-type applications with all of their attendant complexities. Although it may be instructive to develop "full-blown" systems, it is not necessary to do so to realize educational benefits from the cases. An understanding of the process of systems design is of greater benefit than an ability to develop a full set of potential system end products.

PART 1: ANALYSIS AND GENERAL DESIGN REVIEW

The initial assignment is intended as a review of the process and products of systems analysis. Although the data flow, data dictionary, and processing documentation that accompany the cases provide system overviews, these materials do not present sufficient detail for transition into design. The first set of work assignments, therefore, expands the logical system requirements into complete specifications. This part of the case project can begin with the study of *Chapter 1, The Roots of Systems Design*. The following activities and products are required:

- The Diagram 0 provided should be expanded, as appropriate, into leveled data flow diagrams. Particular emphasis should be given to documenting transaction processing routines within which single data streams are broken down and dispatched to separate processes. An example is the maintenance routine that performs additions, changes, or deletions within a master file in response to the type of transaction presented. Another instance requiring further partitioning is an edit routine in which data fields undergo parallel or successive edits.

- For each of the leveled data flow diagrams produced under the previous task, process descriptions should be prepared. These descriptions should expand upon the process narratives presented in the case to provide explicit detail of all business policies and procedures that underlie processing.

- Rough drafts of input, output, and transaction documents or screens should be prepared. These can be drawn from the logical descriptions contained in the data dictionary. It is not necessary, at this point, to specify exact line and column locations of data fields. Rather, the drafts should indicate general layouts for presentation to and discussion with users.

- Determine tentative file organization methods required to support processing. Indicate, for example, whether serial, sequential, direct, or indexed techniques will be needed. If a database management system will be used, provide a general description of the package and modify the data dictionary files to illustrate any new logical views (subschemas) that will be supported.

At the close of this and subsequent parts of the project, it would be beneficial if students conducted walkthrough sessions to review their recommendations. If students are working in project teams, each team could conduct its own session. If students are working individually, the instructor could appoint team members who would review and critique the work of other students.

PART 2: TRANSITION INTO SYSTEM DESIGN

The second assignment is to perform and document a general system design. The logical system specifications that were the products of analysis are given tentative physical specifications. User requirements have been analyzed, and this analysis serves as the point of departure for transition into physical design. The broad hardware/software parameters established during analysis now are elaborated with sufficient detail to verify that the proposed system can be delivered within schedule and budget constraints. Both computerized and manual procedures are brought to a level of detail that provides assurances of technical and operational feasibility. This part of the case project can begin with the study of *Chapter 3, The Technical Environment of Systems Design*.

Materials produced for this part of the case project represent extensions of the specifications prepared during system analysis. The following activities and results are required:

- Computer processing requirements should be defined. For batch processing applications, system flowcharts should be prepared to document job streams. For on-line procedures, system flowcharts should indicate the software product that will drive the procedure, the input and output documents or media that are used, and the files that are needed to support the processes. These flowcharts should be annotated to describe performance requirements, general processing steps, and application timing cycles.

- A set of files to support the application should be defined, and file organization, access methods, and storage media should be specified. Approximate volumes of stored data and anticipated growth patterns should be estimated. If database management systems are to be acquired or used, the specifications should include a summary of modifications necessary to support the new system and an evaluation of the benefits and costs of the software package.

- Performance criteria should be delineated. Included should be required response times for on-line processing, transaction volumes, and other meaningful performance requirements.

- Provide a description of the hardware and system software environment that is assumed. The instructor may establish technical constraints. If not, assume you are working within a shop that allows reasonable flexibility in designs and includes adequate facilities for implementing those designs. If the proposed system involves significant hardware and/or system software changes, however, technical specifications should be prepared. Include a description of requirements for new hardware and software capabilities and any other modifications or additions to the existing computer system.

- Consideration should be given to ways in which the new application will fit within the larger organizational and technical environment. Decisions should be documented concerning methods for data capture and input, the balance between centralized and decentralized or distributed processing, data communication methods, output techniques, and storage methods to support local and remote data bases. Concern should be given to establishing operational controls and controls to ensure the integrity of data within files.

- At least two general designs should be prepared, each representing a cost/benefit trade-off with the other. A brief narrative should be presented explaining the potential economic, operational, budgetary, and scheduling impacts for each alternative. The designs should be presented to the class, which acts as user/managers in consultation with the instructor. The design that is approved and selected then is carried forward through the remaining projects.

PART 3: DETAILED SYSTEM DESIGN

This part of the project adds final technical details to the overall system design developed and selected in Part 2. In particular, consideration is given to human-machine interactions, detailed file design and maintenance, continuity of service in case the computer goes down, and additional security and access controls. The major output of this part is a detailed design specification and can be completed following study of *Chapter 6, Technical Design*. The following activities and products are required:

- Physical file design and layout should be completed. Formal file organization techniques should be defined, and sizes and arrangement of data fields within records should be specified. Data dictionary entries should be updated to include these expanded definitions. Estimates of disk space should be made along with specifications for blocking and physical arrangements of files on storage media.

- Input and output designs should be completed. Formats should be developed for source, input, and output documents and screens based on data dictionary definitions. Interactive patterns of human-machine communication should be specified in terms of menus, "help" displays, and error trapping and recovery routines.

- Program specifications should be prepared. These specifications include structure charts that identify the major processing modules within the software portion of the system and the connections and interfaces among them. The specifications represent the overall software structure rather than complete detailed designs. Process descriptions should be prepared for the modules.

- Consideration should be given to backup and recovery procedures. Necessary routines should be added to incorporate adequate audit trails and logging procedures. Security and control measures aimed at limiting access to equipment and files are documented.

- For batch processing procedures, computer operations documentation should be prepared. The instructions should include brief narratives describing the setup and run procedures for computer operators. For user-driven interactive procedures, tentative outlines of user guides should be prepared and preliminary operating procedures and methods should be written. These guides on how to use particular programs and procedures will be completed in full detail when the system is operational.

- If the project involves moving from one computer system to another, technical specifications describing conversion programs to reformat existing files for new system compatibility should be prepared.

- If appropriate, and if preferred, program and file prototypes can be developed to demonstrate input and output procedures and processing activities. These prototypes should be used to determine effective human-machine interactions. The routines should be coded and refined in connection with the first two activities described above.

PART 4: SOFTWARE DESIGN AND TEST SPECIFICATIONS

This part of the project concerns the design and testing of the software portion of the new system. Inputs to this process are the input, output, and file design specifications completed during the previous part, along with the structure charts that document the overall architectural structure of the software. Activities in this part result in detailed functional designs for all programs that will implement the system and sets of test data to be used in performing operational tests. In addition, user manuals are completed. This part can be completed concurrently with the study of *Chapter 7, Foundations of Software Design* through *Chapter 11, Testing Strategies*. The following activities and products are required:

- Detailed functional designs should be completed for all programs within the system. These designs are expressed as detailed structure charts describing the connections and interfaces among program modules. Design should be carried to a level at which pseudocode can be prepared for each module. Special attention should be given to documenting the data that will be passed between the modules.

- For each module within the system, pseudocode should be written to describe its logic and processing. The pseudocode should be sufficiently detailed to support easy translation into a programming language.

- For each module and for each individual program, test data should be developed. Devise test data that will exercise the logic paths within modules and that will test input and output interfaces. Include additional test data that will be used in testing job streams, subsystems, and the system as a whole.

- Consideration should be given to packaging the design into compilable units. Indicate the modules that will be combined into load modules.

- A plan should be devised for implementing the software. Prepare a procedure for incremental development and testing of modules. Specify designs for stub and/or driver modules and indicate the order of implementation.

- User guides should be completed. These guides should describe complete operating procedures for the novice user. Short tutorials and demonstrations on system operation should be included, as necessary.

All project materials and documentation should be produced with exacting care and neatness. These documents should be put together in a binder, indexed clearly, and cross-referenced to show the design steps and resulting products. There should be clear evidence of logical progression through the parts of the project.

Much of the documentation required for these projects is specified in the case descriptions. As appropriate, data flow diagrams, data dictionary entries, process descriptions, structure charts, pseudocode, and narratives are called for. In other instances, no specific form of documentation is suggested. The student, in consultation with the instructor, should determine the appropriate mechanisms for expressing designs. Although the requirement for clear communication of decisions should not be compromised, the products of analysis and design decisions are less important within case assignments than a clear understanding of the process and rationale for making those decisions. Thus, as appropriate, the student should select documentation techniques that best express those decisions.

CASE STUDY 1: STUDENT RECORDS SYSTEM

SITUATION

Place yourself as project team leader in the CIS department of a large state university. You are to develop designs that will serve to implement a new student records system. At present, virtually all processing is done on a batch basis, although the university does have some on-line capabilities. The plan is to revise the existing system for greater responsiveness in tracking student academic progress.

One of the main reasons for wanting to develop this system is to achieve greater student registration efficiency. The current procedure involves a centralized, walk-in registration center and the passing out of class cards. These procedures have been plagued with difficulties for years. Delays in reporting class closings and new openings have resulted in considerable student and staff confusion, as well as in excessive numbers of drop/add requests at the close of registration. There is a strong demand for some form of on-line system to provide increased currency in reporting class offerings and for allowing an expanded registration session that occurs over several weeks rather than being compressed within one or two hectic days.

At the same time, a new, computerized system could help assist advisors in performing services that are inconvenient to provide under present methods. Currently, advisors are responsible for determining student registration qualifications—in terms of declared majors and completion of prerequisites. It is hoped that the new system will be able to automate registration checks of this type, leaving advisors more time for the substantive parts of their jobs.

Another target for the new system is to improve reporting and record-keeping procedures. Grade reporting and transcript preparation could be integrated into the system. The availability of transcript files could mean that fewer advisors would be needed to monitor student registrations to assure completion of prerequisites. Students, in effect, would gain more responsibility in determining their own courses of study.

CONTEXT DIAGRAM

Over the past several months, the project team has analyzed the current system and developed documentation for the new logical system. This documentation will serve as the starting point for new system design. The scope of the system you are developing is defined by the context diagram shown in Figure A-1. As indicated, the student records system interacts with four external entities:

- The student will use the new system for on-line registration for classes. The ability to support on-line registration capabilities was one of the prime motivations for authorizing development of the new system. The processing of registration requests and requests for changes in classes results in the preparation of student course schedules. Also, the system will support processing of student declarations of major. This process assigns students to academic advisors and ties the student to a particular degree program. A final interaction between the system and the student lies in grade reporting. At the end of each term, the student will receive a report of the grades earned in each class and a summary of credits earned to date.

- The faculty constitutes another key external entity that interacts with the system. Faculty members receive class rosters from the system at the beginning of a term. They submit class grade reports that are used in producing the student grade reports at the close of the term. In addition, grade change requests are submitted for changing grades that were submitted on the class grade reports.

- Academic advisors are responsible for overall tracking of student progress. The advisor receives copies of declarations of majors for students that have chosen a major in the academic area. At the end of each term, advisors receive copies of student grade reports and, as necessary, the advisors can request copies of student transcripts to verify academic progress toward degree requirements.

- The Registrar's Office is responsible for maintaining student academic records. This office, in response to a graduation request, receives a copy of the student's transcript and makes a final check on the meeting of degree requirements. The office also submits class offerings for the upcoming term. Offerings originate in the academic departments and are compiled in the office and released to the student records system.

Of course, as currently conceived, the system does not provide all possible services to the identified entities or to other potential users of the system. This project represents an initial attempt at devising a system that can be expanded later with additional features. For example, future plans call for interfaces with the accounting office and the office of student financial aid along with facilities for providing automatic degree evaluation and checkout. At this time, the interest is in establishing a foundation of procedures and data files that are expandable and can grow into a totally integrated system.

PROCESSING EVENTS

The outputs of systems analysis include a set of data flow diagram (DFD) segments that identify the major processing events in the system and show the data flows and files that interact to transform the inputs into outputs for each process. The data flow diagrams are supported by the data dictionary entries that document the structure and content of input and output data flows and files. Also, process narratives provide detailed explanations of processing events. The individual DFD segments then are combined into a Diagram 0 data flow diagram that describes the completely integrated system.

In studying the context diagram in Figure A-1, the project team identified several major events in the student records system, including:

- Process declaration of major
- Build class file
- Process registration request
- Prepare class rosters
- Process grades
- Produce transcript
- Change student grade.

This list is derived from a review of the inputs and outputs shown on the context diagram. Although every data flow on the diagram is not represented as an event, the list does identify events to cover all of the data flows that are shown.

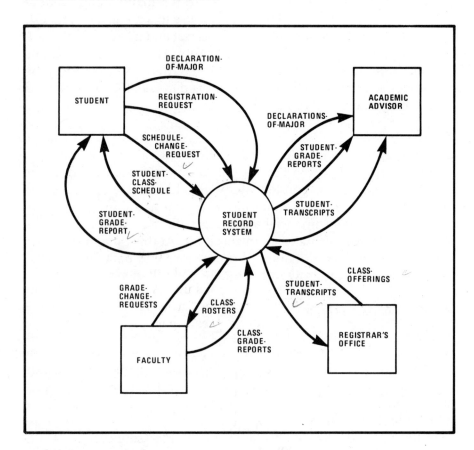

Figure A-1.   Context diagram: STUDENT RECORDS SYSTEM.

SYSTEM DOCUMENTATION

In developing the set of documentation for the new system, each event is described by its own data flow diagram segment. The segment is supported by data dictionary entries for the identified input and output documents and files and by process narratives detailing the processing activities that comprise each processing event. These process narratives are still at a rather high level, describing the business-related activities that are performed to a greater degree than the computer processing steps necessary to implement them.

The entire set of DFD segments is given in Figures A-2 through A-8. The segments are combined into a Diagram 0 in Figure A-9. At this point, the documentation presents a high-level overview of the proposed system. In effect, each bubble represents a suggested subsystem that would be developed to implement the event. As needed, however, lower-level diagrams will have to be developed to partition the processes for further clarification, and diagrams could be combined into integrated software products.

Figure A-10 contains the data dictionary entries for the system's input and output documents. The documents are described logically, in terms of data components and relationships. At this point, little concern is given to the physical layout of the forms. Figure A-11 contains the data dictionary entries for system files. Again, physical design has not yet taken place. File content represents the major entities about which the system maintains information and the keys through which access takes place. Yet to be determined are the file organization techniques and data field arrangements. Reorganization of record keys may be necessary to comply with programming language requirements. The file access diagram for the system is shown in Figure A-12.

Finally, system documentation includes the process narratives for all of the bubbles on Diagram 0. These descriptions are shown in Figures A-13 through A-19. They describe the processing activities that comprise the system processing events, along with the data files and data elements that are affected.

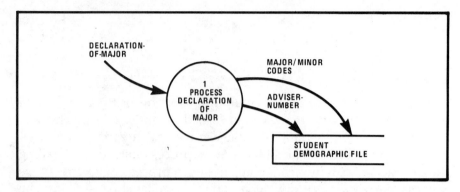

Figure A-2. Data flow diagram. Process 1: PROCESS DECLARATION OF MAJOR.

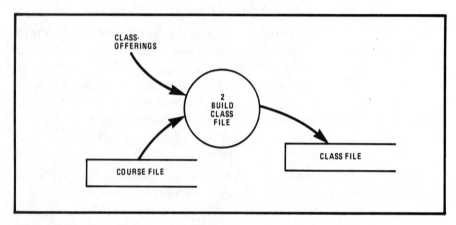

Figure A-3. Data flow diagram. Process 2: BUILD CLASS FILE.

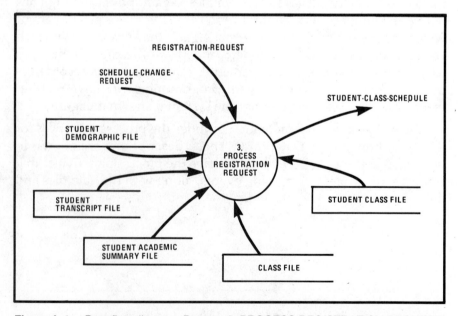

Figure A-4. Data flow diagram. Process 3: PROCESS REGISTRATION REQUEST.

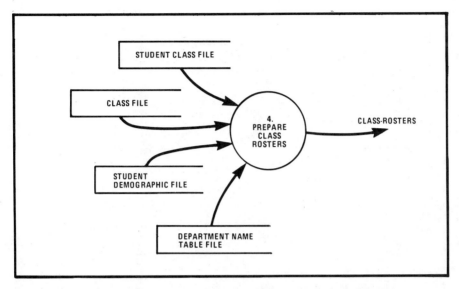

Figure A-5. Data flow diagram. Process 4: PREPARE CLASS ROSTERS.

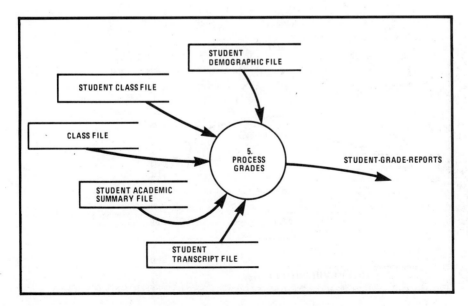

Figure A-6. Data flow diagram. Process 5: PROCESS GRADES.

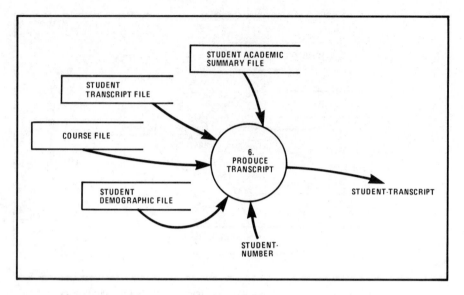

Figure A-7. Data flow diagram. Process 6: PRODUCE TRANSCRIPT.

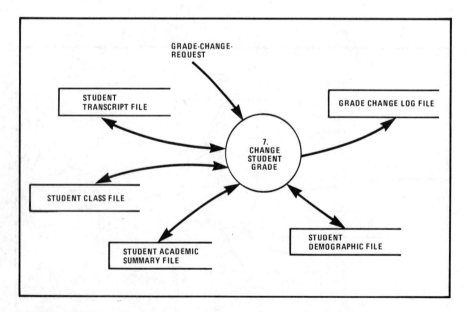

Figure A-8. Data flow diagram. Process 7: CHANGE STUDENT GRADE.

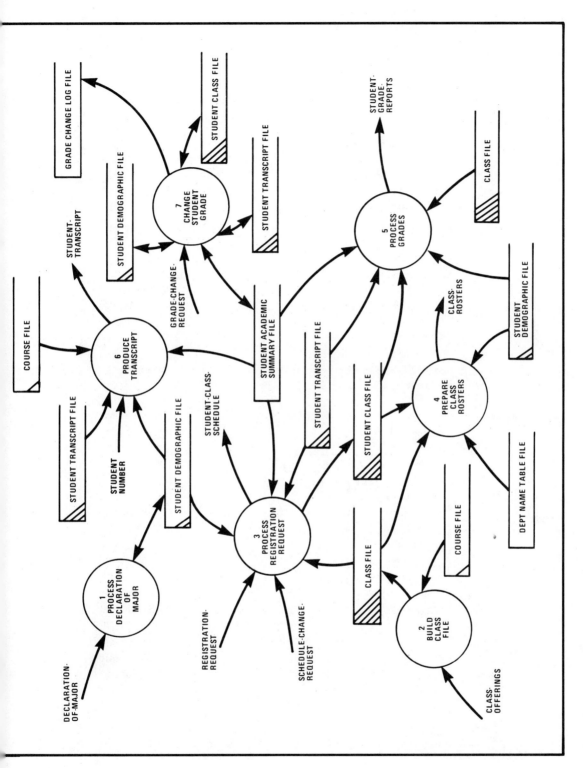

Figure A-9. Diagram 0: STUDENT RECORDS SYSTEM.

Figure A-10. Data dictionary entries for input and output data flows.

```
                              INPUT/OUTPUT DOCUMENTS

CLASS-GRADE-REPORT = DEPT-ID +              CLASS-ROSTER    = DEPT-ID +
                     TERM-ID +                                TERM-ID +
                     REPORT-DATE +                            ROSTER-DATE +
                     DEPT-NAME +                              DEPT-NAME +
                     INSTRUCTOR-NAME +                        INSTRUCTOR-NAME +
                     CLASS-NUMBER +                           CLASS-NUMBER +
                     SECTION-NUMBER +                         SECTION-NUMBER +
                     CREDIT-HOURS +                           CREDIT-HOURS +
                     CLASS-NAME +                             CLASS-NAME +
                     LOCATION +                               LOCATION +
                     DAYS-OFFERED +                           DAYS-OFFERED +
                     TIME-OFFERED +                           TIME-OFFERED +
                    ⎧ STUDENT-NUMBER +⎫                      ⎧ STUDENT-NUMBER +⎫
                    ⎨ STUDENT-NAME +  ⎬                      ⎨ STUDENT-NAME +  ⎬
                    ⎩ STUDENT-GRADE + ⎭                      ⎪ MAJOR-CODE +    ⎪
                     TOTAL-STUDENTS +                         ⎩ ACADEMIC-LEVEL +⎭
                     INSTRUCTOR-SIGNATURE                     TOTAL-STUDENTS


CLASS-OFFERINGS   = DEPT-ID +               DECLARATION-OF-MAJOR = STUDENT-NUMBER +
                    DEPT-NAME +                                    STUDENT-NAME +
                    CURRENT-DATE +                                 STREET-ADDRESS +
                    TERM-ID +                                      CITY +
                   ⎧ CLASS-NUMBER +    ⎫                           STATE +
                   ⎪ SECTION-NUMBER +  ⎪                           ZIP +
                   ⎪ CREDIT-HOURS +    ⎪                           PHONE-NUMBER +
                   ⎪ CLASS-NAME +      ⎪                           MAJOR-NAME +
                   ⎨ LOCATION +        ⎬                           MAJOR-CODE +
                   ⎪ DAYS-OFFERED +    ⎪                           (PREVIOUS-MAJOR-NAME) +
                   ⎪ TIME-OFFERED +    ⎪                           (PREVIOUS-MAJOR-CODE) +
                   ⎪ INSTRUCTOR-NAME + ⎪                           (MINOR-NAME) +
                   ⎩ SEATS-MAXIMUM     ⎭                           (MINOR-CODE) +
                                                                   DECLARATION-DATE +
                                                                   STUDENT-SIGNATURE +
                                                                   ADVISOR-SIGNATURE +
                                                                   ADVISOR-NUMBER
```

```
REGISTRATION-REQUEST      = STUDENT-NUMBER +         STUDENT-CLASS-SCHEDULE = STUDENT-NUMBER +
                            STUDENT-NAME +                                    STUDENT-NAME +
                            STREET-ADDRESS +                                  STREET-ADDRESS +
                            CITY +                                            CITY +
                            STATE +                                           STATE +
                            ZIP +                                             ZIP +
                            PHONE-NUMBER +                                   [REGISTRATION-DATE +    ]
                            REGISTRATION-DATE +                              [SCHEDULE-CHANGE-DATE + ]
                            TERM-ID +                                         (CLASS-NUMBER +    )
                           (CLASS-NUMBER +                                    |SECTION-NUMBER +  |
                           |SECTION-NUMBER +                                  |CLASS-NAME +      |
                           |CLASS-NAME +                                      {CREDIT-HOURS +    }
                           {CREDIT-HOURS +                                    |LOCATION +        |
                           |LOCATION +                                        |DAYS-OFFERED +    |
                           |DAYS-OFFERED +                                    |TIME-OFFERED +    |
                           |TIME-OFFERED +                                    (INSTRUCTOR-NAME + )
                           (INSTRUCTOR-NAME +                                 TOTAL-HOURS
                            TOTAL-HOURS +
                            (ADVISOR-SIGNATURE) +
                            (ADVISOR-NUMBER)               STUDENT-GRADE-REPORT = STUDENT-NUMBER +
                                                                                 STUDENT-NAME +
                                                                                 STREET-ADDRESS +
SCHEDULE-CHANGE-REQUEST = STUDENT-NUMBER +                                        CITY +
                          STUDENT-NAME +                                         STATE +
                          STREET-ADDRESS +                                       ZIP +
                          CITY +                                                (CLASS-NUMBER +  )
                          STATE +                                               |CLASS-NAME +    |
                          ZIP +                                                 {CREDIT-HOURS +  }
                          PHONE-NUMBER +                                        (STUDENT-GRADE + )
                          SCHEDULE-CHANGE-DATE +                                 HOURS-ATTEMPTED +
                          TERM-ID +                                              HOURS-COMPLETED +
                         (ADD-COURSE-NUMBER +                                    CREDITS-EARNED +
                         |ADD-COURSE-SECTION +                                   TERM-GPA +
                         |ADD-COURSE-NAME +                                      CUMULATIVE-HOURS-ATTEMPTED +
                         {ADD-CREDIT-HOURS +                                     CUMULATIVE-HOURS-COMPLETED +
                         |ADD-LOCATION +                                         CUMULATIVE-CREDITS-EARNED +
                         |ADD-DAYS-OFFERED +                                     CUMULATIVE-GPA
                         (ADD-INSTRUCTOR-NAME +
                         (DROP-COURSE-NUMBER +
                         {DROP-COURSE-SECTION +}
                         (DROP-COURSE-NAME +
                          (ADVISOR-SIGNATURE) +
                          (ADVISOR-NUMBER)
```

```
STUDENT-TRANSCRIPT              = TRANSCRIPT-DATE +
                                  STUDENT-NUMBER +
                                  STUDENT-NAME +
                                  STREET-ADDRESS +
                                  CITY +
                                  STATE +
                                  ZIP +
                                  MAJOR-CODE +
                                 ⎧ TERM-ID +                          ⎫
                                 ⎪  ⎛ CLASS-NUMBER + ⎞                ⎪
                                 ⎪  ⎜ CLASS-NAME +   ⎟                ⎪
                                 ⎪  ⎜ CREDIT-HOURS + ⎟                ⎪
                                 ⎪  ⎝ CLASS-GRADE +  ⎠                ⎪
                                 ⎨ HOURS-ATTEMPTED +                  ⎬
                                 ⎪ HOURS-COMPLETED +                  ⎪
                                 ⎪ CREDITS-EARNED +                   ⎪
                                 ⎪ TERM-GPA +                         ⎪
                                 ⎪ CUMULATIVE-HOURS-ATTEMPTED +       ⎪
                                 ⎪ CUMULATIVE-HOURS-COMPLETED +       ⎪
                                 ⎪ CUMULATIVE-CREDITS-EARNED +        ⎪
                                 ⎩ CUMULATIVE-GPA                     ⎭


GRADE-CHANGE-REQUEST            = GRADE-CHANGE-DATE +
                                  STUDENT-NUMBER +
                                  STUDENT-NAME +
                                  TERM-ID +
                                  CLASS-NUMBER +
                                  SECTION-NUMBER +
                                  REPORTED-GRADE +
                                  CHANGED-GRADE +
                                  REASON-FOR-CHANGE +
                                  INSTRUCTOR-SIGNATURE +
                                  CHAIRPERSON-SIGNATURE
```

Figure A-10.   (cont'd).

FILES

```
CLASS-FILE                    = {CLASS-RECORD}
CLASS-RECORD                  = DEPT-ID +
                                CLASS-NUMBER +
                                SECTION-NUMBER +
                                CREDIT-HOURS +
                                CLASS-NAME +
                                LOCATION +
                                DAYS-OFFERED +
                                TIME-OFFERED +
                                INSTRUCTOR-NAME +
                                SEATS-MAXIMUM +
                                SEATS-AVAILABLE +
                                (PREREQUISITE-MAJORS) +
                                (PREREQUISITE-HOURS) +
                                (PREREQUISITE-COURSE-GRADE)


COURSE-DESCRIPTION-FILE       = {COURSE-DESCRIPTION-RECORD}
COURSE-DESCRIPTION-RECORD     = ⎡ CLASS-NUMBER +           ⎤
                                ⎣ {CLASS-DESCRIPTION-TEXT} ⎦


COURSE-FILE                   = {COURSE-RECORD}
COURSE-RECORD                 = DEPT-ID +
                                CLASS-NUMBER +
                                CREDIT-HOURS +
                                CLASS-NAME +
                                (PREREQUISITE-MAJORS) +
                                (PREREQUISITE-HOURS) +
                                (PREREQUISITE-COURSE-GRADE)


DEPT-NAME-TABLE-FILE          = {DEPT-NAME-TABLE-RECORD}
DEPT-NAME-TABLE-RECORD        = DEPT-ID +
                                DEPT-NAME
```

Figure A-11. Data dictionary entries for system files.

511

```
GRADE-CHANGE-LOG-FILE              = {GRADE-CHANGE-LOG-RECORD}
GRADE-CHANGE-LOG-RECORD            = TERM-ID +
                                     STUDENT-NUMBER +
                                     CLASS-NUMBER +
                                     CLASS-GRADE +
                                     CHANGED-GRADE +
                                     GRADE-CHANGE-DATE


STUDENT-ACADEMIC-SUMMARY-FILE      = {STUDENT-ACADEMIC-SUMMARY-RECORD}
STUDENT-ACADEMIC-SUMMARY-RECORD    = STUDENT-NUMBER +
                                     TERM-ID +
                                     HOURS-ATTEMPTED +
                                     HOURS-COMPLETED +
                                     CREDITS-EARNED +
                                     TERM-GPA +
                                     CUMULATIVE-HOURS-ATTEMPTED +
                                     CUMULATIVE-HOURS-COMPLETED +
                                     CUMULATIVE-CREDITS-EARNED +
                                     CUMULATIVE-GPA


STUDENT-CLASS-FILE                 = {STUDENT-CLASS-RECORD}
STUDENT-CLASS-RECORD               = STUDENT-NUMBER +
                                     CLASS-NUMBER +
                                     SECTION-NUMBER +
                                     CLASS-GRADE


STUDENT-DEMOGRAPHIC-FILE           = {STUDENT-DEMOGRAPHIC-RECORD}
STUDENT-DEMOGRAPHIC-RECORD         = STUDENT-NUMBER +
                                     STUDENT-NAME +
                                     STREET-ADDRESS +
                                     CITY +
                                     STATE +
                                     ZIP +
                                     PHONE-NUMBER +
                                     MAJOR-CODE +
                                     (MINOR-CODE) +
                                     ACADEMIC-LEVEL +
                                     ADVISOR-NUMBER


STUDENT-TRANSCRIPT-FILE            = {STUDENT-TRANSCRIPT-RECORD}
STUDENT-TRANSCRIPT-RECORD          = STUDENT-NUMBER +
                                     TERM-ID +
                                     CLASS-NUMBER +
                                     CLASS-GRADE
```

Figure A-11.

(cont'd).

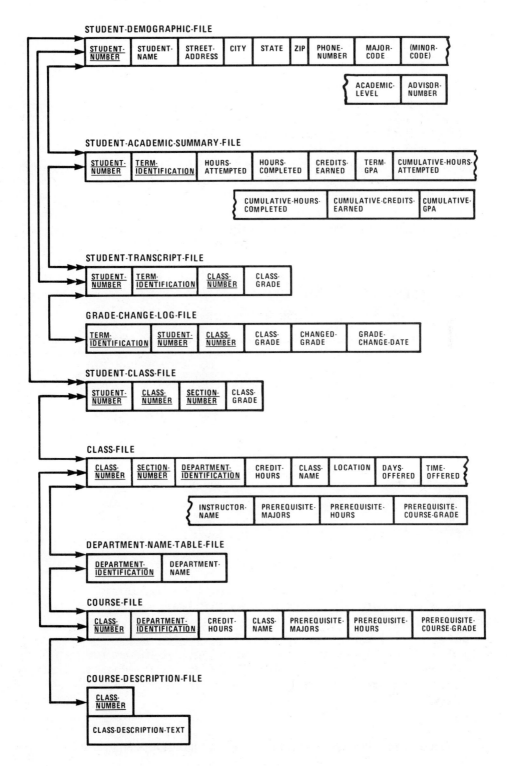

Figure A-12. Data access diagram describing primary associations between files.

513

PROCESS 1: PROCESS DECLARATION OF MAJOR

1. The student returns the original copy of the DECLARATION OF MAJOR form to the Registrar's Office after it has been completed and signed by the student and the academic advisor.

2. Within the Registrar's Office, the new or changed-to major and/or minor are entered into the STUDENT-DEMOGRAPHIC-FILE. The student's record is accessed on STUDENT-NUMBER, and the new major and/or minor codes are added. MAJOR-CODEs are the same as those used for course prefixes in the major academic areas. Each is a three-character alphabetic code. Also, the ADVISOR-NUMBER for the new academic advisor is added to the student record replacing the number for the previous advisor.

3. After the file is updated, the DECLARATION-OF-MAJOR form is filed within the Registrar's Office in sequence by STUDENT-NAME.

Figure A-13.   Process narrative. Process 1: PROCESS DECLARATION OF MAJOR.

PROCESS 2: BUILD CLASS FILE

1. CLASS-OFFERINGS are submitted by academic departments for the current term. These are checked by the Registrar's Office for conformance with style and completeness before being released for data entry.

2. CLASS-FILE is built with information from the CLASS-OFFERINGS and from the COURSE-FILE. The CLASS-NUMBER appearing on the CLASS-OFFERINGS form is used to access the class information from the COURSE-FILE. These data are displayed for verification of the offerings, and are used in building the CLASS-FILE.

Figure A-14.   Process narrative. Process 2: BUILD CLASS FILE.

PROCESS 3:   PROCESS REGISTRATION REQUEST

If Registration Request

1.   Student brings completed REGISTRATION-REQUEST to registration center.

2.   Registration is carried out on-line. Course requests are entered interactively into the computer. CLASS-FILE is accessed to verify course descriptions, to determine if seats are available, and to check on prerequisites. If there are prerequisites for enrolling in a course, the STUDENT-DEMOGRAPHIC-FILE, STUDENT-TRANSCRIPT-FILE, and/or STUDENT-ACADEMIC-SUMMARY-FILE are accessed to verify meeting of prerequisites. Also, DAYS-OFFERED and TIME-OFFERED are checked to make sure that there are no conflicts within the requested schedule. For classes that are filled, alternate sections may be displayed for selection.

3.   After final schedule is verified, the CLASS-FILE is updated by subtracting 1 from the SEATS-AVAILABLE. A record is written to the STUDENT-CLASS-FILE for each class in which the student enrolls. A CLASS-GRADE field is appended to the record for later use in recording the grade earned in the class.

4.   A STUDENT-CLASS-SCHEDULE is prepared and written on the terminal printer. Information from the STUDENT-DEMOGRAPHIC-FILE and CLASS-FILE is combined on the form. A copy of the schedule is retained in the Registrar's Office, filed sequentially by STUDENT-NAME.

If Schedule Change Request

1.   Student brings completed SCHEDULE-CHANGE-REQUEST to registration center.

2.   Schedule change is carried out on-line. STUDENT-CLASS-FILE and CLASS-FILE are accessed to display the student's current schedule. Requests for added courses are verified for availability and prerequisites as is done in original registration request.

3.   Following verification of new schedule, CLASS-FILE is updated by adding to the SEATS-AVAILABLE field. Records are added to or deleted from the STUDENT-CLASS-FILE as necessary.

4.   A STUDENT-CLASS-SCHEDULE is prepared and written on the terminal printer using information from the STUDENT-DEMOGRAPHIC-FILE and CLASS-FILE. A copy of the revised schedule is retained in the Registrar's Office under the STUDENT-NAME.

Figure A-15.   Process narrative. Process 3: PROCESS REGISTRATION REQUEST.

PROCESS 4: PREPARE CLASS ROSTERS

1. CLASS-FILE is accessed in sequence by DEPT-ID (major), CLASS-NUMBER (intermediate), and
    SECTION-NUMBER (minor).

2. For each class, STUDENT-CLASS-FILE is accessed on CLASS-NUMBER + SECTION-NUMBER to
    identify all students within the class. Using the STUDENT-NUMBER recorded in the
    STUDENT-CLASS-FILE, the STUDENT-DEMOGRAPHIC-FILE is accessed to get student information
    for printing on the rosters. This student information is sorted by STUDENT-NAME.

3. Rosters are printed in DEPT-ID/CLASS-NUMBER/SECTION-NUMBER order using information from
    the CLASS-FILE and sorted information from the STUDENT-DEMOGRAPHIC-FILE. A total of the
    number of students in each class is tallied and printed on the rosters. DEPT-NAME is
    pulled from a table of department IDs and corresponding names built from the
    DEPT-NAME-TABLE-FILE.

Figure A-16.   Process narrative. Process 4: PREPARE CLASS ROSTERS.

PROCESS 5: PROCESS GRADES

1. CLASS-GRADE-REPORTs are submitted by each academic department on special marked-sense forms that can be read by a machine that writes the information onto magnetic tape. This tape file is then sorted on CLASS-NUMBER within STUDENT-NUMBER, and the sorted file is used to update the STUDENT-CLASS-FILE by adding the appropriate letter grade within the CLASS-GRADE field.

2. The STUDENT-CLASS-FILE is primary input to this routine. For all class records for a particular student, the CLASS-FILE is accessed on CLASS-NUMBER + SECTION-NUMBER, and the corresponding CLASS-NAME and CREDIT-HOURS are input.

   a. HOURS-ATTEMPTED is calculated by totaling the number of CREDIT-HOURS taken for which a grade was earned.

   b. HOURS-COMPLETED is calculated by totaling the number of CREDIT-HOURS for all classes for which a passing grade was earned.

   c. CREDITS-EARNED is calculated by multiplying CREDIT-HOURS by the numeric grade equivalent (A=4, B=3, C=2, D=1, F=0) for each class, and then totaling the products of the calculations.

   d. TERM-GPA is calculated by dividing CREDITS-EARNED by HOURS-ATTEMPTED.

3. The STUDENT-ACADEMIC-SUMMARY-FILE is accessed on STUDENT-NUMBER + TERM-ID for the previous term, if available. The cumulative hours, credits, and GPA values are used to calculate the cumulative values for the current term. A new summary record for the current term is added to the file.

4. The STUDENT-TRANSCRIPT-FILE is updated by adding a record for each course taken in the current term.

5. The STUDENT-DEMOGRAPHIC-FILE is accessed on STUDENT-NUMBER. The CUMULATIVE-HOURS-COMPLETED through the current term is used in assigning the new ACADEMIC-LEVEL value in the record (0-30 hours = FR, 31-60 hours = SO, 61-90 hours = JR, Over 90 hours = SR). The name and address fields from this file are combined with the information that was calculated and with that from the other files in order to print a STUDENT-GRADE-REPORT.

Figure A-17. Process narrative. Process 5: PROCESS GRADES.

PROCESS 6: PRODUCE TRANSCRIPT

1. STUDENT-DEMOGRAPHIC-FILE is accessed on STUDENT-NUMBER to input student name and address
    information.

2. STUDENT-NUMBER is used to access STUDENT-ACADEMIC-SUMMARY-FILE.   Each record in this file
    contains summary information for the term identified by the TERM-ID along with the
    accumulated hours, credits, and GPA up through that term.   All records for the particular
    student for the multiple terms are input.

3. Using   the   STUDENT-NUMBER + TERM-ID from the STUDENT-ACADEMIC-SUMMARY-FILE,   the
    STUDENT-TRANSCRIPT-FILE is accessed.   CLASS-NUMBERs and CLASS-GRADEs for each of the
    classes taken during each of the terms is input.

4. The CLASS-NUMBERs from the STUDENT-TRANSCRIPT-FILE are used as partial keys for accessing
    the COURSE-FILE.   CREDIT-HOURS and CLASS-NAMEs for the corresponding classes taken are
    input from this file.

5. Information from the four files is combined on the STUDENT-TRANSCRIPT which lists the
    classes taken and grades earned for each term along with the term and cumulative academic
    summary.   The transcript is printed in order by term.

Figure A-18.   Process narrative. Process 6: PRODUCE TRANSCRIPT.

PROCESS 7: CHANGE STUDENT GRADE

1. The STUDENT-NUMBER + TERM-ID + CLASS-NUMBER from the GRADE-CHANGE-REQUEST is used to access the STUDENT-TRANSCRIPT-FILE to input the reported CLASS-GRADE.

2. Information from the STUDENT-TRANSCRIPT-FILE is displayed for verification and the CLASS-GRADE is changed interactively through the computer terminal.

3. The STUDENT-CLASS-FILE for the corresponding term is accessed and the change is reflected in this file as well.

4. The STUDENT-ACADEMIC-SUMMARY-FILE is updated. Recalculations are made for the hours, credits, and GPA values for the current term and cumulatively. Since grade changes are not allowed for previous terms, it is not necessary to provide system features for updating the file for a previous term and successive terms thereafter.

5. If the grade change causes the HOURS-COMPLETED to change the academic level of the student, the STUDENT-DEMOGRAPHIC-FILE will require updating to effect the changed level

6. A record is written to the GRADE-CHANGE-LOG-FILE. This file journalizes changes. The record contains the original CLASS-GRADE, the CHANGED-GRADE, and the CHANGE-DATE.

Figure A-19. Process narrative. Process 7: CHANGE STUDENT GRADE.

SUPPLEMENTARY PROJECT

The student records system described above provides a foundation for expanding processing into several other record-keeping and reporting areas. One of the potential features that could be incorporated within the same system is a graduation checkout procedure. In essence, such a subsystem would provide verification that the student has met all graduation requirements for a particular degree or would indicate progress toward or deficiencies in meeting degree requirements. Such a subsystem, of course, would have to recognize allowable substitute and elective courses and take into account required grade point averages. It would be desirable to make progress reports available on request from a student or advisor.

If required by your instructor, conduct analysis and design for adding this graduation checkout process to the student records system. Add to or modify any of the existing documentation to integrate the necessary procedures within the system before proceeding with design.

CASE STUDY 2: ORDER PROCESSING SYSTEM

SITUATION

You have been appointed project team leader in the CIS department of a large mail-order merchandising company. The company specializes in computer products and supplies that it sells through catalog orders to both individuals and businesses. Merchandise is maintained in an inventory from which all orders are filled. Your task is to produce a new system design for an automated order processing system. This system will permit the acceptance of customer orders, the preparation of shipping orders, and the preparation of customer invoices and statements. The analysis phase of the project is nearly complete. Data flow diagrams, process narratives, and a data dictionary have been developed for the new system. Now, this documentation is to be used to design the procedures and processing activities that will implement this as a computer-based system.

CONTEXT DIAGRAM

The scope of the system you are to develop is defined by the context diagram shown in Figure A-20. As noted, the order processing system interacts with four external entities:

- Customers will use the new system for placing orders with the company. Orders originate either as mail or phone orders. In either case, the order is entered into the system in a standard form, denoting the bill-to and ship-to addresses of the customer

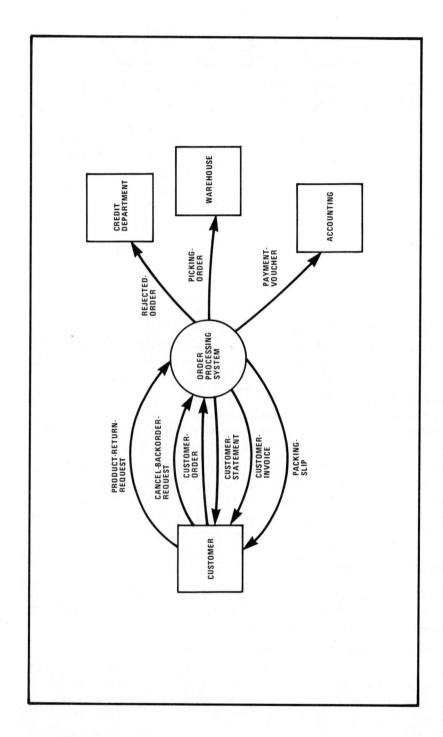

Figure A-20. Context diagram: ORDER PROCESSING SYSTEM.

and one or more order items identifying the products and quantities ordered. During this data entry process, the customer's credit standing is checked, along with the availability of the products ordered. It is assumed that all ordering is done on account. That is, the customer must have established an account with the company before ordering merchandise. After credit is approved, the customer is added to the list of approved customers. Items not in stock will be back-ordered. When the merchandise is shipped to the customer, a packing slip is included. This form lists the products that were shipped and provides a count of the number of items shipped and back-ordered. Subsequent to the shipment, an invoice is prepared and sent to the customer. The invoice shows the pricing for each item ordered and the total amount of the order. Later, during the company's regular billing cycle, the customer receives a statement. This document provides the total amount due the company for any and all invoices for the customer. Items for which back-orders exist can be cancelled before shipment. Also, customers may return merchandise and receive either credit toward their accounts or cash payments for the value of the returned merchandise.

- The order processing system provides picking slips to the warehouse. These slips are used by merchandise "pickers" who assemble orders from the stocks in the warehouse. The picking slips indicate the products and quantities and are printed in order by merchandise warehouse location to facilitate picking and assembly of orders. Packing slips also are sent to the warehouse so that they can be included in the order that is sent to the customer.

- If the customer does not have an account with the company, or if the value of the merchandise ordered is over the approved credit limit, the order is rejected and forwarded to the credit department. This department contacts the customer to establish an account or to adjust the order for possible resubmission.

- When items are returned, the customer can elect to apply the value of the merchandise against the account balance. Alternatively, the customer may choose to request a cash refund, in which case a payment voucher is sent to the accounting office as a request for payment.

PROCESSING EVENTS

In studying the context diagram in Figure A-20 and other collected documents, the project team identified several major events in the order processing system. In particular, the following events were identified:

- Write customer order
- Prepare back-orders
- Prepare shipping order
- Prepare customer invoice
- Prepare customer statement
- Cancel customer back-order
- Process returned merchandise.

This list does not necessarily include all possible or necessary procedures in the system. However, it represents a starting point for general design. Additional analysis probably will be necessary to understand and document further details within each process.

The outputs of systems analysis also included a set of data flow diagram (DFD) segments that document the major processing events in the system. These diagrams indicate the data flows and files that interact when inputs are transformed into outputs within each process. In support of the data flow diagrams are the data dictionary entries that document the structure and content of input and output data flows and files. In addition, process narratives have been prepared to provide explanations of the activities that comprise each processing event. In total, the DFD segments, data dictionary entries, and process narratives provide logical descriptions of the system under study. The individual DFD segments then have been combined within a Diagram 0 data flow diagram to describe the integrated system. It will be necessary, however, to develop leveled data flow diagrams for selected processes before detailed design is begun.

SYSTEM DOCUMENTATION

The set of data flow diagram segments for the major processing activities of the order processing system is given in Figures A-21 through A-28. At this point, the documentation presents a high-level overview of the proposed system. As needed, lower-level diagrams will be prepared to partition processes for further description and clarification of activities.

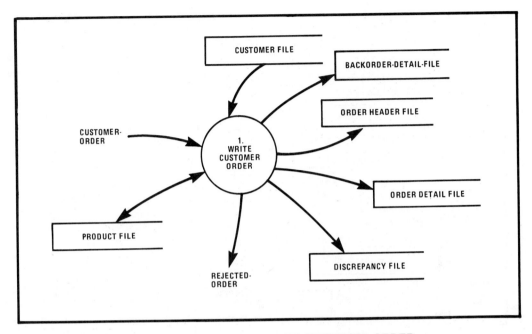

Figure A-21.   Data flow diagram. Process 1: WRITE CUSTOMER ORDER.

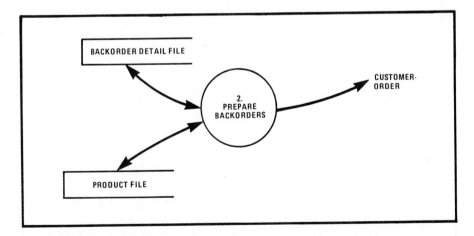

Figure A-22.   Data flow diagram. Process 2: PREPARE BACKORDERS.

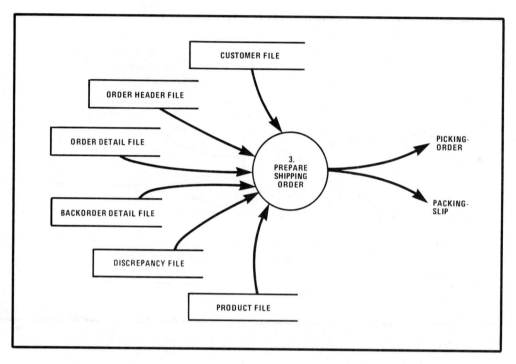

Figure A-23. Data flow diagram. Process 3: PREPARE SHIPPING ORDER.

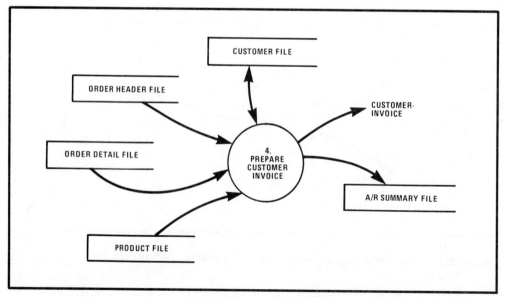

Figure A-24. Data flow diagram. Process 4: PREPARE CUSTOMER INVOICE.

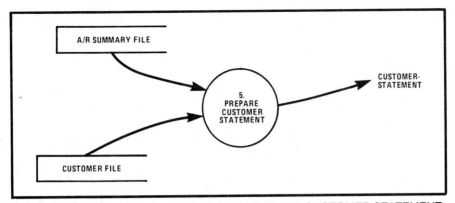

Figure A-25. Data flow diagram. Process 5: PREPARE CUSTOMER STATEMENT.

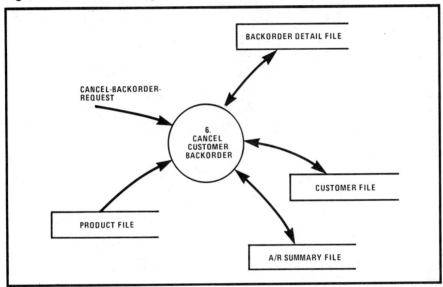

Figure A-26. Data flow diagram. Process 6: CANCEL CUSTOMER BACKORDER.

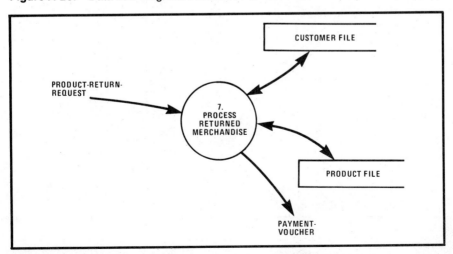

Figure A-27. Data flow diagram. Process 7: PROCESS RETURNED MERCHANDISE.

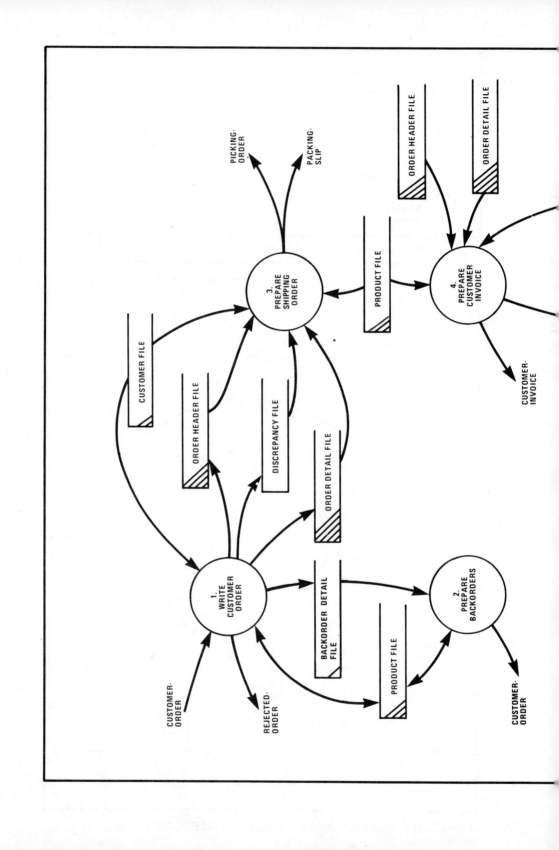

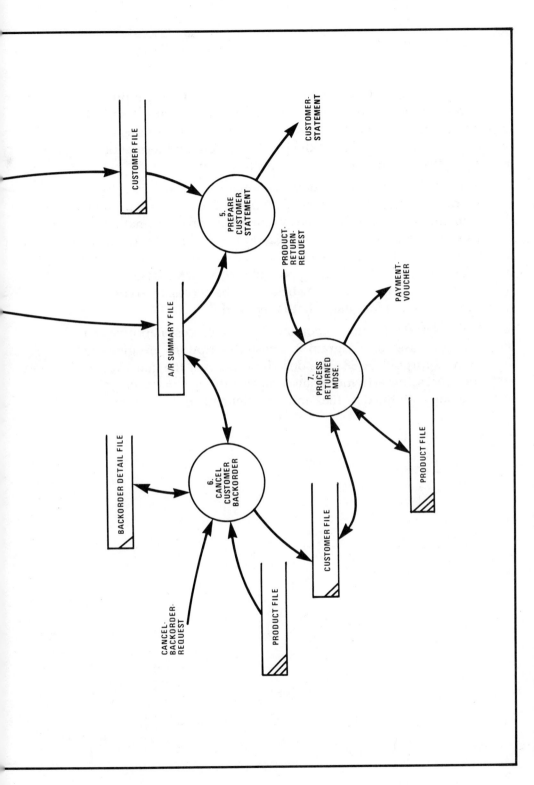

Figure A-28. Diagram 0: ORDER PROCESSING SYSTEM.

Figure A-29 contains the data dictionary entries for the input and output documents of the system. These documents are described logically, in terms of data components and relationships. At this point, little concern has been given to the physical layout and format of the source, input, output, and transaction documents.

Figure A-30 contains the data dictionary entries for the files. Again, physical design has not yet taken place. File content represents the major entities about which the system maintains information, as well as the access keys. File organization techniques and record layouts are still to be determined. Figure A-31 shows the file access diagram for the system, indicating possible access paths based on the record keys. Rearrangement of key fields may be necessary to conform to the programming language that is selected.

Finally, system documentation includes the process narratives for all of the processes on Diagram 0. These descriptions are given in Figures A-32 through A-38. The descriptions cover the general patterns of processing activities that comprise the system processing events, along with the data files and data elements that are involved.

```
CANCEL-BACKORDER-REQUEST = ACCOUNT-NUMBER +          CUSTOMER-ORDER  = ORDER-DATE +
                           ORDER-NUMBER +                             ACCOUNT-NUMBER +
                           CANCEL-DATE +                              CUSTOMER-NAME +
                           CUSTOMER-NAME +                            STREET-ADDRESS +
                           STREET-ADDRESS +                           CITY-STATE-ZIP +
                           CITY-STATE-ZIP +                          / SHIP-TO-CUSTOMER-NAME + \
                          { PRODUCT-NUMBER-CANCEL + }               {  SHIP-TO-STREET-ADDRESS +  }
                          { QUANTITY-ORDERED-CANCEL }                \ SHIP-TO-CITY-STATE-ZIP + /
                                                                     / PRODUCT-NUMBER + \
                                                                    {  QUANTITY-ORDERED +  }
CUSTOMER-INVOICE         = INVOICE-DATE +                           {  UNIT-OF-MEASURE +    }
                           ORDER-DATE +                             {  PRODUCT-DESCRIPTION + }
                           ORDER-NUMBER +                           {  UNIT-PRICE +         }
                           ACCOUNT-NUMBER +                          \ PRODUCT-AMOUNT + /
                           CUSTOMER-NAME +                            ORDER-SUBTOTAL +
                           STREET-ADDRESS +                           SALES-TAX +
                           CITY-STATE-ZIP +                           ORDER-TOTAL
                          / PRODUCT-NUMBER + \
                         {  QUANTITY-ORDERED +  }
                         {  UNIT-OF-MEASURE +    }  PRODUCT-RETURN- = RETURN-DATE +
                         {  PRODUCT-DESCRIPTION + } REQUEST           ACCOUNT-NUMBER +
                         {  QUANTITY-SHIPPED +    }                   CUSTOMER-NAME +
                         {  QUANTITY-BACKORDERED + }                  STREET-ADDRESS +
                         {  UNIT-PRICE +          }                   CITY-STATE-ZIP +
                          \ PRODUCT-AMOUNT + /                        ORDER-DATE +
                           INVOICE-SUBTOTAL +                         ORDER-NUMBER +
                           SALES-TAX +                                PRODUCT-NUMBER +
                           INVOICE-TOTAL                              UNIT-PRICE-PAID +
                                                                    [ CASH-REFUND-REQUEST  ]
                                                                    [ ACCOUNT-CREDIT-REQUEST ]

CUSTOMER-STATEMENT      = BILLING-DATE +
                           ACCOUNT-NUMBER +
                           CUSTOMER-NAME +
                           STREET-ADDRESS +
                           CITY-STATE-ZIP +
                          / ORDER-DATE +   \
                         {  ORDER-NUMBER +  }
                          \ INVOICE-TOTAL + /
                           AMOUNT-DUE
```

Figure A-29   Data dictionary entries for input and output data flows.

```
PACKING-SLIP  = ORDER-DATE +
                ORDER-NUMBER +
                ACCOUNT-NUMBER +
                CUSTOMER-NAME +
                STREET-ADDRESS +
                CITY-STATE-ZIP +
                SHIP-TO-CUSTOMER-NAME +
                SHIP-TO-STREET-ADDRESS +
                SHIP-TO-CITY-STATE-ZIP +
                ⎧ PRODUCT-NUMBER +        ⎫
                ⎪ UNIT-OF-MEASURE +       ⎪
                ⎨ QUANTITY-ORDERED +      ⎬
                ⎪ QUANTITY-SHIPPED +      ⎪
                ⎪ QUANTITY-BACKORDERED +  ⎪
                ⎩ PRODUCT-DESCRIPTION +   ⎭
                {INVALID-PRODUCT-NUMBER}
```

```
PAYMENT-VOUCHER = VOUCHER-DATE +
                  ACCOUNT-NUMBER +
                  CUSTOMER-NAME +
                  STREET-ADDRESS +
                  CITY-STATE-ZIP +
                  PAYMENT-AMOUNT
```

```
REJECTED-ORDER  = * A CUSTOMER-ORDER for
                  which the customer has
                  not established an account
                  or for which the order
                  total exceeds the
                  credit limit *
```

```
PICKING-ORDER = ORDER-DATE +
                ORDER-NUMBER +
                ACCOUNT-NUMBER +
                CUSTOMER-NAME +
                STREET-ADDRESS +
                CITY-STATE-ZIP +
                SHIP-TO-CUSTOMER-NAME +
                SHIP-TO-STREET-ADDRESS +
                SHIP-TO-CITY-STATE-ZIP +
                ⎧ LOCATION-CODE +         ⎫
                ⎪ PRODUCT-NUMBER +        ⎪
                ⎪ UNIT-OF-MEASURE +       ⎪
                ⎨ QUANTITY-ORDERED +      ⎬
                ⎪ QUANTITY-SHIPPED +      ⎪
                ⎪ QUANTITY-BACKORDERED +  ⎪
                ⎩ PRODUCT-DESCRIPTION     ⎭
```

Figure A-29 (cont'd).

FILES

```
A/R-SUMMARY-FILE        = {A/R-SUMMARY-RECORD}      ORDER-HEADER-FILE     = {ORDER-HEADER-RECORD}
A/R-SUMMARY-RECORD      = ACCOUNT-NUMBER +          ORDER-HEADER-RECORD   = ORDER-NUMBER +
                          ORDER-NUMBER +                                    ORDER-DATE +
                          ORDER-DATE +                                      ACCOUNT-NUMBER +
                          INVOICE-TOTAL                                     SHIP-TO-CUSTOMER-NAME +
                                                                           SHIP-TO-STREET-ADDRESS +
                                                                           SHIP-TO-CITY-STATE-ZIP

BACKORDER-DETAIL-FILE   = {BACKORDER-DETAIL-RECORD}
BACKORDER-DETAIL-RECORD = ORDER-NUMBER +
                          PRODUCT-NUMBER +          PRODUCT-FILE          = {PRODUCT-RECORD}
                          BACKORDER-QUANTITY        PRODUCT-RECORD        = PRODUCT-NUMBER +
                                                                           LOCATION-CODE +
                                                                           PRODUCT-DESCRIPTION +
DISCREPANCY-FILE        = {DISCREPANCY-RECORD}                            UNIT-OF-MEASURE +
DISCREPANCY-RECORD      = ORDER-NUMBER +                                  UNIT-PRICE +
                          INVALID-PRODUCT-NUMBER                          QUANTITY-ON-HAND +
                                                                           REORDER-POINT +
                                                                           REORDER-QUANTITY +

CUSTOMER-FILE           = {CUSTOMER-RECORD}
CUSTOMER-RECORD         = ACCOUNT-NUMBER +
                          CUSTOMER-NAME +
                          STREET-ADDRESS +
                          CITY-STATE-ZIP +
                          CREDIT-LIMIT +
                          ACCOUNT-BALANCE


ORDER-DETAIL-FILE       = {ORDER-DETAIL-RECORD}
ORDER-DETAIL-RECORD     = ORDER-NUMBER +
                          PRODUCT-NUMBER +
                          QUANTITY-ORDERED +
                          QUANTITY-SHIPPED +
                          QUANTITY-BACKORDERED
```

Figure A-30.   Data dictionary entries for system files.

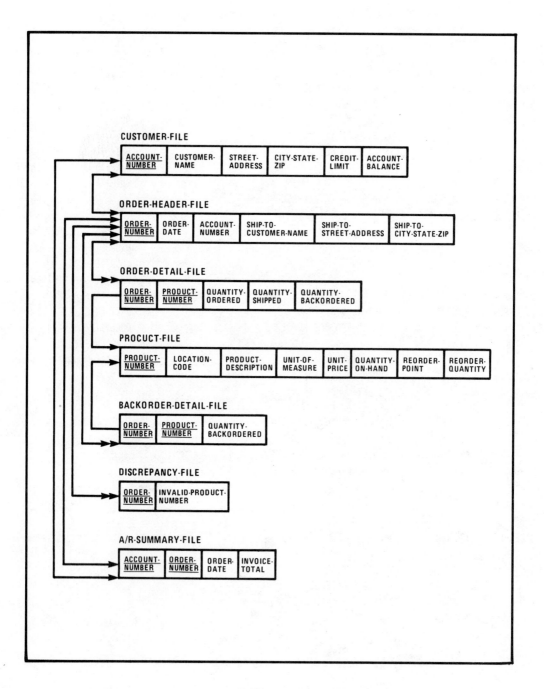

Figure A-31.   Data access diagram for system files.

PROCESS 1: WRITE CUSTOMER ORDER

1. Customers submit orders either on mail-order forms or over the telephone. In either case,
    the order is captured in a standard format. A computer terminal is to be used for entering
    the order into the system.

2. The account number submitted with the order is verified against the ACCOUNT-NUMBER in the
    CUSTOMER-FILE. If the account number is valid, the order is processed. If the customer
    does not have an account, the order is forwarded to the credit department and no further
    processing takes place within the order processing system.

3. The total amount of the order is verified against the CREDIT-LIMIT reported in the
    CUSTOMER-FILE. The limit on the amount of the order is given by subtracting the
    ACCOUNT-BALANCE from the CREDIT-LIMIT. If the ORDER-TOTAL on the CUSTOMER-ORDER is equal
    to or less than the order limit, the order is processed. Otherwise, the order is forwarded
    to the credit department.

4. An ORDER-NUMBER is assigned to the order. If there is no ship-to information on the
    CUSTOMER-ORDER, the customer name and address (the bill-to address) are used as the ship-to
    information. A record is written to the ORDER-HEADER-FILE keyed to the ORDER-NUMBER.

5. Order details are written to the ORDER-DETAIL-FILE. For each PRODUCT-NUMBER on the
    CUSTOMER-ORDER,

    a. The product record is accessed from the PRODUCT-FILE using the PRODUCT-NUMBER.

    b. If the PRODUCT-NUMBER is valid and if there is sufficient QUANTITY-ON-HAND to fill the
        order, the QUANTITY-SHIPPED is equal to the QUANTITY-ORDERED. If there is not
        sufficient merchandise in stock to fill the order, the QUANTITY-SHIPPED becomes the
        QUANTITY-ON-HAND and the QUANTITY-BACKORDERED is the difference between the
        QUANTITY-ORDERED and the QUANTITY-SHIPPED.

    c. The PRODUCT-FILE is updated to show the reduction in merchandise for quantities
        shipped. The QUANTITY-ON-HAND is reduced by the QUANTITY-SHIPPED.

    d. A record is written to the ORDER-DETAIL-FILE.

6. If the PRODUCT-NUMBER reported on the CUSTOMER-ORDER is invalid, a record is written to the
    DISCREPANCY-FILE. When shipping orders are prepared, these numbers will be listed on the
    PACKING-SLIP.

Figure A-32.   Process narrative. Process 1: WRITE CUSTOMER ORDER.

```
PROCESS 2:   PREPARE BACKORDERS

1.   The BACKORDER-DETAIL-FILE is accessed to locate orders for which backordered quantities
     exist.

2.   If there are quantities back-ordered, the PRODUCT-FILE is accessed to determine if there is
     sufficient merchandise in stock to fill the backorder.  If so, the ORDER-DETAIL-FILE and
     PRODUCT-FILE are updated to indicate filling of the order.

3.   A back-order is written as part of the normal procedure for preparing shipping orders.
```

Figure A-33.   Process narrative. Process 2: PREPARE BACKORDERS.

```
PROCESS 3:   PREPARE SHIPPING ORDER

1.   The same procedure is used for preparing shipping orders for both regular orders and
     back-orders.  In the latter case, the BACKORDER-DETAIL-FILE is processed instead of the
     ORDER-DETAIL-FILE.

2.   The ORDER-HEADER-RECORD is accessed on ORDER-NUMBER and the corresponding record from the
     CUSTOMER-FILE is accessed on the ACCOUNT-NUMBER in the header record.

3.   For all matching ORDER-NUMBERs in the ORDER-DETAIL-FILE, the ORDER-DETAIL-RECORDS are
     accessed.

4.   The PRODUCT-RECORD is accessed on PRODUCT-NUMBER for each product ordered.  The products
     are sorted on LOCATION-CODE prior to printing of PICKING-ORDERs and PACKING-SLIPs.

5.   Information from the ORDER-HEADER-RECORD, ORDER-DETAIL-RECORDs, and PRODUCT-RECORDs is
     combined in printing the PICKING-ORDER and PACKING-SLIP.  For the PACKING-SLIP, the
     DISCREPANCY-FILE is accessed on ORDER-NUMBER to identify invalid PRODUCT-NUMBERs.  These
     numbers are printed on the PACKING-SLIP, which is included in the delivered order.  The
     invalid numbers are not included, however, if the shipped order is a back-order.
```

Figure A-34.   Process narrative. Process 3: PREPARE SHIPPING ORDER.

PROCESS 4: PREPARE CUSTOMER INVOICE

1. The ORDER-HEADER-RECORD is accessed using the ORDER-NUMBER.

2. Using the ACCOUNT-NUMBER from the the ORDER-HEADER-RECORD, the customer record is accessed from the CUSTOMER-FILE.

3. For each order item, the ORDER-DETAIL-RECORD is accessed on ORDER-NUMBER and the corresponding PRODUCT-RECORD is accessed on PRODUCT-NUMBER.

4. Individual PRODUCT-AMOUNTs are calculated by multiplying the QUANTITY-ORDERED by the UNIT-PRICE. The INVOICE-SUBTOTAL is an accumulation of all the PRODUCT-AMOUNTs for a particular order. SALES-TAX is calculated at 5% of the INVOICE-SUBTOTAL. INVOICE-TOTAL is then derived by adding the SALES-TAX amount to the INVOICE-SUBTOTAL.

5. CUSTOMER-INVOICE is printed by combining information from the ORDER-HEADER-FILE, CUSTOMER-FILE, ORDER-DETAIL-FILE, and PRODUCT-FILE.

6. The CUSTOMER-FILE is updated by adding the INVOICE-TOTAL to the ACCOUNT-BALANCE. A summary record is written to the A/R-SUMMARY-FILE

Figure A-35. Process narrative. Process 4: PREPARE CUSTOMER INVOICE.

PROCESS 5: PREPARE CUSTOMER STATEMENT

1. All summary records for a particular customer are accessed from the A/R-SUMMARY-FILE.

2. The INVOICE-TOTALs are accumulated.

3. The customer's record is accessed from the CUSTOMER-FILE using the ACCOUNT-NUMBER from the summary file.

4. The statement is written by combining information in the CUSTOMER-FILE and A/R-SUMMARY-FILE. The INVOICE-TOTALS are presented in order by ORDER-DATE.

Figure A-36. Process narrative. Process 5: PREPARE CUSTOMER STATEMENT.

```
PROCESS 6:  CANCEL CUSTOMER BACKORDER

  1.   For each back-ordered product that is cancelled, the BACKORDER-DETAIL-FILE is accessed on
       ORDER-NUMBER + PRODUCT-NUMBER.  The record is deleted from the file.

  2.   The A/R-SUMMARY-FILE is updated by reducing the INVOICE-TOTAL by an amount equal to that
       charged for the product.  The charge is determined from the UNIT-PRICE information in the
       PRODUCT-FILE and the BACKORDER-QUANTITY from the BACKORDER-DETAIL-FILE.

  3.   The ACCOUNT-BALANCE in the CUSTOMER-FILE is reduced by the value of the cancelled order.
```

Figure A-37.   Process narrative. Process 6: CANCEL CUSTOMER BACKORDER.

```
PROCESS 7:  PROCESS RETURNED MERCHANDISE

  1.   If the request is for a credit to the customer's account, the CUSTOMER-RECORD is accessed
       and the amount paid for the product is applied against the ACCOUNT-BALANCE.  If the request
       is for a cash refund, a PAYMENT-VOUCHER is prepared and forwarded to the accounting
       department.

  2.   The PRODUCT-FILE is updated by adding the returned items back onto the QUANTITY-ON-HAND.
```

Figure A-38.   Process narrative. Process 7: PROCESS RETURNED
MERCHANDISE.

SUPPLEMENTARY PROJECT

The order processing system described in this section provides basic services for accepting and filling orders. The system has not yet been expanded to encompass inventory control procedures for maintaining the product file. As a supplementary project, add these necessary features to the system. In particular, provide inventory control procedures for recording data on merchandise received from suppliers within the product file. Also, provide the capability for adding new products to inventory and for changing the prices of merchandise. Establish a procedure for reordering products when the quantity on hand reaches the reorder point.

As an optional supplement, assume that ordering and billing are centralized within the main offices of the company. Order filling and inventory maintenance, however, will be distributed activities. Orders will be filled within the warehouse at the geographic site closest to the customer. These geographically dispersed locations will establish and manage their own inventories.

If required by your instructor, conduct analysis and design for adding inventory control procedures to the system. Add to or modify any of the existing documentation to integrate the necessary processes into the system before proceeding with design.

CASE STUDY 3: PAYROLL SYSTEM

SITUATION

As project team leader in the CIS department of a medium-sized software development company, your job is to head the development of a new payroll system. This system will perform weekly payroll processing for salaried and wage-based employees, print paychecks, pay stubs, payroll reports, maintain employee payroll files, and produce quarterly and yearly governmental reports. The analysis phase of the project is nearly complete. Data flow diagrams, process narratives, and data dictionary entries have been developed for the new system. This documentation will be used to complete any remaining analysis that needs to be done and to design the procedures and processing activities that will implement the payroll process as a computer-based system.

CONTEXT DIAGRAM

The scope of the system that you are responsible for developing is defined by the context diagram in Figure A-39. As shown, the payroll system interacts with four external entities:

- Company employees trigger payroll processing by submitting time slips at the close of each work week. These slips report the number of regular and overtime hours worked. The weekly payroll is calculated on the basis of these figures and the earnings and deductions included in the employee payroll files. At the

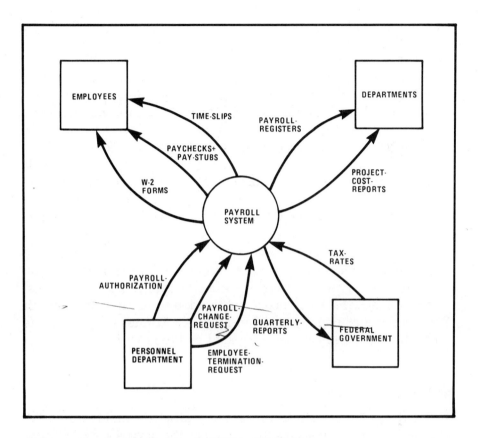

Figure A-39. Context diagram: PAYROLL SYSTEM.

same time, payroll processing is done for salaried employees, who do not report hours worked but are paid on the basis of a set weekly salary. At the end of payroll processing, paychecks and pay stubs are distributed to the employees. Also, at the close of each year, W-2 forms are prepared and distributed to employees for tax reporting purposes.

- The personnel department issues payroll authorizations for new employees. These authorizations are the sources of employee demographic and payroll information used in the payroll files around which processing takes place. The department also provides payroll change requests that authorize changes in the earnings and deductions status of employees. Finally, when an employee is terminated, that information flows through the personnel department into the payroll system.

- Departments within which employees work receive two types of output from the payroll system. A payroll register is printed at the end of each pay period. This report lists the hours worked, earnings, and deductions for the week for all employees within each department. Project cost reports also are prepared for the departments. These reports summarize the hours worked and payroll costs for each project under development within each department. Both reports provide personnel costing information that can be compared with cost projections and budgets developed at the start of projects.

- Finally, the federal government receives reports from the payroll system. Each quarter, the government requires reporting of total federal and F.I.C.A. taxes withheld during the three-month period. Of course, federal and F.I.C.A. tax rates used as inputs to the payroll system are provided by the government.

PROCESSING EVENTS

During the systems analysis process, the project team identified the basic set of processing events that comprise the payroll system. These events do not represent all processing required for an operational computer-based system. They are, however, the major business applications to be implemented. The 10 events are:

- Edit time slips
- Calculate weekly payroll
- Update payroll file
- Update project file
- Print payroll register
- Print employee paychecks
- Print quarterly report
- Print employee W-2 forms
- Maintain employee files
- Print project cost reports.

As noted, it will be impossible to add other processes as processing control measures and to provide additional file maintenance, backup, and reporting procedures.

SYSTEM DOCUMENTATION

The collection of data flow diagram segments for the major processing events listed above is given in Figures A-40 through A-49. A Diagram 0 for the system is shown in Figure A-50. As necessary, leveled data flow diagrams will be needed to elaborate and partition processes for further description and clarification.

Figure A-51 contains the data dictionary entries for the system's input and output documents. These documents are described logically, in terms of data components and relationships. No attention has been given to the physical layout of the source, input, output, and transaction documents.

Data dictionary entries for the files are given in Figure A-52. Again, physical design has not taken place. Yet to be determined are file organizations, record layouts, and field formats and data types. In Figure A-53 is the data access diagram describing the primary access paths among the four data files in the system, based on their record keys.

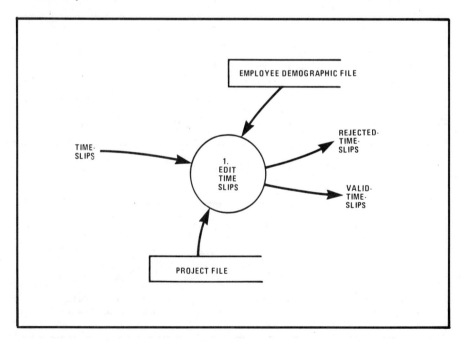

Figure A-40. Data flow diagram. Process 1: EDIT TIME SLIPS.

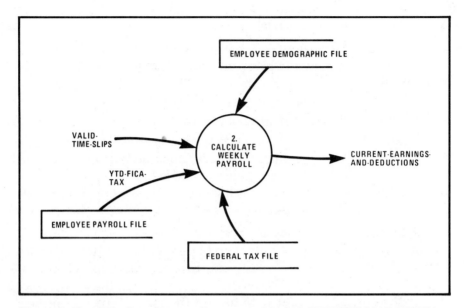

Figure A-41. Data flow diagram. Process 2: CALCULATE WEEKLY PAYROLL.

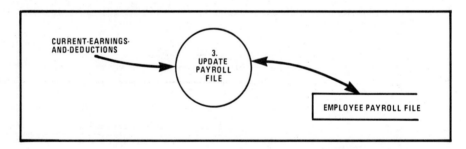

Figure A-42. Data flow diagram. Process 3: UPDATE PAYROLL FILE.

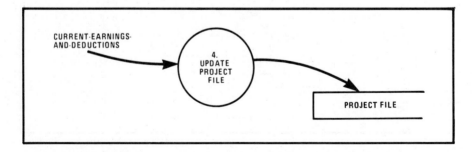

Figure A-43. Data flow diagram. Process 4: UPDATE PROJECT FILE.

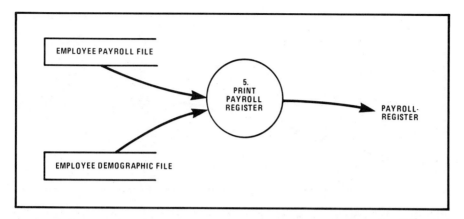

Figure A-44. Data flow diagram. Process 5: PRINT PAYROLL REGISTER.

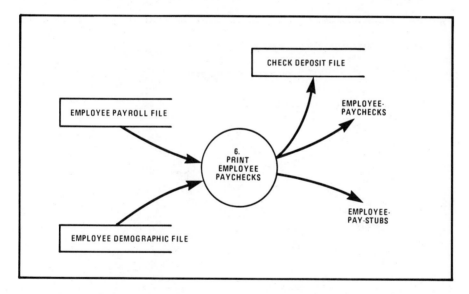

Figure A-45. Data flow diagram. Process 6: PRINT EMPLOYEE PAYCHECKS.

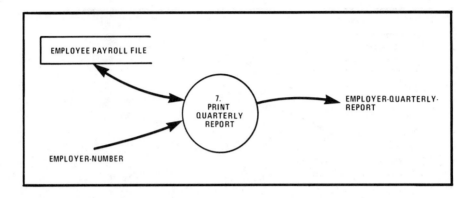

Figure A-46. Data flow diagram. Process 7: PRINT QUARTERLY REPORT.

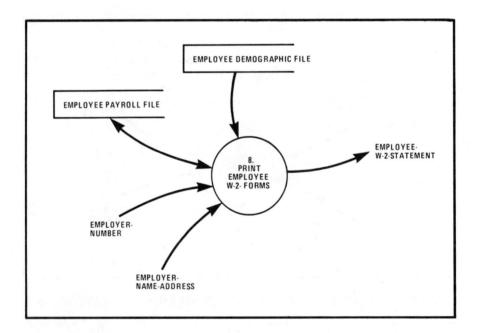

Figure A-47. Data flow diagram. Process 8: PRINT EMPLOYEE W-2 FORMS.

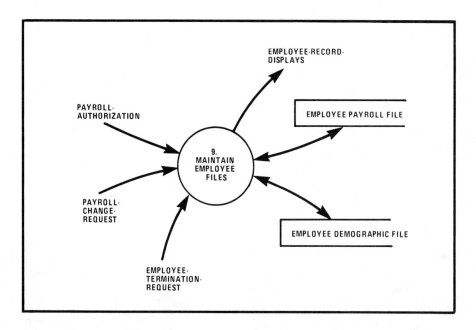

Figure A-48.   Data flow diagram. Process 9: MAINTAIN EMPLOYEE FILES.

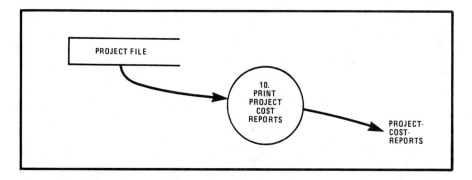

Figure A-49.   Data flow diagram. Process 10: PRINT PROJECT COST REPORTS.

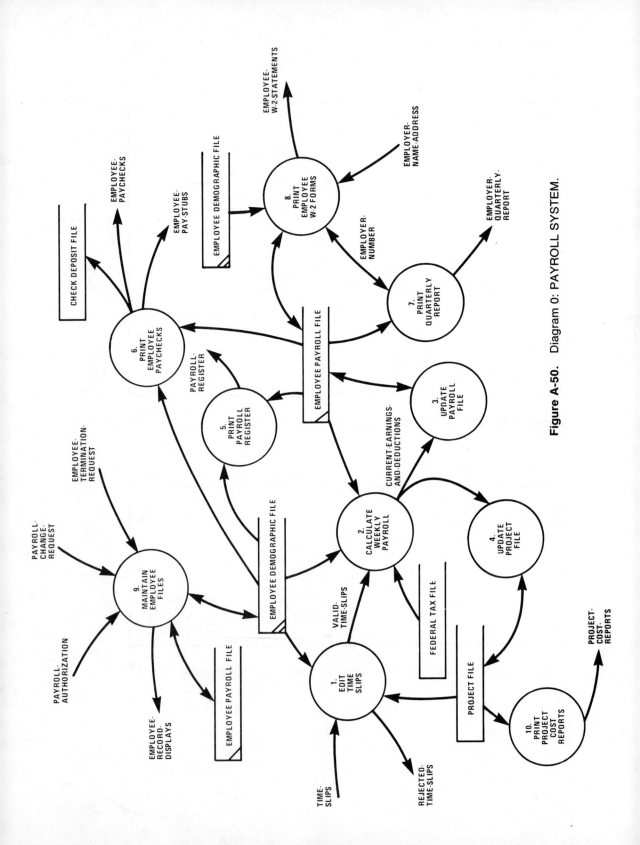

Figure A-50. Diagram 0: PAYROLL SYSTEM.

```
                              INPUT/OUTPUT DOCUMENTS


EMPLOYEE-CHANGE-                                 EMPLOYEE-TERMINATION-
REQUEST           = EMPLOYEE-NUMBER +            REQUEST           = DATE-OF-TERMINATION +
                    (DEPARTMENT-NUMBER-CHANGE) +                     EMPLOYEE-NUMBER +
                    (EMPLOYEE-NAME-CHANGE) +                         DEPARTMENT-NUMBER +
                    (STREET-ADDRESS-CHANGE) +                        EMPLOYEE-NAME +
                    (CITY-STATE-ZIP-CHANGE) +                        STREET-ADDRESS +
                    (DATE-OF-EMPLOYMENT-CHANGE) +                    CITY-STATE-ZIP
                    (DATE-OF-TERMINATION-CHANGE) +
                    (NUMBER-OF-EXEMPTIONS-CHANGE) + EMPLOYEE-W-2-
                    (SALARY-WAGE-CODE-CHANGE) +   STATEMENT         = EMPLOYER-NUMBER +
                    (SALARY-WAGE-RATE-CHANGE) +                      EMPLOYER-NAME-ADDRESS +
                    (OVERTIME-WAGE-RATE-CHANGE) +                    EMPLOYEE-NUMBER +
                    (DEPOSIT-CODE-CHANGE) +                          EMPLOYEE-NAME +
                    (CURRENT-GROSS-PAY-CHANGE) +                     STREET-ADDRESS +
                    (CURRENT-FEDERAL-TAX-CHANGE) +                   CITY-STATE-ZIP +
                    (CURRENT-STATE-TAX-CHANGE) +                     YTD-GROSS-PAY +
                    (CURRENT-FICA-TAX-CHANGE) +                      YTD-FEDERAL-TAX +
                    (CURRENT-NET-PAY-CHANGE)                         YTD-STATE-TAX +
                                                                     YTD-FICA-TAX


EMPLOYEE-PAYCHECK = PAY-PERIOD-END-DATE +        EMPLOYER-QUARTERLY-
                    EMPLOYEE-NAME +              REPORT            = QUARTERLY-REPORTING-DATE +
                    CURRENT-NET-PAY                                  EMPLOYER-NUMBER +
                                                                     EMPLOYER-NAME-ADDRESS +
                                                                     QUARTERLY-GROSS-PAY +
EMPLOYEE-PAY-STUB = PAY-PERIOD-END-DATE +                           QUARTERLY-FEDERAL-TAX-WITHHELD +
                    EMPLOYEE-NUMBER +                                QUARTERLY-STATE-TAX-WITHHELD +
                    CURRENT-GROSS-PAY +                              QUARTERLY-FICA-TAX-WITHHELD +
                    CURRENT-FEDERAL-TAX +                            QUARTERLY-EMPLOYER-FICA-TAX +
                    CURRENT-STATE-TAX +                              QUARTERLY-TOTAL-FICA-TAX
                    CURRENT-FICA-TAX +
                    CURRENT-NET-PAY +
                    YTD-GROSS-PAY +
                    YTD-FEDERAL-TAX +
                    YTD-STATE-TAX +
                    YTD-FICA-TAX +
                    YTD-NET-PAY
```

Figure A-51   Data dictionary entries for input and output data flows.

```
PAYROLL-                              PROJECT-COST-
AUTHORIZATION  = DATE-OF-EMPLOYMENT + REPORT  = CURRENT-DATE +
                 DEPARTMENT-NUMBER +          DEPARTMENT-NUMBER +
                 EMPLOYEE-NUMBER +             PROJECT-NUMBER +
                 EMPLOYEE-NAME +              ⌠ ⌠EMPLOYEE-NUMBER +
                 STREET-ADDRESS +             │ │EMPLOYEE-REGULAR-HOURS-WORKED + ⌡
                 CITY-STATE-ZIP +            │ ⟨EMPLOYEE-OVERTIME-HOURS-WORKED +
                 NUMBER-OF-EXEMPTIONS +       │ │EMPLOYEE-REGULAR-PAY +
                 SALARY-WAGE-CODE +           │ ⌞EMPLOYEE-OVERTIME-PAY +
                 SALARY-WAGE-RATE +           │ PROJECT-REGULAR-HOURS-WORKED +
                 (OVERTIME-WAGE-RATE) +       │ PROJECT-OVERTIME-HOURS-WORKED +
                 DEPOSIT-CODE +               │ PROJECT-REGULAR-PAY +
                ⌠(CURRENT-GROSS-PAY) +        │ PROJECT-OVERTIME-PAY +
                │(CURRENT-FEDERAL-TAX) +      ⌞ PROJECT-TOTAL-PAY
                ⟨(CURRENT-STATE-TAX) +
                │(CURRENT-FICA-TAX) +
                ⌞(CURRENT-NET-PAY)           TIME-SLIP = PAY-PERIOD-END-DATE +
                                                         DEPARTMENT-NUMBER +
                                                         EMPLOYEE-NUMBER +
                                                        ⌠⌠PROJECT-NUMBER +      ⌡
                                                        │⟨REGULAR-HOURS-WORKED + │
PAYROLL-REGISTER = PAY-PERIOD-END-DATE +                │⌞OVERTIME-HOURS-WORKED +⌡
                   DEPARTMENT-NUMBER +                  TOTAL-REGULAR-HOURS-WORKED +
                  ⌠EMPLOYEE-NUMBER +                    TOTAL-OVERTIME-HOURS-WORKED
                  │SALARY-WAGE-CODE +
                  │EMPLOYEE-NAME +
                  │(TOTAL-REGULAR-HOURS-WORKED) + ⌡
                  │(TOTAL-OVERTIME-HOURS-WORKED) +⌡
                  │CURRENT-GROSS-PAY +
                  ⟨CURRENT-FEDERAL-TAX +
                  │CURRENT-STATE-TAX +
                  │CURRENT-FICA-TAX +
                  ⌞CURRENT-NET-PAY +
                   DEPARTMENT-REGULAR-HOURS-WORKED +
                   DEPARTMENT-OVERTIME-HOURS-WORKED +
                   DEPARTMENT-GROSS-PAY +
                   DEPARTMENT-FEDERAL-TAX +
                   DEPARTMENT-STATE-TAX +
                   DEPARTMENT-FICA-TAX +
                   DEPARTMENT-NET-PAY
```

Figure A-51.   (cont'd).

Figure A-52. Data dictionary entries for system files.

```
                                  FILES


CHECK-DEPOSIT-FILE          = {CHECK-DEPOSIT-RECORD}    FEDERAL-TAX-FILE    = {FEDERAL-TAX-RECORD}
CHECK-DEPOSIT-RECORD        = EMPLOYEE-NUMBER +         FEDERAL-TAX-RECORD  = WEEKLY-GROSS-PAY-RANGE +
                              PAY-PERIOD-END-DATE +                           WITHHOLDING-AMOUNT-
                              CURRENT-NET-PAY                                 BY-EXEMPTION



EMPLOYEE-DEMOGRAPHIC-FILE   = {EMPLOYEE-DEMOGRAPHIC-RECORD}   PROJECT-FILE    = {PROJECT-RECORD}
EMPLOYEE-DEMOGRAPHIC-RECORD = EMPLOYEE-NUMBER +              PROJECT-RECORD   = PROJECT-NUMBER +
                              DEPARTMENT-NUMBER +                              DEPARTMENT-NUMBER +
                              EMPLOYEE-NAME +                                  EMPLOYEE-NUMBER +
                              STREET-ADDRESS +                                 PAY-PERIOD-END-DATE +
                              CITY-STATE-ZIP +                                 TOTAL-REGULAR-HOURS-WORKED +
                              DATE-OF-EMPLOYMENT +                             TOTAL-OVERTIME-HOURS-WORKED +
                              DATE-OF-TERMINATION +                            CURRENT-REGULAR-PAY +
                              NUMBER-OF-EXEMPTIONS +                           CURRENT-OVERTIME-PAY
                              SALARY-WAGE-CODE +
                              SALARY-WAGE-RATE +
                              OVERTIME-WAGE-RATE +
                              DEPOSIT-CODE



EMPLOYEE-PAYROLL-FILE       = {EMPLOYEE-PAYROLL-RECORD}
EMPLOYEE-PAYROLL-RECORD     = EMPLOYEE-NUMBER +
                              DEPARTMENT-NUMBER +
                              PAY-PERIOD-END-DATE +
                              TOTAL-REGULAR-HOURS-WORKED +
                              TOTAL-OVERTIME-HOURS-WORKED +
                              CURRENT-REGULAR-PAY +
                              CURRENT-OVERTIME-PAY +
                              CURRENT-GROSS-PAY +
                              CURRENT-FEDERAL-TAX +
                              CURRENT-STATE-TAX +
                              CURRENT-FICA-TAX +
                              CURRENT-NET-PAY +
                              QTD-GROSS-PAY +
                              QTD-FEDERAL-TAX +
                              QTD-STATE-TAX +
                              QTD-FICA-TAX +
                              YTD-GROSS-PAY +
                              YTD-FEDERAL-TAX +
                              YTD-STATE-TAX +
                              YTD-FICA-TAX
```

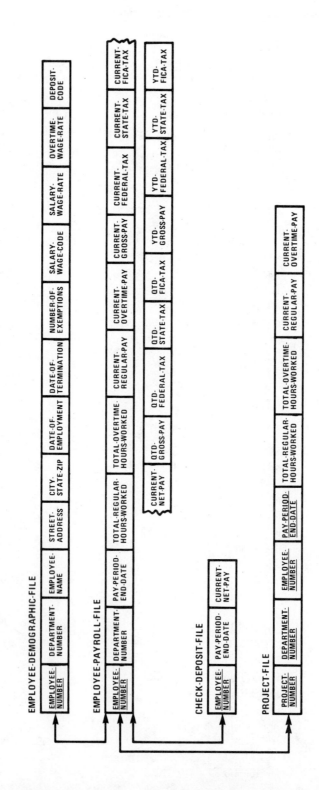

Figure A-53. Data access diagram for payroll system.

System documentation also includes process narratives for all of the processes identified in Diagram 0. These narratives are given in Figures A-54 through A-63. They describe the general patterns of processing activities that comprise the processing events, along with the data files and data elements that are involved in the processing. In most cases, these narratives outline the business functions that are performed rather than specific computer processing tasks. A major part of design, then, will be to specify how the computer will implement the processes that are chosen for automation.

PROCESS 1: EDIT TIME SLIPS

1. For wage-earners--employees who are paid weekly on the basis of an hourly pay rate--TIME-SLIPS are submitted at the end of each week. These slips report the number of regular and overtime hours worked. As appropriate, project numbers are reported for employees assigned to particular projects.

2. The following edits take place:

    a. EMPLOYEE-NUMBER is verified against the number in the EMPLOYEE-DEMOGRAPHIC-FILE.

    b. If present, PROJECT-NUMBERs are checked against the PROJECT-FILE to verify that the employee has been assigned to these projects. The total of REGULAR-HOURS-WORKED and OVERTIME-HOURS-WORKED are verified against the reported totals.

    c. TOTAL-REGULAR-HOURS-WORKED is range-checked to verify that the value falls between 1 and 40.

    d. TOTAL-OVERTIME-HOURS-WORKED is range-checked to verify that the value falls between 1 and 20. No overtime hours should be reported if TOTAL-REGULAR-HOURS-WORKED is less than 40.

3. If the TIME-SLIP fails to pass any of the edits, it is rejected and returned to the originating department for correction.

Figure A-54.   Process narrative. Process 1: EDIT TIME SLIPS.

PROCESS 2:   CALCULATE WEEKLY PAYROLL

1.   This process takes place only for wage earners and not for salaried employees

2.   The SALARY-WAGE-RATE is accesed from the EMPLOYEE-DEMOGRAPHIC-FILE.   The value reported is
     the hourly rate for all regular hours worked.   CURRENT-REGULAR-PAY is calculated by
     multiplying TOTAL-REGULAR-HOURS reported on the TIME-SLIP   by the SALARY-WAGE-RATE.
     CURRENT-OVERTIME-PAY is calculated by multiplying TOTAL-OVERTIME-HOURS-WORKED by the
     OVERTIME-WAGE-RATE.   Then, CURRENT-GROSS-PAY is derived by adding CURRENT-REGULAR-PAY and
     CURRENT-OVERTIME-PAY.

3.   CURRENT-FEDERAL-TAX is extracted from tax tables built from the FEDERAL-TAX-FILE. The
     table contains withholding amounts classified by weekly gross pay range and number of
     exemptions. The employee's CURRENT-GROSS-PAY is used to locate the corresponding row of
     the table; the NUMBER-OF-EXEMPTIONS reported in the EMPLOYEE-DEMOGRAPHIC-FILE is used in
     locating the corresponding column of exemptions.   The withholding amount at the
     intersection of the row and column is the CURRENT-FEDERAL-TAX amount.

4.   CURRENT-STATE-TAX is calculated at 3% of CURRENT-GROSS-PAY.

5.   CURRENT-FICA-TAX is calculated by multiplying CURRENT-GROSS-PAY by 7%. F.I.C.A. tax is
     withheld up to an amount that does not exceed $42,000 a year.   EMPLOYEE-PAYROLL-FILE is
     accessed to verify that the current tax plus year-to-date tax does not exceed this amount.

6.   CURRENT-NET-PAY   is given by subtracting CURRENT-FEDERAL-TAX + CURRENT-STATE-TAX +
     CURRENT-FICA-TAX from CURRENT-GROSS-PAY.

Figure A-55.   Process narrative. Process 2: CALCULATE WEEKLY PAYROLL.

PROCESS 3: UPDATE PAYROLL FILE

1. For wage earners, the current earnings and withholding amounts are added to the quarterly-to-date and year-to-date fields in the EMPLOYEE-PAYROLL-FILE, along with the reported hours worked from the TIME-SLIPs. Records are also updated for employees who did not have hours reported. In this case, zero amounts appear in the current earnings and deductions fields.

2. For salaried employees, the current earnings and withholding amounts remain the same each week. When the employee is hired, these amounts are calculated and placed in the EMPLOYEE-PAYROLL-FILE. File updating, then, involves adding the current values onto the quarterly-to-date and year-to-date figures.

Figure A-56.   Process narrative. Process 3: UPDATE PAYROLL FILE.

PROCESS 4: UPDATE PROJECT FILE

1. For employees assigned to projects, current hours worked and earnings amounts are added to the PROJECT-FILE. A record is added to the file for each employee assigned to each project during each week. An employee will have a PROJECT-RECORD for each project worked on during the week. Records are accumulated within the file until the project is completed.

Figure A-57.   Process narrative. Process 4: UPDATE PROJECT FILE.

PROCESS 5: PRINT PAYROLL REGISTER

1. The EMPLOYEE-PAYROLL-FILE is sorted on EMPLOYEE-NUMBER within DEPARTMENT-NUMBER so that the PAYROLL-REGISTER can be printed by department. Wage or salary codes and employee names are extracted from the EMPLOYEE-DEMOGRAPHIC-FILE for printing on the report.

2. For each department, the report lists the current earnings and deductions for every employee, including those that had no earnings for the week.

3. For each department, the report lists departmental totals of earnings and deductions for the week.

Figure A-58.   Process narrative. Process 5: PRINT PAYROLL REGISTER.

PROCESS 6: PRINT EMPLOYEE PAYCHECKS

1. The EMPLOYEE-PAYROLL-FILE is sorted on EMPLOYEE-NUMBER within DEPARTMENT-NUMBER so that paychecks and pay stubs can be delivered in batch to the departments for distribution to employees.

2. The sorted payroll file is primary input to the check-printing process. The EMPLOYEE-DEMOGRAPHIC-FILE is accessed in order to determine the EMPLOYEE-NAME and CHECK-DEPOSIT-CODE.

3. Both paychekcs and stubs are printed for all employees who will receive pay amounts for the week and who do not choose to have their earnings automatically deposited, as signified by the CHECK-DEPOSIT-CODE. For employees who choose the automatic deposit option, only pay stubs are printed. Instead of paychecks for these employees, records are written to the CHECK-DEPOSIT-FILE. This file is used by the personnel department to authorize release of funds to banks.

4. All paychecks and stubs are forwarded to the personnel department for signature of the payroll authorizing agent before being forwarded to the departments for distribution.

Figure A-59.   Process narrative. Process 6: PRINT EMPLOYEE PAYCHECKS.

PROCESS 7: PRINT QUARTERLY REPORT

1. The quarterly-to-date (QTD) values in the EMPLOYEE-PAYROLL-FILE are totaled for all employees.

2. The employer's contribution to F.I.C.A. taxes is 7% of QUARTERLY-GROSS-PAY, up to an amount not to exceed $42,000 per year for each employee. This amount is added to QUARTERLY-FICA-TAX-WITHHELD to determine the QUARTERLY-TOTAL-FICA-TAX that is due in payment from the company to the federal government.

3. After the report has been printed, all quarterly-to-date fields within the EMPLOYEE-PAYROLL-FILE are reset to zero for accumulation of employee earnings and deductions for the next quarter.

Figure A-60.   Process narrative. Process 7: PRINT QUARTERLY REPORT.

PROCESS 8: PRINT EMPLOYEE W-2 FORMS

1. The year-to-date (YTD) fields within the EMPLOYEE-PAYROLL-FILE are printed on the W-2 forms.

2. Employee names and addresses are accessed from the EMPLOYEE-DEMOGRAPHIC-FILE.

3. After the W-2 forms have been printed, the year-to-date fields within the EMPLOYEE-PAYROLL-FILE are reset to zero.

4. In preparation for the next year's payroll reporting, records are deleted from the employee files for all employees who were terminated during the year. These employees are identified by the date given in the DATE-OF-TERMINATION within the EMPLOYEE-DEMOGRAPHIC-FILE.

Figure A-61. Process narrative. Process 8: PRINT EMPLOYEE W-2 FORMS.

PROCESS 9: MAINTAIN EMPLOYEE FILES

1. New employees are added to the employee files on receipt of a PAYROLL-AUTHORIZATION request. All of the information in the EMPLOYEE-DEMOGRAPHIC-FILE is captured from the authorization form. A new record is also created for the employee in the EMPLOYEE-PAYROLL-FILE, and the earnings and deductions fields are initialized to zero.

2. Changes to the employee files are requested on the PAYROLL-CHANGE-REQUEST. Any field, with the exception of the EMPLOYEE-NUMBER, within the EMPLOYEE-DEMOGRAPHIC-FILE and the EMPLOYEE-PAYROLL-FILE can be changed.

3. Employment terminations are reported on the EMPLOYEE-TERMINATION-REQUEST. The EMPLOYEE-DEMOGRAPHIC-FILE is updated by addition of the DATE-OF-TERMINATION. Records for terminated employees are not removed from the files until the end of the year, after government tax reporting.

4. All maintenance requests should be processed interactively, rather than in batch mode. In addition, the maintenance routines should include inquiry capabilities so that employee records can be displayed selectively when EMPLOYEE-NUMBERs are entered.

Figure A-62. Process narrative. Process 9: MAINTAIN EMPLOYEE FILES.

```
PROCESS 10:  PRINT PROJECT COSTS REPORTS

  1.  The reports are printed by department, showing, for  each  project,  employee  hours and pay
      amounts and total project hours and amounts.

  2.  Hours  worked  and  pay amounts are accumulated for each employee, for each project,  within
      each department, as reported in the PROJECT-FILE.
```

Figure A-63.   Process narrative. Process 10: PRINT PROJECT COST REPORTS.

SUPPLEMENTARY PROJECT

Payroll costs are part of the expenses that can be assigned to projects under development. The total costs of software products developed by the firm also include computer time and other resources. These accumulated costs are used to establish prices and fees charged for the products that are developed. Accounting control, therefore, requires a subsystem for assignment and tracking of costs for all developmental work.

The payroll system should be expanded to include a more comprehensive job costing subsystem. For each project, there should be a way to associate personnel with particular work assignments, to collect information on personnel and other developmental costs related to the project, to record budget and schedule projections for project completion, and to report regularly on deviations from plans. At project completion, the subsystem should provide a complete cost summary.

As required by the instructor, add the necessary procedures to the payroll system. As needed, modify the data collection process to capture the specific payroll and operational information required to assign costs and time charges, establish procedures for collecting budget and schedule projections, and design the routines to compare projections with actual costs and schedules. For completed projects, provide cost summaries.

GLOSSARY

A

absolute position   Specific physical point on a disk surface where a record is located. *See also* relative position.

acceptance review   Session at which project team presents information to a management group on an activity or phase for which approval is necessary.

acceptance testing   A final procedural review to demonstrate a system and secure user approval before a system becomes operational.

access controls   Controls that limit physical access to computer sites, and that limit electronic access to computer systems only to authorized persons.

access diagram   *See* data access diagram.

access path   The correlations or relationships between record keys that establish connections among data items imbedded in file records.

access time   The time necessary to locate a data record on disk and read it into memory. Design consideration when choosing access method.

accuracy   Conformity to a standard, or true value. Consideration when establishing controls throughout a system, especially where input data are entered.

active control   A connection between program modules through conditional or unconditional transfers of processing responsibility, allowing only data to cross the interface.

activity   Within the systems development life cycle, a group of logically related tasks that lead to, and are defined by, the accomplishment of a specific objective.

activity rate   Frequency of record access by an application. Design consideration when choosing access method.

adjustment   Correction or modification generated by feedback within a control process to bring system input or processing back into line with expectations.

administrator (walkthrough)   Experienced system analyst who provides organizational or administrative support for a walkthrough.

afferent   The branches of data flow in which physical inputs are transposed into logical inputs to prepare data for processing.

algorithm   A formula, or series of steps, for defining a problem and describing its solution.

algorithmic approach   A fixed, procedural methodology that provides relatively little flexibility in program design.

alias   An alternate name that can be used to represent an identified data structure within a data dictionary notation.

alphabetic field test   Test to verify that specific data fields contain only alphabetic characters and blank spaces. Used for specific processing control within a computer program.

amplitude modulation (AM)   A method for imparting volume and tone to radio transmissions through variations in signal volume.

analysis   The process of breaking situations or problems down into successively smaller elements for individual study and solution. *See also* systems analysis.

analysis and general design phase   A major segment (phase) of the systems development life cycle. Includes: establishing definitions and descriptions of existing systems, defining requirements for and designing features of a proposed replacement system, and doing a cost/benefit analysis. The report to management at the conclusion of this phase provides the basis for a go/no go decision on implementation of a new system.

application generator   Software that submits prepared program formats to a programmer via a series of menus and prompts. The programmer selects appropriate formats and adds parameter specifications to fit the application requirements. Coding is generated automatically.

application software package   Predesigned software for a specific application, available for purchase and ready for use (possibly with minor modification) in an appropriate CIS; used in place of custom-designed software to reduce overall system costs or shorten development time.

architectural design   The logical structure of processing functions within a software product.

archival file   File that has been processed and retained for special research or historic reference.

archival record   Permanent record of business activity made for legal requirements, historic perspective, and backup security.

archival storage   Storage of archival records in a form that can be easily protected and will not degenerate over time, yet will be accessible when needed.

argument   Data passed to a called module. In a table, a data item.

attribute   Data item that characterizes an object. A consideration in choosing which data structures should be assembled into a composite data structure in the process of normalization.

attribute file   A file that contains data describing or characterizing an entity about which information is maintained in a CIS.

audio output   Data output that is audible and usable—in human language or sound.

audio response   Audio computer output that delivers data as spoken messages generated from digitized voice storage or voice synthesis.

audit trail   Printed documents and computer-maintained records that can be used by auditors in tracing transactions through a system—from input source, to master file update, to output reports—for purposes of verification.

auditability   Degree to which a system is capable of having a successful and complete audit made to evaluate the integrity of data that rely on a system.

author   Initiator or developer of a specific CIS product. Leader of the walkthrough of that product.

B

background   Work that is initiated by programs separate from but complementary to the main processing program.

backup file   Separately retained duplicate physical copy of a transaction file or historic master file, used for reconstruction and recovery of damaged or destroyed files.

balance   Correspondence of amounts between an entered control figure and a computer-developed total. Used as a processing control. Also, correlation between parent and child elements of data flow diagrams in terms of flows in and out and functions accomplished.

balancing   *See* balance.

bar code   Data expressed as a series of bars and spaces printed in a small field on a tag or product label, for capture by optical code reading equipment.

baud rate   A unit of measurement of data transmission indicating the binary units of information transmitted per second.

benefit   Favorable tangible or intangible result that offsets cost; savings or improvements that can be assigned values (either tangible or intangible) that can be balanced against costs as a basis for decision making. *See also* cost.

bi-directional   Capability of a serial printer to print lines of data from left to right, or right to left, in both directions, to eliminate time needed to return to the left side of the paper.

binding   The process of resolving or fixing data values in a program.

bits per second (bps)   A measurement of the rate of data transmission.

black box   Processing entity that produces a predictable output for a given input and whose general function is known, but whose internal processing rules are not known.

black box testing   Monitoring the inputs and outputs of a module in terms of expectations and acceptability, without regard for internal processing logic.

branch   A statement that alters the normal sequential execution of a series of program statements.

bubble   Circular graphic representation within a data flow diagram of a point within a system at which incoming data flows are processed, or transformed, into outgoing data flows.

bytes per inch (bpi)   The number of data bytes that can be recorded in an inch of storage space on magnetic media. Also referred to as bits per inch.

C

capital investment   Financial resources committed by a business for the purchase of equipment or facilities.

case construct   A program module with a selection control structure that executes alternate processing functions chosen on the basis of data content.

category test   Range or reasonableness test applied to non-numeric data that may include table lookup techniques. A processing control.

cathode ray tube (CRT)   *See* CRT terminal.

central transform   The processing branches that lie after the afferent branches and before the efferent branches. These comprise the main processing functions of the application.

channels   Recording patterns on magnetic tape, consisting of bytes recorded next to each other that form rows of aligned bit positions.

check digit   Data bit used for a validity check in which a series of calculations is performed on a numeric value in a certain position within a field. The result must equal one of the digits in the field. A processing control.

check point   A verification step, usually applied through use of periodic output reports of sampled transactions to verify that processing is proceeding to acceptable standards. A processing control.

child diagram   Exploded version of a parent bubble, showing processing or transformation in greater detail. *See also* parent.

cohesion   Degree to which a process has a singular business purpose.

coincidental cohesion   The random existence of relationships between the elements of a module.

collector   Symbol for a point within an information system where separate streams of data are merged, repackaged, and forwarded. Indicated on data flow diagram by a half circle.

common coupling   Shared access to data in a common pool by two or more modules.

communicational cohesion   The lowest level at which processes are related within a module through the sharing of inputs or a data pool.

completeness   Control requiring that all appropriate data to be gathered appear on the source transaction. Also, condition of possessing adequate and appropriate data for the processing at hand.

complexity   A measure of the processing complications or amount of data to be passed from one module to another.

computer-aided-instruction (CAI)   Educational and instructional methods using a computer to guide a learner through an instructional program.

computer information system (CIS)   A total, coordinated information system that includes computers, people, procedures, and all the resources necessary to handle input, processing, output, and storage of data useful to an aspect of the organization.

computer operations documentation   Written instructions for the setting up and execution of computer programs including—but not limited to—data control, back-up procedures, outputs, and any special functions.

computer output to microfilm (COM)   The recording of system outputs on microfilm, usually for archival storage.

concatenate   To link two or more keys together to form a new, combination key. Used to allow unique identification of records and, at the same time, to permit access to related records in a file.

concatenated key   A series of linked keys used for record identification and access.

concentrator   A computer processor that monitors and logs the transmission traffic of several peripheral processors or other user devices, forming a coherent data stream for delivery of messages to their destinations.

conditional   An exchange of control to and from a subordinate module during the execution of a program on the basis of the results of condition tests.

confidentiality controls   Controls designed to protect rights of privacy of persons or organizations described by, or represented in, data records.

connection   A reference by an element within one module to the identifier of another module.

context diagram   Graphic model of an information system that shows a flow of data and information between the system and external entities with which it interacts, to establish the context, or setting, of the system.

continuous value   Data element whose value can vary over a range of optional values. *See also* discrete value.

control   Any method or function that monitors input, checks processing, or evaluates feedback to determine if system performance meets expectations.

control break   A sensed change in record content that triggers the writing of summary data and the resetting of accumulators in the generation of summary reports.

control coupling   The interface that occurs if a subordinate module is passed information that directs its processing.

control (systems development)   The organizational activities that govern the systems development process to monitor functions, budgets, schedules, and quality.

control totals   Numeric totals used for comparison to assure keyboarding accuracy and completeness of records. Includes count of number of documents or records in a batch, hash totals, and monetary or quantity totals.

conversion program   Special program that reformats files of one system for use with a different system.

coordination   Information passed by a module to direct the processing of another module.

corporate culture   The totality of the beliefs, philosophies, and goals of an organization.

correlation   Special identifying relationship between objects and composite data structures.

correlative file   A special file of relationships between record keys appearing in two separate files. Used to establish access paths among physically separate files.

cost   Tangible or intangible expense associated with any system function; encompasses any out-of-book expense associated with any function within a system, as well as human-related intangible costs. *See also* benefit.

cost/benefit analysis   Study and evaluation of a course of action, or proposed solution to a problem or need, that compares projected savings and other benefits to projected costs.

cost-effective   Course of action that produces maximum relative benefit at minimum relative cost.

coupling   Interface on data flow diagram between two higher-level processes, represented by the number of data flows connecting them. Processes with minimum coupling are more independent and more easily maintained. *See also* cohesion.

critical activities   Necessary and essential activities that must be performed individually and that, together, account for the total elapsed time of a systems development project. *See also* critical path, critical path method.

critical path   Sequence representing minimum amount of time necessary for project completion; represented on a critical path method (CPM) visual representation by the longest path through the activities.

critical path algorithm   Mathematical formula used to help identify the longest sequence of activities that will lead to a completed project. *See also* critical path, critical path method.

critical path method (CPM)   Planning and scheduling method for predicting and measuring trade-off relationships between relative costs and alternative completion dates for a project; presented visually on a project graph. *See also* critical activities, critical path.

CRT terminal   Unit that contains a video (cathode ray tube) display screen and a keyboard for entry of data. The data may go into a recording device, or directly into a computer.

cumulative documentation   Relevant documentation generated during CIS project analysis and design phases to support later developmental stages.

D

data access diagram Graphic representation of data files showing formats of files and corresponding relationships, or access paths, between files.

data capture Procedures for recording and putting data into a system through keyboarding or other methods.

data coupling The interface of modules connected only through active control and passing only essential data items.

data dictionary Listing of terms and their definitions for all data items and data stores within an information system.

data element Basic unit of data that has a specific meaning for the system in which it is used.

data entry Converting or transcribing source data into a form acceptable for computer processing.

data flow Movement of data through a system, from an identified point of origin to a specific destination; indicated on a data flow diagram by an arrow.

data flow diagram (DFD) Graphic representation and analysis of data movement, processing functions (transformations), and the files (data stores) that are used to support processing in an information system. Used to improve present utilization or to plan future changes in the system.

data input Transmission of data into a computer, especially by machine. *See also* data capture, data entry.

data processing system (DPS) Collection of methods, procedures, and resources designed to accept inputs, process data, deliver information, and maintain files. Provides direct support for an organization's basic transactions and operations.

data store Storage area for collections of data input or generated during processing; indicated on data flow diagram by open rectangle.

data structure Packet of logically related data that can be decomposed into subordinate data components or data elements.

data structure diagram Graphic representation of relationships among attribute data structures. Indicates: access keys, access paths, access through correlative structures, and relationships among attribute structures sharing the same key.

database Data organized so that multiple files can be accessed through a single reference, based upon relationships among records on the various files rather than through key values or physical position. Also, all data resources needed to support a system.

database management Direction or control of a database through special software that identifies relational values for records, then executes access commands through sequential, direct, or indexed-sequential reference methods, whichever is appropriate to define the relationship specified by the user.

debugging Finding and correcting improperly written statements or logic errors in a program.

decision support system (DSS) Type of computer information system that assists management in formulating policies and plans by projecting the likely consequences of decisions.

decision table Representation of decision-making process showing a multidimensional array of conditions and outcomes with points of correspondence at the intersections of these vertical and horizontal elements. Used for description and/or analysis of processing alternatives.

decision tree Graphic representation of conditions or processing alternatives and outcomes that resembles the branches of a tree.

decompose *See* partitioning.

decomposition The process of partitioning a system into increasingly detailed functions that can be studied separately in relative isolation.

demodulation The conversion of information structured in ASCII binary analog code into a digital signal for use by a computer communications system.

density Average number of data bits per unit of storage space.

design A representation of an object to be constructed.

desk checking *See* walkthrough.

detail report Report of data content of file records.

detailed design and implementation phase The portion (phase) during the systems development life cycle that refines hardware and software specifications, establishes programming plans, trains users, and implements extensive testing procedures, to evaluate design and operating specifications and/or provide the basis for further modification.

detailed design specification The output of technical design that serves as the framework for the implementation and operation of a computer system.

developmental benefit One-time benefit resulting from undertaking a systems development project; includes economic benefits, as well as increased experience and competence for systems developers.

developmental costs Costs of establishing a new system and bringing it into use. Depreciable as a capital investment over the anticipated useful life of the system.

Diagram 0 (Zero) Graphic system documentation and specification model that uses a symbol vocabulary to identify main processing functions, data flows, external entities, and data storage points.

dial-up service A simple telephone service that can be adapted to point-to-point computer communications.

differentiation The process of identifying a company's unique characteristics in order to design an information system oriented toward fulfilling specific needs.

digitizer Pen-like device moved along a graphic shape whose movements are assigned digital values by a computer. Used for entering drawings and graphics as data.

direct file A file organized directly, by location key, and also relatively, by position of a record within the entire field. Data can be accessed randomly from a direct file. Serial or sequential access is also possible.

discrete value Noncontinuous, distinct value. Refers to data element that has only specific options, rather than a range of options, for its value. *See also* continuous value.

disk pack Multi-surface recording device that consists of a set of magnetic disks on which data can be written and read at random, or directly.

diskette A small, flexible, circular magnetic recording medium on a plastic base, enclosed in a paper envelope. Most often used as a storage medium with microcomputers. Also called a floppy disk.

documentation controls   Control procedures used to assure that correct, updated copies of current processing procedures are available to users and that all previous versions of documentation are maintained.

down time   Unexpected interruptions in service in a data processing system.

download   The periodic transfer of information files from a central maintenance facility to distributed computer locations.

driver   A test program routine used in bottom-up program design to call subprograms, pass information, and serve as a temporary superordinate program.

drum-type plotter   Graphic device using a round drum to hold the paper on which lines are plotted.

dump   A procedure in which the contents of memory are recorded onto an output medium.

E

early finish (EF)   Earliest time at which a project activity can be finished, determined by adding estimated completion time to early start time. Used in critical path method (CPM).

early start (ES)   Earliest possible time (date) at which project activity can begin. Used in critical path method (CPM).

efferent   Branches of data flow that convert logical outputs into usable physical outputs.

80/20 rule   Guideline for systems development costs stating that 80 percent of the benefits of a system can be achieved for 20 percent of the cost of the total system; the remaining 80 percent of the cost provides only an additional 20 percent of benefits. Used as guideline in evaluating system features and capabilities.

electrostatic (laser) printer   Highest speed nonimpact printing device. Forms images on a copier drum, then transfers outputs to paper.

encryption   Alteration or encoding of signals representing data. Used when processing involves transmission over communication lines or networks. Also known as signal scrambling.

ergonomics   Study of human factors related to job performance and use of equipment.

exception   Condition outside of the range defined as normal.

exception report   Specially produced report indicating exceptions. Used to identify conditions that require human decision, items that cannot be processed, or out-of-balance situations.

execution structure   The specific processing statements and sequences needed to implement an application on a computer.

execution trace   Routine used in troubleshooting programs. Causes a computer to document a sequential log of processing events.

explode   To expand a unit of a Diagram 0 (Zero) representation to a more detailed level for further scrutiny.

external entity   Person, organization, or system that supplies data to or receives output from a system being modeled. Indicated on data flow diagram by a rectangle.

external output   Documents or reports produced expressly for use outside an organization; includes reports to governmental agencies, documents sent to customers, communications with stockholders, paychecks, etc. *See also* internal output.

F

face validity   Appearance of underlying authenticity and purposefulness in an information-gathering questionnaire.

fan-in   A measure of the number of higher-level modules that call upon a lower-level module. The higher the value, the greater the overall usefulness of the lower-level function.

fan-out   The number of modules that are immediately subordinate to a given module, indicating the degree to which hierarchical partitioning has occurred within a system.

father file   *See* generation.

feasibility report   End result of a feasibility study. Includes recommendation for a specific course of action, description of the existing problem and anticipated changes, preliminary estimate of costs and benefits, impact statement detailing needed changes in equipment and facilities, proposed schedule for completion, and a list of policy level decisions to be resolved by management.

feasibility study   Study that, when completed, will have evaluated initially the relevant factors involved in a problem or need, considered preliminary alternative solutions, recommended a definite course of action, and projected estimated costs and benefits to be derived from the recommended solution.

feedback   A specially designed output used for verification, quality control, and evaluation of the results of data processing.

fiber optics wand   Hand-held device used with optical character reading equipment to "read," capture, and input data recorded in bar code or a special character set from a printed document or label.

fiche   Flat multi-image film sheet. Used with computer output to microfilm (COM) device.

file   A collection of records relevant to an application under development.

file controls   Procedures and methods used to assure proper and authorized handling, storage, use, and backup duplication of files.

file conversion   Process of changing master and transaction files to meet specifications of new system processing requirements.

fill-in-the-blank   Questionnaire item that seeks specific, finite, factual answers not restricted to a given set of choices.

final documentation   Detailed report of systems development project after completion. Included are documentation of programs, processing, procedures, forms, and files to assist in solving day-to-day system operational problems or questions when the system is in operation.

financial feasibility   Evaluation that results from consideration of the economics of a proposed course of action, to determine potential profitability.

finish time (T)   Time at which a project will be completed. Identified on project graph by the symbol T. *See also* critical path method, project graph.

finite   Having a definite or definable beginning and a specific ending point.

first-level factoring   The process of identifying the major processing functions that must be performed to accomplish a specified program function.

first normal form Preliminary partitioning of data structures containing repeating groups into two or more relations without repeating groups that accomplish the same purpose. *See also* hierarchical partitioning.

fixed costs Continuing costs involved in assuring the ongoing existence of a business enterprise that must be considered in any proposed systems development plan. *See also* variable costs.

fixed-type printer Impact printer device that uses a rotating circular printing element in front of a striking device to imprint characters.

flag In programming, a data item used to signal the occurrence of an expected event or special condition that arises during processing.

flatbed plotter Graphic output device that uses a flat area to hold the paper on which lines are plotted.

floppy disk *See* diskette.

font Format that gives a printed character set its particular "look."

fourth-generation language A nonprocedural programming language such as those used in application generators.

frequency modulation (FM) A method for imparting volume and tone to a radio transmission through variations in the signal frequency.

function The transformation that takes place when a program module is executed. A specific computer processing operation.

function testing A procedure used to identify discrepancies between the results of module execution and the expectations established in specifications.

functional cohesion Performance capability of a module to apply a single function to a given data item, producing a predictable output.

functional decomposition A method of system and software design that breaks a complex problem into a set of individual, solvable subproblems.

functionally dependent Describes the relation between nonkey data elements and the primary key in the second normal form. Uniquely identified only by a complete concatenated key, rather than by just a partial key.

G

Gantt chart   Graphic representation of a work project showing start, elapsed time, and completion relations of work units in a project. Used to control schedules as part of project management. *See also* critical path method, project graph, project planning sheet.

generation   Version of master file produced by processing a transaction file against a master file; the previous master file becomes a backup file. Three generations, known as the son file (most current), father file (previous master file), and grandfather file (predecessor of father file), are typically maintained.

global   Encompassing all aspects of a system.

grandfather file   *See* generation.

H

hard copy   Output in the form of permanent records such as paper documents or microfilm.

hard wire   To interconnect computer devices through directly attached cables.

hash function   A formula applied to a record key to determine the storage location for the record in a direct file organization.

hash total   Summation of a numeric field that does not contain quantities or values normally added together. Used only to verify data entry.

header record   Record indicating number of documents in a batch, batch identification number, and date of processing. Input control.

heuristic   Method providing aid or direction in the solution of a problem; a "rule of thumb."

hierarchical   An ordering and division of problems or functions into successively smaller increments, according to logical and/or functional sequence.

hierarchical partitioning   Breaking down a large problem or project into a series of structured, related, manageable parts through iteration, for the purpose of understanding clearly the functions and requirements of individual system parts.

hit rate   *See* activity rate.

human engineering   *See* ergonomics.

human factors feasibility   Evaluation that results from consideration of human reactions to a proposed course of action, to determine whether such reactions might impede or obstruct systems development or implementation.

hypothesis testing   Program troubleshooting technique that attempts to isolate a processing error by devising specific test data that will retrigger the mistake.

I

identifier   An assigned name. Applied to each module within a program.

impact printer   Printing device that creates impressions by striking a ribbon that transfers images to paper. *See also* line printer and serial printer.

incremental   Step by step.

incremental step   Implementation and installation of a larger new system with reasonably independent components in increasingly complete stages. Allows users to learn to use the final system effectively, in stages. Makes it possible to develop the final system with good control over schedules and budgets.

indexed sequential file   File arranged in sequential order, according to key, and also containing an index, or table, to identify the physical location of each key within the file. File can be searched in ascending order according to key, or a single record can be randomly accessed by reference to a physical location in the index.

information   Meaningful data transformed through processing, or knowledge that has resulted from the processing of data.

information center   A specialized computer facility that uses sophisticated software tools to generate functional computer applications in direct response to user service requests.

information hiding   Isolating functions not directly related to the problem within a module so that changes in the processing environment will not require changes in the structure or logic of the program.

information system   The methods, procedures, and resources for developing and delivering information.

initial investigation Activity to handle and evaluate requests for new or improved CIS services. End result is an understanding of the request at a level sufficent to make a preliminary recommendation as to the course of action to be followed.

initial investigation report Report documenting the initial investigation activity, findings, and recommendations.

ink jet printer Nonimpact printing device that sprays microscopic ink particles onto paper to form characters.

input Data that serve as the raw material for system processing or that trigger processing steps. Also, to access data and place them into a computer system. Input tasks include data capture, data entry, and input processing.

input controls Controls used to assure that only correct, complete input data are entered into the system. Encompasses control totals for batch processing, video display, and maintenance of a transaction log to produce control totals for on-line systems.

installation phase Portion (phase) in the systems development life cycle during which the new CIS is installed, the conversion to new procedures is fully implemented, and the potential of the new system is explored.

instrumental input Data recorded directly by a machine, without human interpretation; examples are supermarket bar code reading devices and optical character recognition devices.

intangible Real, but not easily assessable. Describes business costs or benefits not easily quantifiable in monetary terms. *See also* tangible.

intangible benefit Delivered, identifiable improvement that must be identified, and for which a value that is not easily quantified must be ascribed.

intangible cost Cost, in most cases readily identified, but not easily quantified, usually attributable to human reactions to changes in the work environment.

integration testing Procedures that examine the interfaces between system components or modules, certifying that information and control are passed correctly and that output is as expected.

integrity Completeness and unimpairedness. Integrity controls assure that: data files processed represent the actual, current status or condition; materials and mechanisms will exist to reconstruct destroyed files and recover processing capabilities in the event of loss; only authorized transactions will be admitted to a system.

intelligence   Built-in electronic processing capability within a CRT terminal. May include microprocessors, memory units, printing and document originating capabilities.

interface   A connection between modules. The hardware or software that must be used to interconnect systems or devices within a system.

interim documentation   Documents generated during the analysis and development phase of the systems development life cycle to provide orderly, cumulative records of the development process. *See also* cumulative documentation.

internal output   Documents or reports produced for use within an organization, as distinct from documents for use outside of the organization. Includes reports to management, job tickets or production schedules, employee time cards, etc. *See also* external output.

interview   Planned interactive meeting between a data gatherer and one or more subjects for the purpose of identifying information sources and collecting information.

investigation phase   Portion (phase) at the inception of the systems development life cycle to determine whether a full systems development effort or another course of action is appropriate.

iteration   Repetition; indicated on data flow diagram by braces { . . . }. Also, partitioning a problem repeatedly to reach increasing levels of understanding. *See also* hierarchical partitioning and partitioning.

J

job   A grouping, or packaging, of processes into a single processing unit.

job control language (JCL)   The operating system software tool used to identify programs being submitted and the necessary software and equipment support requirements for the processing of application programs.

job step   Within a batch processing environment, an independent segment of a job that has been subdivided into separate processing units, or steps.

job step boundary   The limits of an independent segment of a job stream containing inputs, processing steps, and outputs.

job stream   A sequence of programs or steps that make up a single processing job.

job stream test specification   An extensive, detailed list of all conditions to be tested and the test files to be used.

journal   A log or record kept on a daily or regular basis. *See* transaction log file.

K

key   Access control field that uniquely identifies a record or classifies it as a member of a category of records within a file.

key attribute   Primary key to other data structures and the attributes of those other data structures.

keypunch   A machine that punches holes in cards to make a physical data entry. The cards are machine- or human-readable.

key-to-disk machine   Keyboard entry device that usually includes a CRT terminal and a recording system that processes entries and places them on disk packs.

key-to-diskette machine   Keyboarding device, with or without a CRT, that enters machine-readable data directly onto a diskette.

key-to-tape machine   Keyboarding device that enters machine-readable data directly onto magnetic tape.

L

laser   A high-powered coherent beam of light. *See also* electrostatic (laser) printer.

late finish (LF)   Latest completion of an activity. Determined by adding the activity duration to the late start time. Used in critical path method (CPM).

late start (LS)   Latest time at which an activity can begin without extending the total project completion time. Determined by deducting elapsed time from late finish time for an activity. Used in critical path method (CPM).

layering   Iteration of systems analysis studies to produce additional knowledge and/or understanding of problems and system operations.

leased line service   A fixed, dedicated communications link.

level of abstraction   One of a series of isolated units, or levels, within a top-down process of breaking a problem into increasingly detailed subproblems.

library routine   Software modules that preserve programs for later use in different applications.

light pen input   Input device, resembling a pen, that allows users to manipulate data on the face of CRT screens. Used chiefly for engineering and design applications.

line item   Data represented on a single line of a report or document, such as a single item in an extensive order.

line printer   Printing device that prints documents a full line at a time.

local area network (LAN)   A series of computer processors that share files and peripherals.

log in   A procedure under which a user establishes interaction with a computer system, including entry of codes that authorize access for that individual.

logical cohesion   The degree of correspondence of module elements when a single module performs more than one type of logically related function or class of functions.

logical data structure   A model indicating the content of files.

logical model   Model of a CIS showing only logically necessary data content and handling to aid in documenting and/or analyzing a system. *See also* physical model.

longest path   Minimum time required for project completion, as indicated on a project graph network. *See also* critical path.

lookup table   Program table searched to find entries to match input data. May be used in a category test for processing control.

M

magnetic ink character recognition (MICR)   Input method developed and used by banking industry to identify checks, deposit slips, and other documents preprinted with a special magnetic ink.

maintenance   Altering or replacing software or hardware of a CIS to meet new or changing processing requirements.

management information system (MIS)   Type of computer information system that provides meaningful summarization of data to support organizational management control functions and highlights exception conditions requiring attention or corrective action.

management summary   Summary report prepared for management. Recommends a course of action to solve a problem.

mark sensing   Optical or electrical document reading method that uses the position of marks to indicate the meaning of data.

master file   File containing permanent or semipermanent basic information to be maintained over an extended lifespan. Contains one record for each entity covered.

master-to-slave   A computer configuration system that includes a master, or central, processor and one or more peripheral ''slave'' processors.

materiality   A measure of the relative importance (significance) of a data item within an application.

matrix printing element   Impact printing device containing a series of points that are projected forward to cause printing impressions, thus forming characters.

memory dump   Printouts showing the status of a memory at a given moment. Often used in program troubleshooting.

metric   A formal unit of measurement that can be applied to a software design as opposed to heuristics, simple rules of thumb.

minimodel   Individual changes in a proposed CIS, modeled separately.

mnemonic   An assigned name or value that serves as a memory aid.

model   Mathematical or logical representation of a system that can be manipulated intellectually to assess hypothetical changes. Also, to make graphic or written representations of an information system and its functions, to help people understand the system.

modem   From MODulator-DEModulator, a device that translates analog code into digital code, and digital to analog, allowing computers to utilize telephone lines for direct communication.

modulate To vary the amplitude, frequency, or phase of a transmitted signal.

module In programming, a solution component representing a processing function that will be carried out by a computer.

module testing Procedure used to determine the soundness of logic and functions within a processing step.

monetary total *See* quantity total.

monolithic Programs that behave as single, interrelated blocks of code. These programs are generally large and difficult to implement, maintain, and modify.

most probable time estimate ''Best guess'' of the time that will be required to complete an activity, assuming a normal number of problems or delays. Used in project evaluation and review technique (PERT).

multiple-choice Questionnaire item that provides the respondent with a series of finite, specific choices.

mutually independent State when it is verified that each nonkey data element is independent of every other nonkey element in the relation; test for third normal form.

N

net present value (NPV) Present value of benefits, minus present value of investments; can be positive, zero, or negative. Used to compare alternative investment opportunities with a stated benchmark, or standard. *See also* present value.

network Graphic flow diagram relating the sequence of activities to the sequence of occurrence. Used in project evaluation and review technique (PERT) and critical path method (CPM). *See also* project graph.

networking Linking of multiple devices through communications lines for distribution of processing and/or the transmission of data.

new system design specification Comprehensive proposal for a new CIS, encompassing both user specification and all updated and/or additional detailing of hardware, software, procedures, and documentation needed for actual implementation. Presented to both users and CIS design group for signoff.

node   Beginning or ending point of an activity, represented on a project graph by a circle. A control point processor within a network that monitors and logs data transmission traffic. *See also* network.

noise   Electronic interference on a transmission channel that degrades or distorts the signal.

nonimpact printer   Printing device that causes images to be imprinted without actual contact between print mechanism and paper. *See also* electrostatic (laser) printer, ink jet printer, and thermal printer.

nonincremental   A program design technique that creates all modules before testing them simultaneously as a program.

nonparameterized   Conditional transfers of control that gather information from a global data pool rather than from passed argument lists.

nonredundancy   Criterion for logical data design, characterized by avoiding inclusion of the same data component within two or more data stores, and/or avoiding inclusion of the same data in different forms within the same data store.

nonredundant   Components of data files that appear only once or in only a single form in the entirety of those files that model the data structure of an organization.

normalization   Process of replacing existing files with their logical equivalents, thereby deriving a set of simple files containing no redundant elements.

numeric field test   Test to verify that a given field contains only numeric characters. Used for processing control.

O

object   Entity, or thing, described by or represented in a data structure. *See also* attribute.

observation   Method of gathering information utilizing a highly trained, qualified person who watches firsthand the actual processing associated with a system and records information and impressions of the process.

open-ended question   Questionnaire item offering no response directions or specified options. Used to allow a wide variety of potential responses.

operational benefit   Recurring benefit that results from the day-to-day use of a system, such as reduced operational costs.

operational control   Procedures within computer information systems that deal with access, authorization, and verification.

operational costs   Variable costs that are associated with the use and maintenance of a system.

operational feasibility   Evaluation that results from consideration of manual processing needs and overhead costs of a given systems operation by an organization.

optical character recognition (OCR)   Data input technique that uses reflected light to ''recognize'' printed patterns.

optimistic time estimate   ''Best guess'' estimate of minimum time required to complete a project, assuming all conditions will be ideal. Used in project evaluation and review technique (PERT).

optimum   Most favorable in terms of cost/benefit analysis. Describes business option that produces greatest benefit for the least relative cost.

organizational controls   Methods and techniques for protecting the integrity and reliability of data within a system through patterns of job responsibility. *See also* separation of duties.

organizational structure   A formal recognition by the management of a business of the subsystems that make up the business organization. Reflects fundamental strategy for achievement of the organizations's goals. Often represented on an organization chart.

output   A product, or result, of data processing.

owner (system)   Upper-level personnel who manage lower-level users of a CIS. *See also* user.

P

packet   A block of data, identified by length and destination prefixes, that moves along a network until it reaches its designated user.

parameterized   A type of control structure that draws specific information from argument lists.

parameters   Specifications given to a program generator that define program limits and processes, allowing the generator to create source code.

parent   Single bubble in high-level data flow diagram that can be exploded to produce a more detailed version. *See also* child diagram.

partition   To reduce the complexities of a problem or situation into smaller elements that can be approached as individual, soluble items.

partitioning   Division of a complex problem or situation into smaller separate elements for ease of understanding, and/or solution. *See also* hierarchical.

patch   A program coding subset used to fix or update an application module.

payback   *See* payback period.

payback analysis   Method for determining period necessary for a new system to generate savings great enough to cover developmental costs.

payback period   Length of time necessary to earn an amount equal to the amount required for acquisition of a capital investment.

peer-to-peer   A relationship within a distributed computer network in which separate processors function as equals; there is no central or dominant processor.

percentage completion   Indication on a Gantt chart of the proportion of a project that has been finished.

performance testing   Procedures used to verify the quality of system processing before the system becomes operational.

pessimistic time estimate   Maximum completion time of a project, assuming that everything that can go wrong will go wrong. Used in project evaluation and review technique (PERT).

phase   Set of activities and tasks that, when completed, delimits a significant portion of a systems development project.

physical model   Graphic representation of the processing activities in an information system, shown in sequence and reflecting all data transformations, file alterations, and outputs.

planning   Study and development of projected courses of action for meeting goals or dealing with anticipated problems.

plotter   Computer-driven graphic output device that creates images on paper by guiding a pen-like stylus.

point-of-sale terminal Electronic cash register that transmits sales entries into a recording device or computer.

point-to-point A communications channel, usually a telephone line, that establishes a direct connection between two users.

pointer *See* key.

poll A transmission initiation procedure in which the central processor contacts each network point sequentially, allowing each point time to respond.

population Total group of persons with a commonality of identification. Information providers identified as potential respondents for a questionnaire.

post-implementation maintenance list List of change requests from users, made during system implementation, and noncritical changes to be made after system test procedures, that are to be handled as maintenance after full system implementation.

post-implementation review report Report prepared for CIS, user departments and steering committee. Covers review, conducted after a new system has been in operation for some time, to evaluate actual system performance against original expectations and projections for cost/benefit improvements. Also identifies maintenance projects to enhance or improve the system.

preliminary detailed design and implementation plan Planning document used as a basis for detailed planning, and also to update estimates of development costs before new system design is completed. Encompasses: activities down to major task level, working days required, proposed staffing plan, and dependable planning schedule for activity and task completions.

preliminary installation plan Document prepared during implementation and installation planning. Contains: file conversion and system installation approaches; preliminary list of major files to be created or converted and forms to collect new data; identification of necessary computerized file conversion programs; and preliminary list of installation tasks for the new systems, including any special coordination considerations.

preliminary system test plan Document prepared during implementation and installation planning that establishes expectations of results to be delivered in each system area. Identifies major system products, or functions and interrelationships, and modules to be tested. Also specifies system, program, and user procedures tests.

present value   Current value of money. To determine the value of money in constant dollars, future economic values are discounted backward in time to the present.

present value factor (pvf)   Multiplicand used to determine the present value of a sum of money to be received at a certain time in the future.

prime data area   In an indexed-sequential filing scheme, the disk area that holds data records.

primitive   A basic, simple function that can be executed routinely by a computer.

printing device   Output device that produces printed documents.

private network   A multiple-user data communications system created by an organization for its own use.

procedural cohesion   The grouping of elements within a module on the basis of the flow of control through that module.

procedures manual   Instructional document written to aid people in performing manual procedures within a computer-based system. *See also* training manual.

process   To transform input data into useful information through performance of certain functions: record, classify, sort, calculate, summarize, compare, communicate, store, retrieve. Indicated on data flow diagram by a circle, or bubble.

process description   Set of rules, policies, and procedures specifying the transformation of input data flows into output data flows.

processing controls   Controls designed to assure accuracy and completeness of records each time a file is processed. *See also* exception report and trailer record.

program folder   Documentation within the cumulative project file that supports the certification of each program within a system.

program integration   The procedures used to connect modules to form programs.

program test log Document describing problems noticed as system was tested and brought into use. Log is updated to provide current information as changes are made to individual program modules and programs themselves.

program test specification Extensive, detailed listing of criteria for testing a program.

programming and testing Detailed design and implementation phase activity encompassing actual development, writing, and testing of program units or modules.

project Extensive job involving activities that are finite, nonrepetitive, partitionable, complex, and predictable.

project evaluation and review technique (PERT) Project scheduling and control methodology that provides graphic displays to: identify project activities; order activities in time sequence; estimate completion time for each activity, relationships among activities, and time required for the entire project; and identify critical activities and noncritical activities. *See also* critical path method (CPM).

project graph Graphic network that represents activities as paths between beginning and ending points. Used in project evaluation and review technique (PERT) and critical path method (CPM). *See also* network, node.

project management Method or combination of techniques that facilitates planning, scheduling, and control.

project management review Meeting at which technical or general reports by members of project team are reviewed by team leaders or project managers.

project plan Detailed account of scheduling and staffing—to task level—for the second and succeeding phases of a systems development life cycle.

project planning sheet Worksheet used to identify work units, make personnel assignments, and keep track of planned and actual hours worked and dates of completion. Used for project management. *See also* Gantt chart.

project team A team brought together to carry out a systems development project, representing all user needs and perspectives, usually headed by a senior systems analyst, and including other information system specialists and representatives from each of the functional areas impacted by the system.

protocol   An informal set of rules that governs the exchange of data transmission between processing systems.

protocol emulation   A type of software within data communications systems that interfaces different data formats or binary structures to implement accurate, meaningful data exchange between dissimilar systems.

prototype   A working system that can be developed quickly and inexpensively, given the necessary software tools, to evaluate processing alternatives and specify desired results.

prototyping   Specialized systems development technique using powerful application software development tools that make it possible to create all of the files and processing programs needed for a business application in a matter of days or hours for evaluation purposes.

Q

Quality total   *See* monetary total.

query   A single inquiry sentence that, with a database reference, would seek out and organize all relevant, related records, and present them in a sequence stipulated in the query.

questionnaire   A special-purpose document requesting specific information that can be quantitatively tabulated, usually from large populations of source respondents. Used by systems analysts to gather information relating to potential CIS development.

R

random access   Disk access technique in which records can be read from, and written directly to, disk media without regard for the order of their record keys.

randomizing routine   Algorithm applied for assigning record locations for applications in which keys cannot be used directly as locators.

range test   Test to verify that values of entries in a given field fall between the upper and lower limits established by a program. Used for processing control.

ranking scales Questionnaire item that asks the respondent to order a response in terms of preference or importance.

rating scales Questionnaire multiple-choice item that offers a range of responses along a single dimension. Used to assess responses to a given item or situation.

readability The output of processed information that is delivered to users in a format that is usable and meaningful in the context of the application.

reasonableness test Test applied to determine whether data in a given field fall within a range defined as reasonable, compared with a specified standard. Used for processing control.

reference file File containing constant data to be used each time an application program is run. Used, in conjunction with data from transaction files, to update master files.

reference manual Procedural documentation on use of a system.

referent Identifiers or calls that are placed within modules to define paths for sending and returning unconditional transfers of control.

relational value The comparison, or ordering, of one record relative to another. Used in database management to identify a record to be accessed through sequential, direct, or indexed-sequential reference methods.

relative position Record position on disk media identified relative to the basing point, or first record, in a given file.

reliability Description of level of confidence that can be placed on probability of performance as expected for a function or device.

repetition *See* iteration.

report Data output from a file in a format that is easily readable and understandable.

reprographic system System that forms graphic images for typesetting, printing page makeup, or displays.

requirements specification *See* user specification.

respondent Person selected as potential information source, who receives and answers a questionnaire.

review phase Portion (phase) during systems development life cycle that include two activities: the first to evaluate the successes and failures during a systems development project, and the second to measure the results of a new CIS system in terms of benefits and savings projected at the start of the project.

reviewer (walkthrough)   Member of a team appointed to review quality.

ring structure   A circular, decentralized distributed-processing system in which messages are passed continuously around the circuit.

robustness   The ability of software to respond to and deal with unexpected situations encountered by nontechnical users.

router   Point in an information system where a cumulative flow of data are broken down into a series of individual data streams. Indicated on data flow diagram by reverse-facing half-circle. *See also* collector.

run time   The amount of time required to complete a processing function.

S

sample   A subset of a population of respondents chosen to represent accurately the population as a whole in an information-gathering process.

sampling   Method used to gather information about a large population of people, events, or transactions by studying a subset of the total population that accurately represents the population as a whole. Statistical methods are used to infer characteristics of the entire population.

schedule feasibility   Evaluation that results from consideration of time available to complete a proposed course of action, to determine whether or not it can be implemented in the time available.

scheduling   Relating project activities that must be completed in a time sequence. *See also* planning.

scope of control   The range of effect of a decision on its respective module as well as all modules subordinate to that module.

scope of effect   The range of effect of a decision on the total group of modules encompassing conditional processing based on that decision.

second normal form Second step in normalization, when it is verified that each nonkey data element in a relation is functionally dependent on a primary key.

secondary storage device Equipment used to write data to, and read data from, magnetic media.

secretary (walkthrough) Member of quality review team who produces a technical report listing identified errors or problems noted.

security controls Controls applied to protect data resources from physical damage, and from intentional misuse or fraudulent use.

selection Group of data structures or data elements out of which one, and only one, item may be selected for use.

separation of duties Policy that no one individual should have access to, or know enough about, a system to process data in an unauthorized way, either during development stages or during ongoing use of the system. Major technique of organizational control.

sequence Linking together of data elements or data structures; indicated by ''+'' sign between units.

sequential access Access technique to read from and write to records and files in an order determined by a logical identifier, or key, that is generally a data field within the record.

sequential cohesion A process in which multiple functions, comprising successive data transformations, are performed on the same data element.

sequential file File in which the physical and logical sequences of records match. Records are accessed in an order determined by a key, usually numeric.

serial access Access technique to read from and write to records and files in the same chronological order in which the records were initially recorded.

serial file File in which records are recorded in chronological order, as transactions are entered into a computer.

serial printer Impact printing device that prints one character at a time to produce documents.

service function Function or activity that is initiated in response to, guided by, and aimed at satisfying, user need for information.

sign off To agree formally and commit to a proposed course of action, for the purpose of proceeding with a project.

sign on *See* log in.

sign test Test to identify and verify presence of positive or negative values in fields. Used for processing control.

significance The relative impact of a data item on its respective application. *See also* materiality.

simplicity A requirement that all components of data stores within logical data structures be fixed-length records accessible only by primary keys.

simulation An imitative representation of the functioning of a system or process. *See also* model.

skeleton programming Prerecorded modules, usually from existing applications, that serve as frameworks for additional applications. These range from simple formats to sophisticated, high-level modules.

slack Without tight constraints. Used to describe time spent on subsidiary projects not affecting duration of an entire project.

software design A model of a program outlining its structural and functional characteristics.

software engineering The application of formal procedures and disciplines to the construction of computer programs.

software package *See* application software package.

software testing Executing a program under conditions designed to cause and discover errors.

son file *See* generation.

source document control Authorization measure that must be applied before data are accepted for input to a system; *See also* input controls.

space (blank) test Test to check whether a given field contains some data value or is totally blank. Used for processing control.

span of control *See* fan-out.

speech synthesizer A sound-generating device that can produce sounds understandable by humans as language.

staffing plan   Detailed account of personnel assignments, and days or hours to be worked, for a systems development project.

stamp coupling   The passing of superfluous data elements to a module; increasing the complexity of the intermodular interface and reducing the independence of both modules.

star network   A centralized, distributed data processing system in which the central processor acts as the control point in logging and monitoring information transmission.

start time   Time at which a project begins, indicated on a project graph by the symbol S. *See also* critical path method (CPM).

starving the process   Showing, in a logical model, only the logically necessary elements or steps needed. Distinguished from a physical model's representation of an actual processing sequence.

statement   In programming, a single command that directs a computer to carry out a processing operation.

status review   Meeting held to keep user management informed on progress of a project. Participants include project leader, key user manager, and possibly project team members who can make special contributions.

steering committee   A committee that sets organizational priorities and policies concerning CIS support. Composed of top management personnel representing all user areas.

stepwise refinement   Top-down abstractions of problems to create workable submodules that can be coded, related, and implemented. *See also* hierarchical partitioning.

structure chart   Graphic representation of overall organization, and control logic of processing functions (modules) in a program or system.

structure clash   A noncongruence of input and output data structures that must be resolved before a program structure can be designed.

structured English   Formal English statements using a small, strong, selected vocabulary to communicate processing rules and to represent the structure of a program or system.

structured specification   *See* user specification.

stub   A routine applied in top-down program testing that simulates the processing that will take place in modules still to be coded.

stylus   Electromechanically driven writing device used on a plotter to produce lines.

subsystem   A secondary or subordinate small system within a large system.

subsystem test specification   *See* systems test.

summary report   Report showing accumulated totals for specific groups of detail records. Used by middle-level managers for review of business activity.

synergistic   The way that a system's parts function together, producing results with a greater value than would be produced by the system's separate parts working alone.

synthesis   The process of bringing information system component parts together into a remodeled system in which previously existing problems have been eliminated.

system   A set of interrelated, interacting components that function together as an entity to achieve specific results.

system flowchart   Graphic representation of a system showing flow of control in computer processing at the job level. Represents transition from a physical model of computer processing to a set of program specifications that will be prepared at the start of the detailed design and implementation phase.

system life cycle   Activities or conditions common to all computer information systems from inception to replacement: recognition of need, systems development, installation, system operation, maintenance and/or enhancement, and obsolescence.

systemic control   Measures taken to deal with hardware configurations and software updating to establish full compatibility.

systems analysis   The application of a systems approach to the study and solution of problems, usually involving application of computers.

systems analyst   A problem solving specialist who analyzes functions and problems, using a systems approach, to produce a more efficient and functional system, usually involving application of computers.

systems approach   Way of identifying and viewing component parts and functions as integral elements of a whole system.

systems design   The technical plans and methods for implementing a computer information system.

systems development   Process that includes identifying information needs, designing information systems that meet those needs, and putting those systems into practical operation.

systems development life cycle (SDLC)   Organized, structured methodology for developing, implementing, and installing a new or revised CIS. Standard phases presented in this book include investigation, analysis and general design, detailed design and implementation, installation, and review.

systems development recap report   In-depth review document prepared for CIS management covering completed systems development project. Aimed at enhancing or improving individual members' and the organization's performance on future projects.

systems test   Extensive test of full system. Conducted chiefly by users after all programs and major subsystems have been tested. Assures that data resources handled by the system will be processed correctly and protected fully. Careful documentation is maintained through program test logs and system test logs.

systems testing   Test procedures conducted by CIS personnel to view the entire system as an operational, functional entity. Provides an overview of the implemented system and all procedures relevant to user training.

T

table   An index that records the physical location of each key within an indexed-sequential file, making possible random access to individual records.

tangible   Cost or benefit readily quantifiable in monetary terms. *See also* intangible.

tangible benefit   Benefit realized when a new system makes or saves money for its organization.

tangible cost   Cost of equipment or human factors associated with the operation of a system.

tape drive   Peripheral storage unit that performs input and output of data on magnetic tape. Also called a tape unit.

task Smallest unit of work that can be assigned and controlled through normal project management techniques; normally performed by an individual person, usually in a matter of days. *See also* activity.

technical design Activity within detailed design and implementation phase that builds upon specifications produced during new system design, adding detailed technical specifications and documentation.

technical feasibility Evaluation that results from technical consideration of available computer hardware and software capability to carry out a proposed course of action.

teleprocessing The utilization of telephone lines for data communication between devices in a computer system.

temporal cohesion Also called classical cohesion, this term describes related functions that are specified to occur in the same time frame.

test data file High-quality information to be input for module testing. Care must be taken that each file is specific to the function of its respective module.

test plan Documents listing the schedules, objectives, criteria, and procedures for system testing. An overview document encompassing all plans for integrating the tasks of program and system testing.

test specifications and planning Activity during detailed design and implementation phase to prepare detailed test specifications for individual modules and programs, job streams, subsystems, and for the system as a whole.

thermal printer Nonimpact printing device that develops images through exposure of special paper to heat.

third normal form Third stage of normalization process, during which duplicate data elements or elements that can be derived from other elements are removed. *See also* mutually independent.

time reporting Accounting procedure for reporting work completed and still to be done. Controls are applied at the task level.

time scale Horizontal axis on a Gantt chart reading from left to right, indicating passage of time.

time value Changing value of money as time goes by, assuming inflationary devaluation or investment growth. Money invested at a percentage return will have a value equal to principal plus interest; money left uninvested loses purchasing power as inflation occurs.

timeliness Quality factor. Meeting needs of users or process for delivering results when needed to meet service requirements.

top-down Partitioning of functions into successive levels of detail from the top-level module, representing the general system or program function as a whole, down through to lower-level modules that perform actual processing.

top-down testing Procedures applied to test a program from its highest, most general module down through more specific modules.

total slack Time difference between the early start and late start dates, or early finish and late finish dates, for a noncritical activity. *See also* critical path method (CPM).

touch-screen input Method of inputting data directly through touch contact with specially sensitized locations on the face of CRT terminal video display screens.

track *See* channels.

trade-off Term referring to decision-making consideration that weighs advantages and disadvantages of alternatives as a basis for selection.

trailer record Last record in a file, containing totals for all numeric fields in all records in the file. Compared with field totals each time the file is processed. *See also* processing controls.

training manual Easy-to-use reference manual that teaches operators how to learn to perform procedures within a computer-based information system. *See also* procedures manual.

transaction A basic act of doing business. The exchange of value for goods or services received.

transaction analysis A data-flow-oriented design process that organizes system design around the transactions in the higher-level modules and recaptures processing similarities using lower-level modules with high fan-in.

transaction document Form upon which data generated by transactions are recorded. Used to capture data at its source to report on results of transactions, control business activity, and for historic purposes.

transaction file   Collection of records containing specific, timely data pertaining to current business activity. Used to update master files.

transaction log file   Continuously updated master accounting record that records all transactions of an on-line processing system chronologically. Serves as starting point for an audit trail and can be used for recovery purposes if master or transaction file data are damaged or lost.

transform   To process data for conversion (transformation) into information.

transform analysis   Modeling a problem structure by decomposing the problem into data flow and transformation requirements.

turnaround document   Computer output document that also serves as input document for a follow-up processing activity.

turnaround time   The amount of time required for a system to accept input, process data, and output meaningful information.

U

unconditional   A transfer of control in which the receiving module then determines the next assignment of control rather than routinely returning control to the originating module.

unit record   A single keypunched card containing an entire data record that may be broken into several fields.

unit testing   The inspection of individual program modules to determine the soundness of their logic and functions.

universal product code (UPC)   Bar code used extensively in supermarkets and other retail outlets for optical sensing of product identification.

user   Term referring both to lower-level personnel who use, and upper-level personnel who own, a CIS. *See also* owner (system).

user concurrence   Agreement by user that capabilities described in the user specification contain a full and complete statement of user needs and that the solution is feasible from operational and human factors standpoints.

user-friendly   A phrase used to indicate a high degree of user convenience.

user procedures manual   *See* procedures manual, training manual.

user specification   User-oriented report presenting a complete model of a new CIS for user evaluation and approval. Can include data flow diagrams, description of system inputs and outputs, performance requirements, security and control requirements, design and implementation constraints, and unresolved policy considerations that must be dealt with before the system can be implemented.

user training   Activity during detailed design and implementation phase of the systems development life cycle. Encompasses: writing user procedure manuals, preparation of user training materials, conducting training programs, and testing manual procedures.

user training outline   Specification document prepared during implementation and installation planning that includes: content outlines for user training manuals, details for preparation of manuals to cover user procedures to be installed, and list of proposed activities and assignments for users and analysts who will write these manuals.

utility   Programs that provide standard functions such as sorting, collating, and report writing.

V

validity   Description of transaction or data to indicate they are authorized, that transactions actually took place, and that data really exist.

variable costs   Costs incurred only when a system is used. *See also* fixed costs, operational costs.

version   *See* incremental step.

version installation   Technique of installing a new system as a series of functional areas or incremental steps.

video display   Visual data display device using a CRT (cathode ray tube).

voice input   A method of inputting data directly through voice commands.

volatility   Rate of change and expansion of a file. Factor to be considered in determining file organization.

W

walkthrough   Technical quality review of a CIS product that can be identified as a separate unit capable of introducing errors into the system.

white box   The exposure of the logical functions of a module to clarify its functions and operations for systems analysts and designers.

white box testing   Complete tracing of the decision path(s) of data through a module.

wide area telephone service (WATS)   A contracted communications service that reduces the costs of heavy telephone utility use.

work station   The physical area in which people and computers interact.

working papers   Documents accumulated during work completion that are useful for project review or for guiding the performance of ongoing work.

INDEX

Page numbers in *italics* refer to figures in the text.